4th International Conference on Environmental Systems Research (ICESR 2017)

IOP Conference Series: Earth and Environmental Science
Volume 178

Singapore
14 – 16 December 2017

ISBN: 978-1-5108-6873-1
ISSN: 1755-1307

Printed from e-media with permission by:

Curran Associates, Inc.
57 Morehouse Lane
Red Hook, NY 12571

Some format issues inherent in the e-media version may also appear in this print version.

Printed by Curran Associates, Inc. (2018)

For permission requests, please contact the Institute of Physics
at the address below.

Institute of Physics
Dirac House, Temple Back
Bristol BS1 6BE UK

Phone: 44 1 17 929 7481
Fax: 44 1 17 920 0979

techtracking@iop.org

Additional copies of this publication are available from:

Curran Associates, Inc.
57 Morehouse Lane
Red Hook, NY 12571 USA
Phone: 845-758-0400
Fax: 845-758-2633
Email: curran@proceedings.com
Web: www.proceedings.com

Table of contents

Volume 178

2017 4th International Conference on Environmental Systems Research (ICESR 2017)

14–16 December 2017, Singapore,

Accepted papers received: 20 July 2018
Published online: 16 August 2018

Preface

Papers

Water Resources Management

Resources and Environmental Economics

Urban and Agricultural Engineering

IOP Conf. Series: Earth and Environmental Science **178** (2018) 011001 doi:10.1088/1755-1315/178/1/011001

PREFACE

It is our great pleasure to welcome you to the 2017 4th International Conference on Environmental Systems Research (ICESR 2017) which was held in Nanyang Executive Centre in Nanyang Technological University, Singapore during December 14-16, 2017. ICESR 2017 was dedicated to issues related to Environmental Systems Research.

ICESR 2017 was held successfully in Nanyang Executive Centre in Nanyang Technological University, Singapore during December 14-16, 2017. Prof. James T. Anderson, from West Virginia University, USA, did the keynote speech with the title "Pollution Effects on Fish and Wildlife Communities". Prof. Ki-Hyun Kim, from Hanyang University, South Korea did the keynote speech with the title "The need for VOC removal in ambient air and advances in sorbent technology". Assoc. Prof. YU Liya E., from National University of Singapore, Singapore, did the keynote speech with the title "Urban Airborne Pollutants in Southeast Asia: What do we know and what can we do?". There were five sessions for all of the presentations in which Eiji Haramoto, Haoyang Cai, Yu Fang, Anlin Liu, Xiaowei Yang and Ova Candra Dewi received the best presentations awards. In addition, many researchers, engineers, academicians as well as industrial professionals from all over the world have presented their research results and development activities.

The major goal and feature of the conference was to bring academic scientists, engineers, industry researchers together to exchange and share their experiences and research results, and discuss the practical challenges encountered and the solutions adopted. Professors from Korea, Singapore, and USA were invited to deliver keynote speeches regarding the latest information in their respective expertise areas. It has been a golden opportunity for the students, researchers and engineers to interact with the experts and specialists to get their advice or consultation on technical matters, sales and marketing strategies.

The proceedings present a selection of papers submitted to the conference from universities, research institutes and industries. All of the papers were subjected to peer-review by conference committee members and international reviewers. The papers selected depended on their quality and their relevancy to the conference. The volume tends to present to the readers the recent advances in the field of Environmental Systems Research and various related areas such as Air Pollution, Air Quality Management, Biodiversity Conservation, Carbon Sequestrations, Clean Fuel and Clean Energy, Clean/Green Production, etc..

We would like to thank all the authors who have contributed to this volume and also to the organizing committee, reviewers, speakers, chairpersons, sponsors and all the conference participants for their support to ICESR 2017.

Prof. Ki-Hyun Kim
Hanyang University, South Korea
March 12, 2018

Peer review statement

All papers published in this volume of *IOP Conference Series: Earth and Environmental Science* have been peer reviewed through processes administered by the proceedings Editors. Reviews were conducted by expert referees to the professional and scientific standards expected of a proceedings journal published by IOP Publishing.

IOP Conf. Series: Earth and Environmental Science **178** (2018) 012001 doi:10.1088/1755-1315/178/1/012001

Mine Water Resource Utilization

Wei Bo Sun[1,2], Yan Wang[1,2], Lang Liu[1,2], Lei Ming Zhang[1], Xin Wang[1] and Yan Da Liu[1]

1 School of Energy Engineering, Xi'an University of Science and Technology, Xi'an, China
2 Key Laboratory of Western Mine Exploitation and Hazard Prevention, Ministry of Education, Xi'an, China
E-mail: sweibo@163.com

Abstract. In view of the influence and destruction of coal mining on water resources, from the concept of sustainable development and green mining, this paper explores the ways and means of mine water utilization in underground mining of coal mine. By using the storage of underground space formed by mining, people can purify mine water and combine mine heat source with mine water. By discussing the development trend of mine water, the key core technologies, methods and development of mine water resources utilization in the future are prospected.

1. Introduction
China ranks third in the world's coal reserves, and it has been the first largest coal producer in the world since the 21st century. China is a land composed of multiple tectonic plates that are spliced together by multiple geological structures. Geological conditions are very complicated. The complex geological conditions have made China one of the countries with the most serious water damage in coal mines in the world. With the continuous increase of mining depth, the reserves of coal resources in shallow and upper groups are gradually exhausted, and the problem of water damage of coal resource exploitation in the deep and lower coal groups is becoming increasingly serious. At the same time, the hydrogeological conditions of mine water filling are becoming more and more complex, and the control factors of water inrush are increasing, and the mechanism and type of water inrush are becoming more complex and changeable. According to the survey and statistics of the National Coal Mine Safety Supervision Bureau in 2012, in recent years, the actual drainage of coal mines is 7.17 billion m^3 per year, and the water inflow of 61 coal mines is more than $1000m^3/h$ in China [1]. Because of coal mining, the destruction of groundwater is about 8 billion m^3 each year, but the utilization rate of mine water is only about 25% [2]-[3]. The loss of mine water resources is equivalent to 60% of the annual industrial and living water shortage in China (10 billion t) [4]. How to keep the balance between coal mine exploitation and water resources protection and make good use of mine well water is one of the major problems in the development of coal science, and also one of the core contents of ecological civilization construction in coal mine area.

2. Underground Reservoir of Coal Mine
Underground reservoirs in coal mines (storage and utilization under well water well), that is, to store ground water in the space of the rock mass in the goaf formed by coal mining. The safety coal pillar is joined together by artificial dam, and the reservoir dam is formed [5]. Meanwhile, mine water storage facilities and water intake facilities are built at the right position. Make full use of natural purification

IOP Conf. Series: Earth and Environmental Science **178** (2018) 012001 doi:10.1088/1755-1315/178/1/012001

of goaf rock mass for the mine water, construct the underground coal mine reservoir project. This technology breaks through the original "interception method" water conservation concept, and adopts the idea of "guiding and storing". The method of dredging the mine water to the underground goaf for storage and utilization. This method avoids the efflux of evaporation loss, the high cost of surface water treatment plant construction and operation (Figure 1). It opens up a technical approach to coordinate the exploitation of coal with the protection and utilization of water resources.

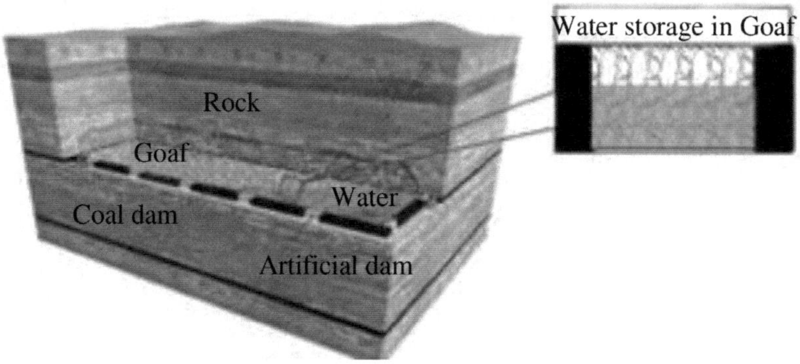

Figure 1. Underground reservoir of coal mine.

For multilayer coal, people can set up a number of underground reservoir, the reservoir can be water diversion. The reservoir should establish the monitoring and control system of water level and water quantity, realizing the water transfer between the reservoirs, ensuring the safety of the reservoir. (Figure 2).

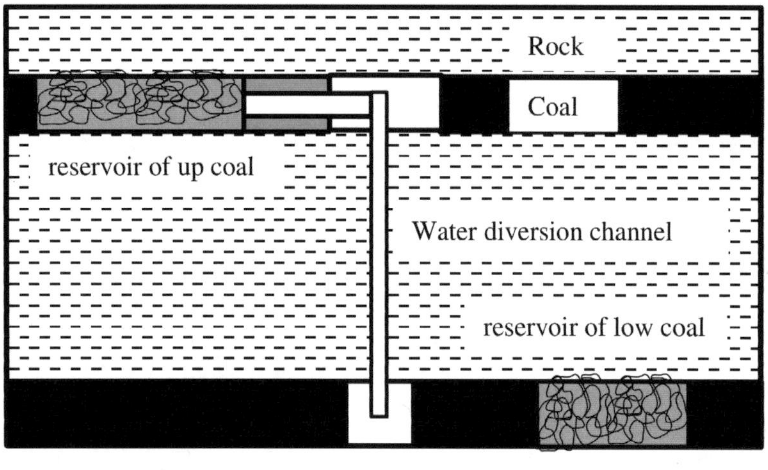

Figure 2. Water transfer of multi-layer underground reservoir

The coal mine underground reservoir technology has been used in Shendong mining area completing a total of 32 underground reservoirs of coal mine. In Shendong mining area, storage capacity of 31 million m^3, is currently the world's only coal mine underground reservoir. The underground reservoir of coal mine supply more than 95% of the water for the mining area, which ensures the sustainable development of the mining area. The technology of coal mine underground reservoir not only provides production, life and ecological water of Shendong mine, the only 2 billion ton mining area in the world, but also lets the vegetation coverage of the mining area increased by about 30%.

3. Utilization of Mine Water Heat Energy

The coal mining in the Middle Eastern part of China is mostly in the deep area of the underground. The depth of mines is generally above 500m [6]. With the increase of mining depth, there is "high temperature" mine water in the deep area, and the temperature of mine water in many coal mines is above 30 degrees, and even there is heat damage. Harnessing the heat damage of the mine is also becoming an important part of the workers' occupational health protection.

By using the technology of water source heat pump unit, the utilization of the "hot water" in the mine is combined with the heating of the mining area. On the one hand, the "waste water" of mine is fully utilized, and on the other hand, the problem of daily bathing heat and winter heating in coal mines is solved. It is a scientific and technological project, and it is also a project of environmental engineering, benefit engineering and sustainable development.

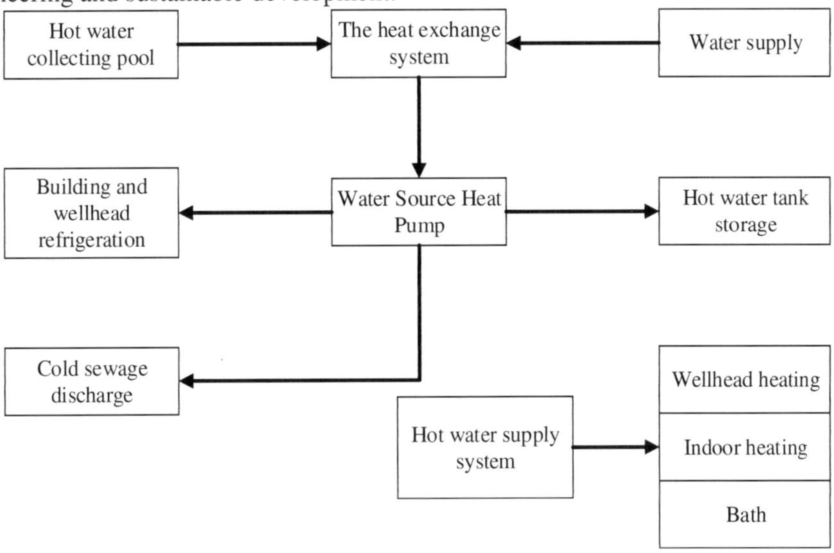

Figure 3. Water source heat pump equipment and system

For example, by using the technology of water source heat pump, the original boiler system is reformed by "hot water" in Xin Yi coal mine, solved the heating problem of mining area. After the use of hot water in the mine, the heat exchanged can not only meet the workers' bath, but also basically meet the winter heating and the wellhead heat preservation in the mining area. (Figure 3). In the Zhao Lou coal mine, through the implementation of the cooling water of the downhole refrigerator and the waste heat utilization of mine water, the collection and utilization of the waste heat of the underground mine are realized. Zhao Lou coal mine therefore reduced 3 coal-fired boilers of 10t/h, and save standard coal more than 3300t/a.

4. Mine Pumped Storage Power Generation

The principle of drop type underground pumped storage power generation is consistent with the principle of conventional pumped storage. The difference is that the water storage facilities are placed underground, such as abandoned coal mines or metal mines in roadways and mined out areas (Figure. 4) [7]. Compared with conventional pumped storage facilities, the water storage space has already existed, only need to do corresponding transformation, and does not consume the land resources of the surface and damage the environment of the earth's surface.

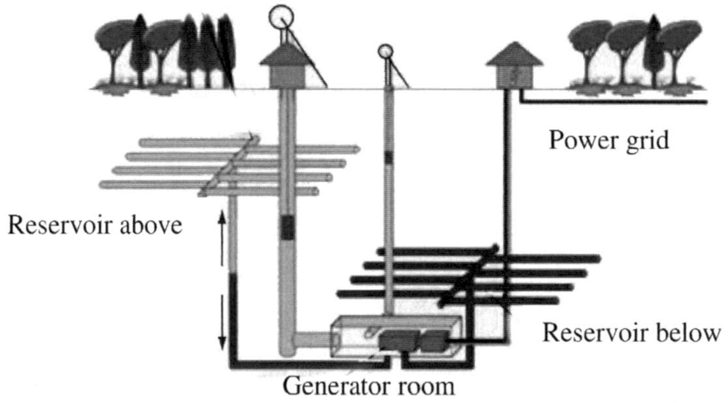

Figure 4. Sketch of underground storage facilities in metal and coal mines

In the northern plain area, it is difficult to find the natural high drop terrain conditions for building surface pumped storage power station. So underground pumped storage power plants are a good choice. Especially, the construction of pumped storage power generation facilities by using the mine or abandoned mine space can not only reduce the cost of construction, but also produce good economic, social and environmental benefits, and shorten the period of building the database. Many large and medium-sized coal mines and abandoned coal mines are distributed in Northeast China, North China, Inner Mongolia and Xinjiang province. If some suitable coal mines or abandoned coal mines can be converted into pumped storage power stations, they can not only provide necessary conditions for the development of wind energy and solar energy, but also make use of the existing underground space. At the same time, through increased stability, the modified underground space can avoid the large surface subsidence and collapse of the ground, and the follow-up problem of water and air pollution.

5. Conclusion

It is the basis and guarantee of mine water resources utilization that we should further strength the "coal water" dual resource type mine construction, development of theory and technology research, as well as "water control coal mining" core technology research and development. The inevitable way to realize the harmonious and sustainable development of the modern coal industry includes:

1 According to the characteristics of high water pressure and high water temperature in the conditions of deep coal mine, the mine resources are comprehensively utilized.

2 The mine water, underground temperature, which is regarded as damages several years ago, can be pre pumped, or be utilized by the ground source heat pump.

3 The development of coal resources and the utilization of water and heat resources are studied as a comprehensive system to build a coal mine-water-heat multi resource mine.

6. References

[1] Qiang W. Progress, problems and prospects of prevention and control technology of mine water and reutilization in China *J. Journal of China Coal Society,* 2014, **39**(5):795-805.

[2] Yuqun X. Present situation and prospect of groundwater numerical simulation in China *J. Geological Journal of China Universities*, 2010, **16**(1):1-6.

[3] Qiang W and Zhiqiang W. A research on an optimized five-in-one combination of mine water control, treatment, utilization, back-filling and environment friendly treatment. *J. Journal of China Coal Society,* 2010, **36**(2):109-112.

[4] Qiang W and Duo L. Research of "Coal-water" double-resources mine construction and development. *J. Coal Geology of China,* 2009, **21**(3):32-36.

[5] Dazhao G. Theory framework and technological system of coal mine underground reservoir *J. Journal of China Coal Society*, 2015, **40**(2):239-246.

[6] Heping X, Zhengmeng H, Feng G, Lei Z and Yanan G. A new technology of pumped-storage power in underground coal mine: Principles, present situation and future. *J. Journal of China Coal Society*, 2015, **40**(5):965-972.

[7] Lin M, Chaofeng D and Dengtao A. Application of mine water heat energy utilization system in large mining depth coal mine. *J Energy Technology and Management,* 2017, **40**(1):141-143.

Acknowledgments

The authors are grateful for the support provided by the National Natural Science Foundation of China (No.51674188, 51504182), the Natural Science Basic Research Plan of Shaanxi Province of China (No.2015JQ5187), the Scientific Research Program funded by the Shaanxi Provincial Education Department (No.15JK1466), the Project funded by China Postdoctoral Science Foundation (No.2015M582685), and the Xi'an University of Science and Technology Research Fund Program for Young Scholars (No.2014-NY-018), Training Fund of Xi'an University of Science And Technology (NO.2014059, NO.2014054), Dr. Start Foundation of Xi'an University of Science And Technology (NO.2014QDJ021, NO.2014QDJ040).

Evaluation of Plain River Network Hydrologic Connectivity Based on Improved Graph Theory

Yuqin Gao[1], Xuan Xiao[1], Mingming Ding[2], Yuqiang Tang[1] and Hongyu Chen[1]

1 College of Water Conservancy & Hydropower Engineering, Hohai University, Nanjing 210000, China;
2 Water Conservancy Bureau of Nanjing, Nanjing 210036,China

Abstract. Based on graph theory, the river network was generalized by a graph model. Considering the influence of water quantity on the hydrologic connectivity of river network, the connectivity factor reflecting water quantity index was taken as the weighted value, the quantitative evaluation model of hydrologic connectivity degree was setted up. By calculating the river network weighted connectivity degree with MATLAB, the quantitative evaluation of river network hydrologic connectivity was realized. Taking the Qinhuai River Basin as an example, the HEC-HMS hydrological model was established for the simulation of short-term and long-term flood. The connectivity level of the river network in the basin under different scale floods was graded by the quantitative evaluation results of connectivity degree. The results showed that the constructed evaluation model has good applicability and reliability.

1. Introduction

Changes in hydrologic connectivity pattern will affect the water cycle path and regulatory capacity of the river, thus affect the play of hydrologic connectivity functions[1]. The evaluation method of hydrologic connectivity has undergone a process from qualitative analysis to quantitative analysis. More and more attention has been paid to the research of water system connectivity evaluation in our country with the increasing demand of flood control and disaster reduction, water resource allocation and water environment improvement.

However, the study in this aspect is still in its infancy. Guanglai Xu[2] constructs the degree of patency matrix by taking the reciprocal of the hydraulic resistance as the weight of the edge and took the average of the degree of patency at all the vertexes as the weighted connectivity degree of the river network; Xing Chen[3] applies the graph theory separately to quantitatively evaluating the structural connectivity and hydraulic connectivity of the Yanjingwei Plain River Network, in Changshu City before and after river system planning; Erős[4] searches key river sections for protection and environmental management based on network analysis function of the graph theory, and using graph theory to quantify the connectivity of freshwater ecosystem habitats further, proving that graph theory can be used for the protection of freshwater ecological resources; Segurado[5] uses the relevant theory of graph theory to determine the obstacles that affect the structural connectivity of the basin, and then judges which obstacles should be cleared firstly to improve the overall connectivity more effectively. In this paper, the connectivity degree reflects the degree of connectivity of plain river network based on the improved graph theory.

2. Establishment of Evaluation Model

2.1. Fundamentals

Based on the graph theory, the river network graph model $G(V,E)$, which can be represented by an adjacency matrix $A(a_{ij})_{n\times n}$, a_{ij} is the number of edges directly connected from V_i to V_j. According to the judgement criteria, if all of the elements of the judgment matrix S are not zero, the graph G is a connected graph, otherwise it is a non-connected graph.

$$S(s_{ij})_{n\times n}=\sum_{k=1}^{n-1}A^k\left(a_{ij}^{(k)}\right) \tag{1}$$

2.2. Establishment of the Evaluation Model

The existing graph theory can only reflect whether the river channels are connected with each other, it can not reflect the dynamic transfer capacity of water quantity among river networks, and can not analyze the impact of water quantity on hydrologic connectivity either. To make up the above deficiencies, the improved graph theory was proposed in the paper (Figure 1).

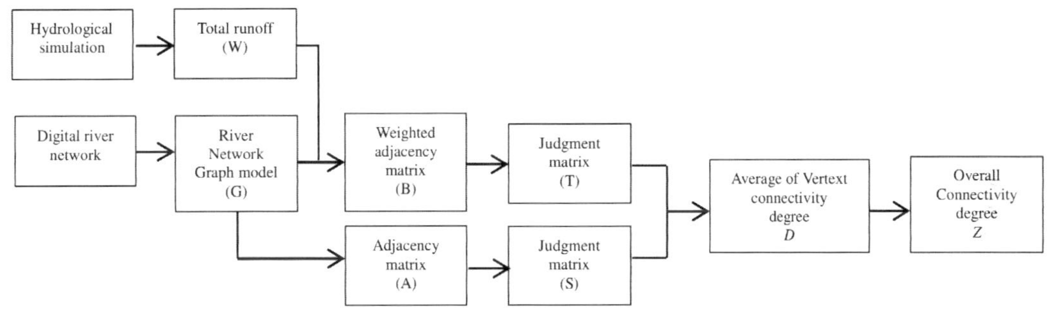

Figure 1. Calculation process of the improved graph theory.

For the graph with the weight, we can use the weighted adjacency matrix $B(b_{ij})_{n\times n}$ to represent. The element t_{ij} belongs to the judgment matrix T , which represents the sum of the connected factors whose length is $k(1,2,\ldots n-1)$ from the vertex V_i to the vertex V_j.

$$T(t_{ij})_{n\times n}=\sum_{k=1}^{n-1}B^k\left(b_{ij}^{(k)}\right) \tag{2}$$

2.2.1. Selection of the water quantity index. There are many indexes to characterize the water quantity in a river channel, such as runoff Q, total runoff volume W, runoff modulus M, runoff depth R, runoff coefficient α and so on[6]. Runoff coefficient α can indicate the relative relationship between rainfall and runoff, and can comprehensively reflect the impact of various natural factors on the runoff formation. Therefore, it is more practical to select runoff coefficient α as the index to reflect water quantity in a river.

2.2.2. Weight of the edge. It can avoid the problem that the complicated calculation process and the long calculation time by using runoff coefficient α , which can be used as the connectivity factor to characterize the weight of the river network model. Equation (3)~(5) are the calculation process of the connectivity factor b_{ij}.

$$W_i=\int_0^{\Delta t}Q_i dt \tag{3}$$

$$R_i=\frac{W_i}{A} \tag{4}$$

$$b_{ij} = \frac{R_i}{P_i} \tag{5}$$

Where: W_i is total runoff volume at the node of the river network, m^3; Δt is the study period, h; Q_i is the instantaneous runoff, $m^3 \cdot s^{-1}$; R_i is runoff depth, mm; A is the area of the study area, km^2; P_i is rainfall depth over the period, mm. It is determined by Thiessen polygons method, which assumes that precipitation in each sub-basin is represented by the nearest rainfall station.

2.2.3 Calculation of weight connectivity degree. The connectivity degree of each vertex is expressed by the average of all connected paths between V_i and V_j, and the equation is:

$$D_{ij} = \frac{t_{ij}}{s_{ij}} \tag{6}$$

Where: s_{ij} is the sum of the connected paths whose lengths are 1, 2, ... n-1 from V_i to V_j; t_{ij} is the sum of the connected factors whose lengths are 1, 2, ... n-1 from V_i to V_j.

The overall hydrologic connectivity degree is calculated as the mean of the average connectivity degree of all the vertices. The equation is:

$$Z = \frac{\sum_{i=1}^{n}\sum_{j=1}^{n} D_{ij}}{n^2} \tag{7}$$

3. Case Study

3.1. Study Area

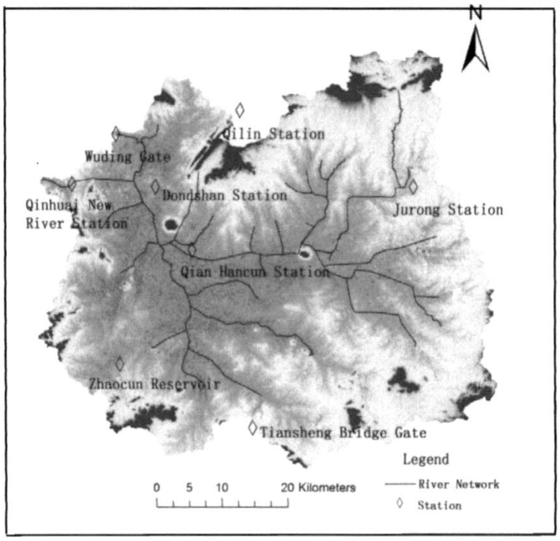

Figure 2. River distribution of Qinhuai River Basin

Qinhuai River Basin is located in the downstream of the Yangtze River in Jiangsu Province with a area of $2631km^2$. There are two sources of Qinhuai River, north source is the Jurong River, and south source originates the Lishui River in the Dong Lu mountian. The two rivers converge in the northwestern village of Jiangning District and become the main stream of the Qinhuai River. When flowing to Dongshan Town, they flow into two streams. The north branch flows along the main river channel in Nanjing, outflowing from Wuding Gate. The west branch is Qinhuai New River, who outflows from the Qinhuai New River Gate. Both of them flow into Yangtze River[7] (Figure 2).

3.2. Data Processing

The water system was extracted and the study area was divided into 18 sub-basins based on GIS. The Thiessen polygon was established by the distribution of rainfall stations in the basin (Figure 3). Due to the deviation between the water system extracted based on GIS and the actual river network, the extracted river network was imported into Google Earth for correction, the digital water system graph was obtained and further generalized to 25 main points river network topology graph (Figure 4).

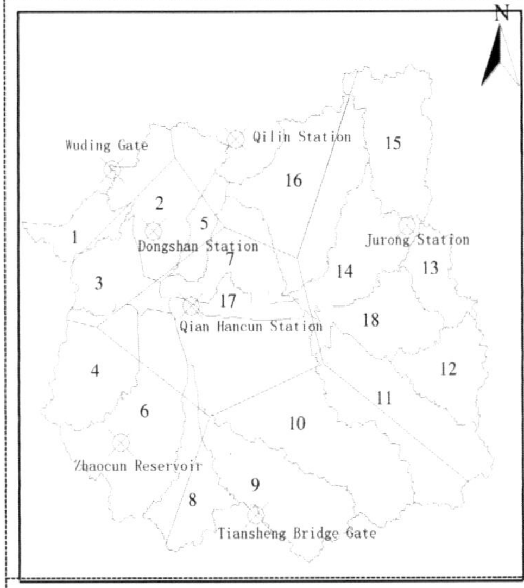

Figure 3. Qinhuai River sub-basin division and rainfall station Thiessen polygons

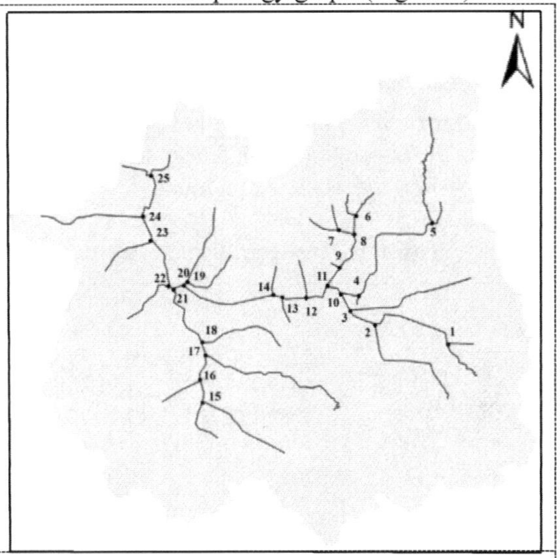

Figure 4. The main river network digital water system graph of the study area

3.3. Flood Process Simulation

HEC-HMS hydrological model can be used to simulate the short-term and long-term flood processes in the Qinhuai River basin. The relative error of peak discharges and total runoff volumes were both less than 20%, the Nash coefficient and correlation coefficient were both greater than 0.8, and the Nash coefficient of long-term flood processes were all above 0.7, so this model can meet accuracy requirements[8]. The comparison between simulated values and measured values of some flood processes were shown in Figure 5.

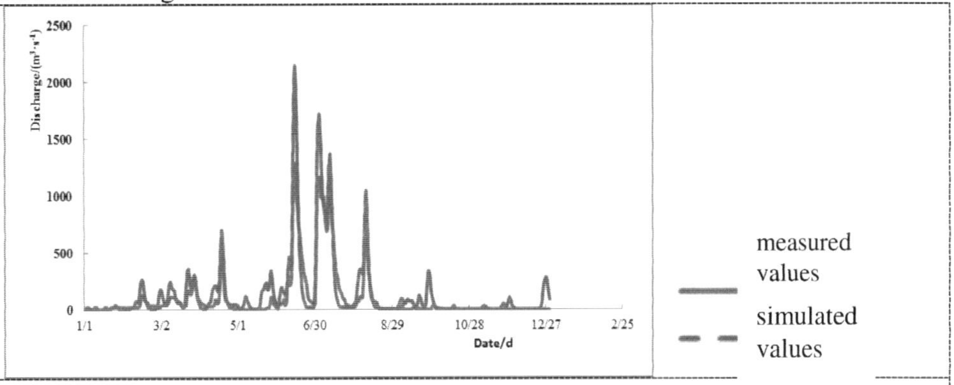

(a) 1991

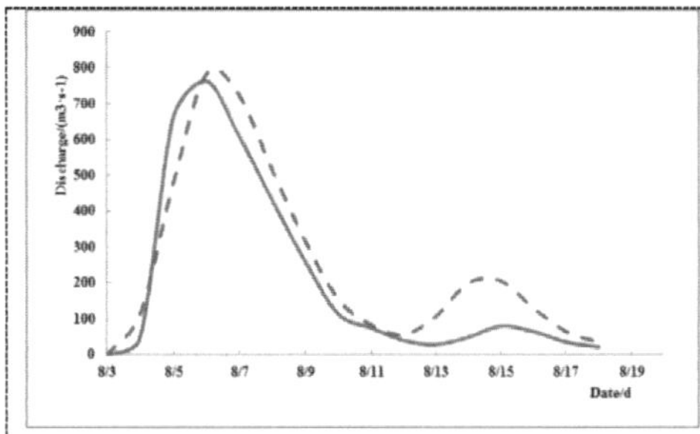

(b) 198908

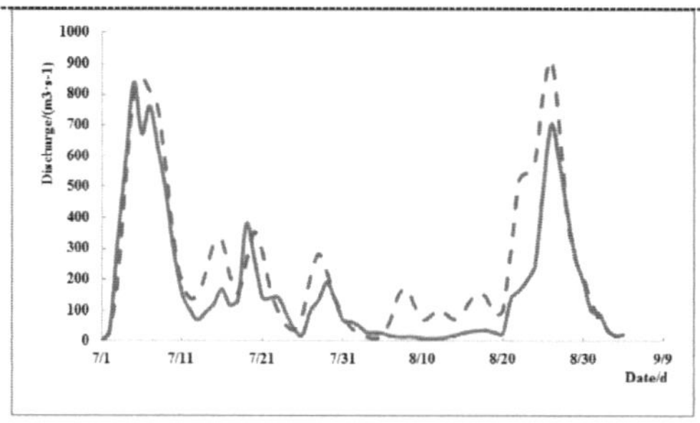

(c) 198707

Figure 5. Comparisons between simulated and measured values of some flood processes

Based on the improved graph theory method, the overall connectivity degree of the basin during some rainfall periods are as shown in Table 1.

Table 1. The overall connectivity degree of the basin during some rainfall periods

Flood Processes		Simulated Values of Peak Discharge ($m^3 \cdot s^{-1}$)	Simulated Values of Total Runoff Volume (mm)	Nash	Degree
Short-term Floods	199603	169.4	19.15	0.758	0.0013
	198908	782.7	138.04	0.888	0.0066
	199106	1336.9	400.53	0.751	0.0083
Long-term Floods	198707	910.2	564.72	0.705	0.0195
	200306	1376	803.46	0.784	0.0327
	199105	2058.7	1314.92	0.753	0.0545
	1991	2140.6	1827.39	0.702	0.0617

4. Results and Discussions

4.1. Relationship between Total Runoff Volume and the Connectivity Degree

As shown from Table 1, the connectivity degree tends to increase with the increase of volume, and a quantitative relationship in line with the two can be studied further. Taking total runoff volume as independent variable, connectivity degree as dependent variable, they were basically in line with the linear relationship .If confidence level α is 0.05 and degree of freedom is 5, the critical value $r_{critical}=0.754$, $r > r_{critical}$,indicating that there is a significant linear relationship between the two variables[9].

4.2. Classification of Hydrologic Connectivity

The level of hydrologic connectivity is divided into four classes, which are better, good, general and bad respectively in this paper. The better class indicates the level of hydrologic connectivity is high, which can meet demands of industry, agriculture and ecology. Other classes' level of connectivity are low relatively.

During the study period, a flood disaster broke out in 1991, taking the connectivity degree as the highest and using it as the standard value[10]. The evaluation criteria corresponding to different classes are shown in Table 2.

Table 2. The evaluation criteria corresponding to different classes

Classes	Evaluation Criteria
better	the standard value×70%≤the overall connectivity degree≤the standard value
good	the standard value×50%≤the overall connectivity degree < the standard

	value×70%
general	the standard value×25%≤the overall connectivity degree < the standard value×50%
bad	the overall connectivity degree < the standard value×25%

According to the evaluation criteria proposed in this paper, the classification of simulated short-term floods are bad. The reason may be that although peak discharge is large, the water quantity is small due to the short period of time, which can not meet the water needs of industry and agriculture. Therefore, the evaluation results are reasonable.

4.3. The Reliability Provement of the Constructed Evaluation Model

It can be seen from the calculation results that the river network of Qinhuai River Basin was connected within the study period and the connectivity degree was 1 based on the traditional graph theory, indicating that the river network can play normal connectivity function and can promote the economic development in this area. Considering the effect of water quantity on the connectivity, it can be seen from Table 1 that the overall connectivity degree $Z \in (0, 0.0617]$ and the connectivity degree of 1991 flood was 0.0617, which was the maximum value within the research period, and the connectivity effect was better. Under this flood, the connectivity factor was 0.984 at node 16, and it was 0.978 at node 15 in Figure 4, indicating that the water flew from node 15 to 16. The other nodes had the same characteristics, which were in line with the actual flow path of rivers and reflected the dynamic effect of water quantity transferrig between rivers.

5. Conclusions

Based on the evaluation model proposed in this paper, it proved that volume was positively correlated with the connectivity degree. Under the premise of meeting the needs of urban water supply, agricultural irrigation and ecological replenishment, the river channel can ensure the smooth flow if it can storage water to an extreme. Therefore, this model was of great guiding significance for the optimal dispatch of water in a basin.

Based on the corresponding evaluation criteria, it was of great significance to popularize the model to evaluate hydrologic connectivity of different scales floods in other basins. The corresponding standard values can be determined according to the hydrological characteristics in the basin, which was convenient and applicable.

The standard value of connectivity was determined at the annual scale in this paper. So, we can only evaluate the hydrologic connectivity under the annual scale of the river basin. The connectivity degree and evaluation of hydrologic connectivity under the longer time scale needs to be further studied.

6. References

[1] Biao Ru, Xing Chen, Qicheng Zhang, Weiping Wang and Qin Xu 2013 *Water Resources and Power* 31(5) 9-12
[2] Guanglai Xu, Youpeng Xu and Liuyan Wang 2012 *Advances in Water Science* 23(6) 776-781
[3] Xing Chen, Wei Xu, Kunpeng Li, Qilin Zhang and Song Lin 2016 *Water Resources Protection* 32(2) 26-29
[4] Tibor E, Dénes S and Robert S S. 2011 *Biological Conservation* 144(1) 184-192
[5] Pedro S, Paulo B and Maria T F 2013 *Ecology* 28(7) 1231-38
[6] Yuting Liang 2014 Study on the factors of rainfall-runoff coefficient. (Changsha: Hunan University) 26-42
[7] Huaizhi Wang, Yuqin Gao, Yu Yuan and Yuxue Guo 2017 *Journal of Water Resources & Water Engineering* 28(1) 81-87
[8] Yu Yuan, Yuqin Gao, Xi Wu. 2015 *J of China Three Gorges Univ.(Natural Sciences)*, 37(5) 34-39
[9] Yingzi Nan 2011 *Statistics and Decision* 2 26-28
[10] Qian Li 2012 Changing of spatial pattern and the hydrological response of urbanization in qinhuai river basin (Nanjing: Nanjing University) 37-64

Acknowledgments

The study was supported by the National Natural Science Foundation of China (No.51309076), the Fundamental Research Funds for the Central Universities (No.2014B05814). The Project was also funded by the Priority Academic Program Development of Jiangsu Higher Education Institutions.

IOP Conf. Series: Earth and Environmental Science **178** (2018) 012003 doi:10.1088/1755-1315/178/1/012003

The Evaluation of PPP Mode of Environmental Pollution Prevention and Control Based on the View of the Perspective of Sustainable Development

Zhu Rong[1] and Jiang Yalong[2]

1 School of Civil Environmental Engineering, Anhui Xinhua University, Hefei, China
2 School of Civil Environmental Engineering, Anhui Xinhua University, Hefei, China
E-mail:eternitylollipop@qq.com

Abstract: The reality of the water environment pollution has seriously affected people's production and life, and even threaten people's life and living environment, it is urgent that we should curb water pollution increasing trend. To promote the prevention and control of water pollution in the field of PPP (Public - Private - Partnership, a public-private partnership) project implementation, and to guide social capital to enter the field of water pollution control, can effectively solve the problems of water pollution, such as the shortage of funds, low operating efficiency. This paper first describes the current situation of PPP project research, then analyzes the feasibility of PPP project in water pollution prevention and treatment, and analyzes the existing problems in China, finally puts forward some targeted suggestions and measures.

1. Introduction

With the development of economy in our country, the pollution situation of domestic environment is becoming more and more serious. Although China's government financial investment is increasing, there is still a big gap between pollution control goals and environmental protection needs. The establishment of a market mechanism to attract social capital into the ecological environment protection and to carry out the innovation of environmental protection investment and financing, PPP mode has become a new type of cooperative organization relationship model of ecological environment governance. Water pollution control involves many factors such as resources, technology and economy, make the evaluation and analysis on the PPP mode of water environmental governance from the angle of sustainable development, will make a strong complement of the existing mode of PPP project in the depth.

2. Research Significance

From the perspective of sustainable development, the evaluation and analysis of PPP mode of water pollution prevention and control will help to improve the efficiency of resource allocation and pollution control, improve the enthusiasm of social capital, to form effective competition, and to reduce costs. It is helpful to alleviate the main problems that individual project is difficult to disperse and co-ordinate; it is conducive to the overall regional project collaboration and sharing of resources.

In addition, due to the large coverage of China's environmental problems, similar financing projects will be further increased with the development of society economy, the contradiction between the project financing and governance effect in the next period of time will exist[1]. Therefore, in the near future and long-term the PPP project of environmental pollution control under the premise of a wide range of promotion, to evaluate the research and optimize countermeasures is particularly important from the view of perspective of sustainable development.

3. Research Status and Trends in China and abroad

Since 2014, the government and social capital cooperation (PPP) model has received widespread attention[2]. The State Council and the Ministry of finance, development and Reform Commission and other relevant ministries have launched more than 20 documents, in energy, transportation, water conservancy, environmental protection, municipal engineering and other public services, to encourage the use of government and social capital cooperation mode, to attract the participation of social capital. PPP model has been developed for decades in China, but it is still a new thing in the field of environment[3], especially in the field of municipal sewage and garbage. Promoting the PPP mode in the environmental field will be beneficial to broaden the channels of environmental protection investment and financing, and have a greater impact on the environment of investment and financing pattern.

At present, as an effective means of pollution control, promoting environmental governance by the government and social cooperation mode, has been widely used in developed countries, some developed regions in China also began to promote and carry out exploration.

Now, the existing researches are mainly related to these subjects:

3.1. Research on PPP Mode

The PPP (Public-Private-Partnership, PPP) mode is that the public sector and the private sector provide public services, which is established by way of contract and based on long-term cooperation mechanism of risk sharing and benefit sharing (Deng Xiaopeng, 2007). Some foreign scholars had discussed the mode of mutual cooperation between the public sector and the private sector very early, and established a relatively complete theoretical framework[4]-[6]. In recent years, with the introduction of PPP mode, many public facilities in cities have achieved a series of successes through this method. More and more problems need to be paid attention to in the application of PPP mode. For example, in 2003 Zhang Tongjiang, Liu Erlie deeply researches and elaborates the system of PPP mode, pointed out that with the rapid rise of private enterprises, capital accumulation, will be introduced into the PPP mode of public facilities. In recent years, many domestic scholars have studied the PPP mode, including the theoretical basis, application methods, legal framework, risk sharing, pricing model and so on.

3.2. Research on PPP Mode of Water Environment Prevention and Control in Domestic

From the existing research data, at present, the research on PPP mode project are more, but related to environmental pollution, it is less, the reason is mainly because that PPP project in China is too late to start in the field of environmental protection (Yuan Tang Lu, 2015). However, with the promotion of PPP mode of water environment prevention and control in recent years, the relevant content has gradually attracted the attention of scholars. Existing research data focused on the feasibility of PPP mode in environmental governance (Mo Lu Group, 2016; Lin Donghai, 2016; Liu Jun, 2016). With the policy oriented in nearly a year, it has gradually started to increase PPP project application in environmental pollution control efforts in domestic. According to the "13th Five-Year" plan, PPP mode is expected application in all fields in 2017, it will cover the construction of infrastructure and public services (Tang Chuan, 2017).

3.3. Evaluation on PPP Mode of Water Environment Prevention and Control from the View of Perspective of Sustainable Development

The relationship between environmental protection and economic development is a game, which is bound to exist between the contradictory subjects. Environmental protection is inseparable from economic development, economic development is bound to require environmental protection, the establishment of a market based cooperation mechanism is an effective way to solve environmental protection and sustainable economic development (Lei Qinli, 2010). From the point of sustainable development, the analysis of environmental pollution control is an effective means to evaluate the effectiveness of social development and environmental governance. The evaluation index system of sustainable development and its index are more and more recognized as an effective evaluation tool in practice, and also an effective tool for policy formulation and public communication (Singh R K, 2009). Most of the current researches focus on the application and review of PPP mode in environmental pollution control. On the basis of the feasibility study of environmental pollution control by the third party, the existing projects are discussed (such as Liu Jun, 2016; Zhao Baoqing, 2016; Yin Tailing, 2016).

Overall, due to a late start, at present the evaluation research on PPP mode of water pollution prevention, especially from the angle of sustainable development, are very few. In applied research, the research on water environment pollution control is less and stays on the conceptual framework level, and there is no further analysis on the sustainability of water environment pollution control. Therefore, in-depth study of practical applications may be particularly important and urgent.

4. Feasibility of Introducing PPP Mode into Water Environment Prevention and Control Project

4.1. Economic Growth

After thirty years of development, China's economic development has made great achievements, China's folk has accumulated enormous wealth of society. by the end of 2014, China's savings rate is as high as 50%, of which private savings accounted for the proportion of total savings reached 90%, the savings amount reached 48 trillion yuan. In the government limited financial resources, private capital investment potential does not effectively guide the social development, the participation of private capital investment to the prevention and control of water pollution is an effective way to promote private investment. Research Report by the world bank research institutions show that when the water pollution control investment accounted for the proportion of fiscal expenditure 0.5%-0.7% interval, water pollution can be prevented, when the time reaches 0.7%-1% water environment quality can be improved significantly. In 2015 April, "water pollution prevention action plan" is formally introduced, opening a water pollution control market with 2 trillion investment feast.

4.2. The Government Has Created a Favorable Institutional Environment for the Implementation of the PPP Mode for Water Pollution Control Projects

The government has been encouraging private capital to enter the field of public utilities, and has issued a series of laws and regulations to create a good system and policy environment for the introduction of PPP mode. In 2015 May, the Ministry of finance, Ministry of environmental protection jointly issued "Opinions on the implementation of government and social capital cooperation (PPP) in the field of water pollution prevention and control", put forward to perfect the relevant rules and norms, optimize the public-private cooperation mechanism design, supply mode transformation, improve management mode, accelerate the prevention and control of water pollution. The PPP mode should be popularized in the field of water pollution prevention and control, and the PPP mode should be taken into the first place in the areas where the key water pollution basins and the capital gap are prominent[7].

4.3. Technological Progress Has Created a Competitive Environment for Private Capital to Enter the Field of Public Goods Supply

The progress of science and technology to reduce the monopoly of public utilities, natural monopoly supply of public goods and services is relative, with the progress of technology, the market structure is changing, the natural monopoly of the strength of the trend will change. The supply of public goods is broad and the demand for funds is huge. With the progress of technology and the establishment of property right system, it is possible to introduce the PPP mode into the public goods supply field and introduce the market competition mechanism[8]. The field of water pollution control through the introduction of the PPP model, the advantages of the public and private sectors, to solve the water pollution prevention and control of the single investment structure, low operation efficiency of water pollution control projects, high operating costs, such as the poor quality of service.

5. Main Problems in PPP Mode of Water Pollution Control Project

5.1. The Laws and Regulations are Imperfect and Not Unified

1. A nationally consistent organization which is responsible for PPP mode application is lacking

In developed countries, such as Europe and America, they set up a professional organization to be responsible for the improvement of the laws and regulations of the PPP mode. For example, the United States set up public-private partnerships Commission, which is responsible for the problems of system, laws when it encountered in the process of implementation of the PPP project and the regulations to conduct a comprehensive, scientific, authoritative interpretation[9].

2. A national uniform law which is applicable to the PPP model is lacking

The Lacking of a national level of unified regulations, easily lead to contradictions between the PPP mode of water pollution control project and other laws or regulations. This is common in china. The general solution is to report to the NDRC and the State Council, which greatly increases the cost of cooperation between the two sides.

5.2. Risk Sharing is Unreasonable

Because the PPP project construction and operation cycle is long, the amount of investment is large, and the components of the participants are more complex, it will cause many uncertainties, namely risk in PPP project construction and operation process as shown in Figure 1.

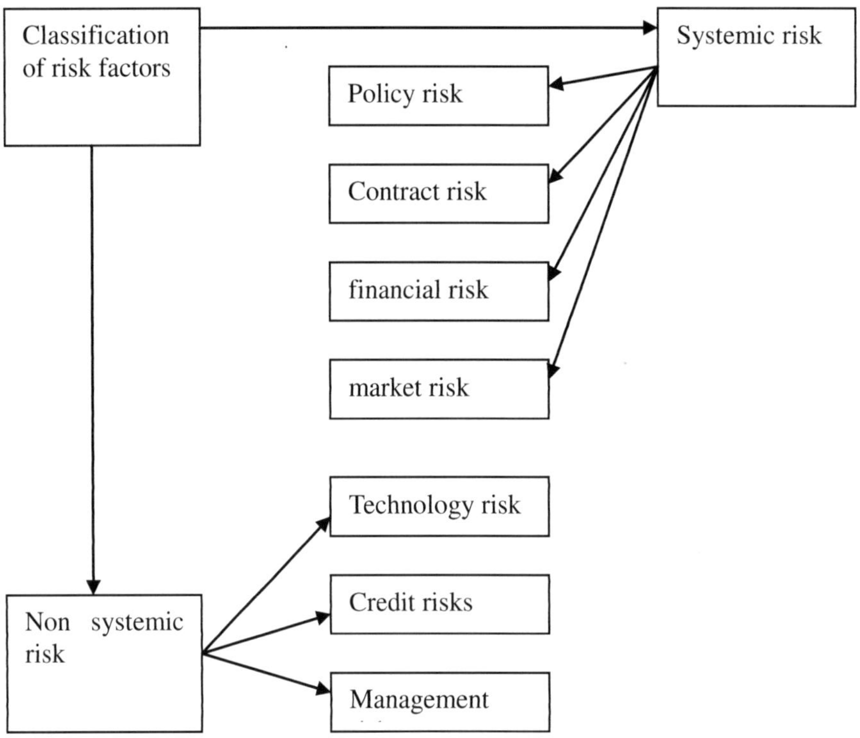

Figure 1. Classification of risk factors of PPP

5.3. The Role Change of Government is Difficult

The Chinese government is always the role of the production government. In the past, the government has carried out all round intervention in financing, construction and operation of water pollution control projects. But in the PPP mode, the government needs to change their roles, the role of the government is not exhaustive, but provides the key implementation of policies and regulations and system guidance and support. From the direct investment, construction and operation management of the water pollution prevention project to the supervision of the project, the government changed from "paddle" to "steering". For the government, it is not only the change of the system, but also the change of the long-term idea.

6. Policy Recommendations for Improving PPP Mode of Water Pollution Control Project

6.1. Improve the Relevant Laws and Regulations System

A scientific, complete, clear, effective and unified system of laws and regulations is the necessary condition for the successful implementation of the PPP mode. Because of water pollution in the life cycle of PPP project faces the risk of the change of policy change, it is necessary to establish a perfect system of laws and regulations, in order to avoid the financial difficulties which are brought by inconstant policy[10].

6.2. Change the Government's Role Indeed

In the PPP mode, the role of the government should be divided into two perspectives:
(1) Macro role
a. Political risk control.
Political risk mainly includes two aspects: first, the government change will bring the validity of the commitment.
The two is the stability risk of the government economic policy, and the change of the government's fiscal and tax policies. The government should control the risk, first they can set the system mechanism to ensure the effectiveness of government commitment, and then establish a compensation mechanism to deal with the risk due to policy changes brought to the PPP project.
b. Improve relevant laws and regulations
As a new mode of public-private partnership, the PPP mode involves complex aspects, such as guarantee, finance, taxation, contract, entrustment, etc.. The government should improve the relevant laws and regulations to lay the foundation and premise for the operation of PPP project[11].
c. The government should cultivate the industry and guidance the policy
The PPP mode has the advantage that other cooperative models do not have. For our country, it is still in a stage of introduction and absorption. The government should assume the guidance and promotion of PPP model in china.
(2) Micro role
a. Providing support
From a micro perspective, supporting PPP projects is one of the responsibilities of the local government, including economic support, political support and other support.
b. Cooperation and supervision
The government has changed from the role of former project leader to supervisor and collaborator, and the role of cooperation is that the government should take the management work to the private sector[12]. At the same time, another important role of the government is supervision, government supervision includes two aspects: first, the supervision of water pollution prevention and control of PPP project financing, construction, and the use of funds. Two is for water pollution prevention and control PPP project itself, the government should to supervise the effect of pollution control.

6.3. Reasonable Determination of Price

For the prevention and control of water pollution PPP project, if the water pollution treatment cost and reclaimed water price are not reasonable, it will be a serious threat to the existence of PPP project. It the price is too low, it will affect the enthusiasm of private investment; if the price is too high, it will damage the interests of the public.

6.4. Encourage Public Participation

Water pollution prevention and control PPP project has a strong utility spillover, water pollution prevention and control is in line with the interests of the public. People are demanding the clean water. Public participation is to a great extent play a supervisory role, although the government to some extent represents the interests of the public, but they are subject to government management, some policy bottlenecks. the participation of social groups which represent the interests of public are lack, it is difficult to guarantee the interests of the public to make people get good maintenance in operation management process.

7. Reference

[1] E.S.Savas. Privatization and Pubic Private Partnerships.2000. *New York: Chatham House,* 187-191

[2] Dennis D C,Erik D. An ex ante bidding model to assess the incentive creation capability of a public–private partnership pipeline. 2016.*International Journal of Project Management*, 341

[3] Yanli Z, Rui Z,Shengfeng G. A new PPP algorithm for deformation monitoring with single-frequency receiver.2014. *Journal of Earth System Science*, 12:38

[4] Po-Han C,Robert L.K. T. Analysing the structure of public–private partnership projects using network theory.2011. *Abu Naser Chowdhury, Construction Management and Economics* . 3 L 123

[5] Shuibo Z,Ying G,Zhuo F,Weizhuo S. PPP application in infrastructure development in China: Institutional analysis and implications.2014. *International Journal of Project Management* 234

[6] Rui M,Martin d J, Joop K. The rise and fall of Public–Private Partnerships in China: a path-dependent approach.2010. *Journal of Transport Geography*. 4 L527

[7] Matti S. Public-Private Partnership Networks: Exploring Business-Government Relationships in United Kingdom Transportation Projects.2011. *Economic Geography* .

[8] Zhijia T.Capacity and toll choice of an add-on toll road under various ownership regimes.2012. *Transportation Research Part E* . 6 L 121-123

[9] Xiao H J,Guo Z. Modelling optimal risk allocation in PPP projects using artificial neural networks.2010. *International Journal of Project Management* . 5 L 342

[10] Li Y. Tang Q S, Eddie W.L. A review of studies on Public–Private Partnership projects in the construction industry.2009 . *International Journal of Project Management* . 7 L 134

[11] Jeffrey K. Pinto, Dennis P. Slevin, Brent E. Trust in projects: An empirical assessment of owner/contractor relationships.2008. *International Journal of Project Management* . 6 L 427

[12] Abu N C,Chotchai C. Factors influencing finance on IPP projects in Asia: A legal framework to reach the goal.2008.*International Journal of Project Management* . 1L 228

IOP Conf. Series: Earth and Environmental Science **178** (2018) 012004 doi:10.1088/1755-1315/178/1/012004

Summary Study on the Development of Water Assessment

Sudi GAO[1] and Rong ZHU[2]

1 Associate professor, Anhui Xinhua University : School of Civil and Environmental Engineering, China
2 Lecturer, Anhui Xinhua University : School of Civil and Environmental Engineering, China
E-mail: gaosudi@163.com

Abstract: Throughout the development process of the water quality evaluation method, the paper outlines several water quality evaluation methods, summarizes them and points out the advantages and disadvantages. Finally the paper points out that water quality evaluation method will be further improved with the wide application of computer technology combined with various subjects of mutual cross-penetration.

1. Methods Introduction

Water quality assessment evaluates qualitatively or quantitatively the water quality elements on some water environment monitoring data and in accordance with the evaluation criteria and evaluation methods. The water quality assessment is the basis for the calculation of the capacity of the water environment and water resources planning and management. It is also a prerequisite for the water control planning. Water quality assessment results can determine the water quality model and provide parameters and information for model identification and parameter identification[1]. It can accurately reflect the water quality situation, understand and master the water pollution impact and development trends and provide a scientific basis for the protection of the water environment and water resources planning and management. Water quality assessment is not only the water quality of the scientific description of reality, but also points out the objectives and direction for water pollution control programs. We can build the appropriate water quality model for water pollution control programs by the information provided by the water quality assessment.

In the 1960s, Jacobs proposed the water quality index of concepts and formulas of the water quality evaluation. Because it is simple and easy to operate, until today people still use it in the domestic water quality evaluation practice. Later, water quality experts at home and abroad also raised various comprehensive water quality index methods. Which the Nemerow mode, mean square error mode[2] and vector mode[3] are more typical. Since the 1980s, Some uncertainty, the emergence of the nonlinear theory method and the promotion and application of computer technology provides a theoretical basis and technical support for the new model of water quality assessment[4]. The WQA-SPA model couples fuzzy logic theory and set pair analysis theory, combining identity、discrepancy and contrary(IDC). According to this model, the water level recognition results and the membership of various levels can be calculated by using the river water quality data.Some of the commonly used methods can be summarized into the following categories throughout the development process of the water quality evaluation method,

2. Index Evaluation Method

Index evaluation method can be divided into single-factor evaluation and comprehensive evaluation method according to the number of evaluation projects selected by the monitoring data. Single factor evaluation method is also known as "single-indicator evaluation method" and "one vote veto method". Index evaluation is an evaluation method based on monitoring data of the physical and chemical parameters of water quality, monitoring data divided by the evaluation criteria data results the sub-indices, the water pollution index can be obtained through mathematical computation and we can use it as a water quality evaluation scale. The standard index is a representative of the single factor evaluation methods, it can be clearly pointed out where the problem is water quality but the shortcoming is that the quality status of water environment can not be given. In order to overcome the deficiencies of the method, domestic and international water quality expert propose various comprehensive water quality index methods, among them, the Nemerow index method and the weighted-type method are more typical[1].

Nemerow index method which is proposed by U.S. scholars Nemerow. Now this method is used more often.It is a balanced extreme value or highlight the maximum weighted multi-factor environmental quality index. The Nemerow Index specifically considers the most polluted factor and avoids the influence of subjective factors in the weight coefficient.

3. Fuzzy Theory Evaluation Method

Because of the vagueness and the randomness of the river water environment systems, fuzzy theory evaluation method classifies objective things through the establishment of fuzzy similarity relations. In the classification of water quality assessment, research methods commonly use the clustering analysis based on equivalent fuzzy relations or fuzzy graph theory tree clustering method[5]-[7]. The improved fuzzy hybrid cluster analysis method has been successfully used to finish the lower reach's water quality assessment of the Fengman hydropower station. Basing on fuzzy mathematics and fuzzy relation composite theory, fuzzy comprehensive evaluation quantifies some ill-defined, difficult to quantitative factors, and thus to achieve a comprehensive evaluation. In the fuzzy evaluation methods, fuzzy pattern recognition method is a fuzzy set theory's evaluation method of which develops rapidly in the field of environmental assessment in recent years [8] [9]. At present, this method has been widely used in various types of water quality assessment.

With the large-scale application of gray theory in environmental systems, the gray relational analysis method of the evaluation of water quality is used. The concept of Gray relational analysis is clear and the evaluation method is convenient and operable. But its evaluation of the value tends to be homogenized, the resolution is not high. For this shortfall, on this basis, by calculating the conditional extremum, the eutrophication gray evaluation model, the water quality assessment of gray pattern recognition model, the water quality assessment of gray situation decision-making model and the water quality assessment of gray relation projection model are established.

4. Mathematical Statistics Evaluation Method

In recent years, constantly domestic researchers use mathematical statistics methods to evaluate water quality conditions. However, mathematical statistics often requires a large amount of measured data to support the result, it does not apply to the smaller capacity situation of the water samples.

The 25 surface water quality conditions was evaluated by using principal component analysis method. Rivers and lakes water quality was evaluated by using profiling techniques method. To promote and deepen the principal component analysis, factor analysis method also joined the water quality assessment[10]. In the comprehensive assessment of water pollution situation of the Huangshui river valley, several statistical methods were synthetically used[11]. First using the factor analysis method to evaluate, sort, and analyze the degree of water pollution of the water to be evaluated, studying each section of water pollution similarity by using cluster analysis, then the classification processing results were obtained.

The water quality assessment model based on the set pair analysis theory (WQA-SPA) for evaluating water quality of the river mainstream is proposed[12]-[14]. The WQA-SPA model couples fuzzy logic theory and set pair analysis theory, combining identity、discrepancy and contrary(IDC). According to this model, the water level recognition results and the membership of various levels can be calculated by using the river water quality data.Using the index weight vector W and contact matrix, the water quality of the sampling points evaluated by the weighted composite method. The following formula:

$$\mu_k = \sum_{k=1}^{4} W_k \mu_{rk} \qquad (1)$$

The water level recognition results of the sampling points can be calculated by equation. Therefore, this evaluation model is the water quality assessment model based on the set pair analysis theory (WQA-SPA)[15].

5. Artificial Neural Network Analysis Method

The artificial neural network is a non-linear science which rises suddenly in the 1980s. In the environmental assessment, common methods in the evaluation study for river water quality and Lake water eutrophication are feedforward BP neural network model based on error back propagation algorithm and the feedback neural network model with association function-Hopfield model[12]. It tries to simulate the human brain's characteristics, such as self-adaptability, self-organization and fault tolerance, a better effect has been got in the field of medical diagnostics and automatic control by Analyzing and judging things. For the standard BP algorithm is slow convergence, poor stability, and easy to fall into local minima and other shortcomings, the multi-layer feedforward BP neural network constructed with radial basis function (RBF) was used to evaluate the water quality and achieved good results[16] [17]. In addition, the combination of artificial neural networks and fuzzy mathematical theory may constitute a fuzzy neural network model of the lake water eutrophication assessment. The artificial neural network simulates the human way of thinking and intuitively infers the judgment simply the essential characteristics of things.

6. Conclusion

Rivers, lakes and other surface water bodies and groundwater is constantly contaminated with the increase of our population, economic development, and the accelerating process of urbanization. Pollution control of rivers has attracted wide attention. Water quality evaluation plays an important role in the prevention of water pollution. Its requirements are getting higher and higher. With the development of science and technology and the wide application of computer technology, combined with various subjects of mutual cross-penetration, water quality evaluation method will be further improved and innovative.

7. References

[1] Peng Wenqi. *Theories and methods of modern water quality evaluation.* Beijing: Chemical Industry Press, 2005,33~39.(in Chinese)

[2] Xu Hongkai. *Half set of variance of water quality evaluation model.* Environmental Science, 1985, 6(4): 55~57(in Chinese)

[3] Shen Baocheng, Xu Guoyi. *Hilbert space theory in the comprehensive evaluation of the environmental pollution.* Environmental Science,1981 ,2(4):57~60(in Chinese)

[4] Li Ruzhong. *Progress and trend analysis of theoretical models of water quality assessment.* Journal of Hefei University of Technology, 2005,28(4):369~373(in Chinese)

[5] Wu Kaiya, Jin Juliang, Pan Zhengwei. *Urban flood vulnerability contact number assessment model based on triangular fuzzy number cut set*[J]. Journal of Hydraulic Engineering, 2010, 41(6) : 712-718.

[6] Li Xiangzhong. *The application of cluster analysis in the classification of water quality.* Fuzzy Mathematics, 1984,(1):85~90(in Chinese)

[7] Fu Yanpeng, GaoJiarui. *Fuzzy Mathematics in the Water Quality Assessment.* Wuhan: Huazhong Institute of Technology Press, 1986:35~70(in Chinese)

[8] Zhao Wujing, Wu Kaiya, Jin Juliang. *Flood control engineering safety assessment model based on set pair analysis and variable fuzzy set*[J]. Water Resources and Power ,2007, 25(2) : 5 - 7, 13.

[9] Bai Ximeng, Man Chunsheng. *Fuzzy graph tree clustering method in the evaluation of environmental quality.* China Environmental Science, 1985,5(6):38~42(in Chinese)

[10] Chen Shouyi, Xiong Deqi. *Fuzzy set theory and model of lake eutrophication.* Journal of Lake Sciences, 1993,5(2):144~152(in Chinese)

[11] ChenSY, XiongDQ. *Fuzzy set theories and methods for municipal environmental quality assessment[*A].Proceeding of International Conference on City Development[C].Beijing: China Environmental Science Press,1993:24~30
[12] Chen Dongjing,Ma Anqing,Xu Dongmin. *Factor analysis in water quality assessment.* Hydrological ,2002,22(3):29~31(in Chinese)
[13] Wang Xiaopeng. *The application of multivariate statistical analysis in the evaluation of river pollution.*Systems Engineering Theory and Practice, 2001,21(9):118~123(in Chinese)
[14] Li Zuoyong. *Quality and nutritional status evaluation based on BP network model and the effect of test.*Acta Scientiae Circustantiae, 1995,15(2):186~191(in Chinese)
[15] Zhao Keqin. *Set pair analysis and its preliminary application*[M].HangZhou: Zhejiang science and Technology Press,2000: 10-40.
[16] Liu Guodong,Huang Chuanyou. *The artificial neural network model of comprehensive evaluation of water quality.* China Environmental Science, 1998,18(6):514~517(in Chinese)
[17] Dong Manling,Huang Shengwei. *Radial basis function neural networks in water quality assessment.* Environmental Science and Technology,2003,26(1):23~25(in Chinese)

Acknowledgements:

This work was financially supported by the quality engineering project of Anhui Xinhua University (NO.2015xqjdx01) and the Natural Science Research Projects of Anhui Educational Commission (NO.KJ2016A302).

IOP Conf. Series: Earth and Environmental Science **178** (2018) 012005 doi:10.1088/1755-1315/178/1/012005

A Water Trading Model Based on the Construction of Water Bank with Considering Multiple Uncertain Information

Xueting Zeng[12], Tienan Li[1]*, Shoujie Zhang[1], Zhenjiang Si[1], He Dong[1]

1 Heilongjiang Provincial Hydraulic Research Institute, Heilongjiang, 150080, China;
E-mail: zxt1231@sina.com
2 Capital University of Economics and Business, Beijing 10070, China;

Abstract. On the context of population growth and industrialization development, water demands of China present an increasing tendency, which results in conflicts between human activities and water resources. Particular in an arid region, the water deficit has attracted policymakers' attentions, which requires a more effective water resources allocation method to satisfy the harmony between human being and natural resources. Thus, a "quasi-market" approach (mixed market approach or government method) is developed for water allocation in an arid region; meanwhile, water bank would be introduced to improve water trading. An interval stochastic-fuzzy programming (ISFP) is introduced to tackle uncertainty by probability distribution and interval number; meanwhile it can hedge the risk by the simulation process of water trading. Based on analysis of feature of arid region, a sustainable water resources strategy and mode can be provided to coordinate the human activity and resources utilization.

1. Introduction

The high speed of population growth and economic development can lead the shrinking of available water resources in China, which would motivate water crisis today. The decision-makers have confronted big challenge for maintaining sustainable between human development and limited water resources, particularly in some arid regions. More efficient schemes for water resources management such as water trading have been required, with aim to allocate water resources optimally [1]. Although trading mechanism can increases the economic productivity of water by encouraging its movement from low to high valued, water is a quasi-good, which can not be allocated by market value completely [2]. Therefore, a "quasi-market" approach (mixed market approach or government method) can be established for water allocation in an arid region; meanwhile, water bank would be introduced to improve water trading [3]. However, multiple uncertainties and corresponding complexities in a water trading system would increase the risk of decision-making, which have essentially placed them beyond the conventional deterministic optimization methods.

System analysis and optimization method can be deemed as effective tools to support decision-making processes in water trading with considering multiple uncertainties [4]. Among these, two-stage stochastic mathematic programming (TSP) is an effective manner to handle randomness characterized by probability distribution (such as water flow), where the expected target can be recoursed by an examination of random water availability [3]. Meanwhile, through the construction of linkage between policies and the economic penalties, TSP can reflect complexities of system uncertainties as well as analyzing policy scenarios when the pre-regulated targets are violated [4], [5]. However, the TSP method has difficulty in identifying the probability distribution of coefficient, which should be traduced an interval parameter mathematic programming into TSP optimization framework [6]. Moreover, many uncertainties (e.g., economic data and water availabilities) are often not obtained as probability distributions, which make deterministic parameters of water availability mislead or bias the decision makers easily. Therefore, credibility constrained programming (CCP) can be introduced to measure in fuzzy water trading system with the confidence levels, which would be useful in the presence of weaker sources of information [6], [7]. Previously, few reports in the consideration of multiple uncertainties previously, which requires a hybrid method to deal with uncertainties and economic penalties in a water trading system based on trading mechanism.

Therefore, an interval stochastic-fuzzy programming (ISFP) for water trading is developed for water resources management in arid regions of China. The proposed method can not only reflect the tradeoffs between resource benefits and penalties, but also deal with uncertainties expressed as intervals, probability distributions and fuzzy membership functions. The ISFP method will be applied to one of the arid regions in Northwest China. The model can help releasing more water resources to join into water trading based on bank, which can reduce the loss of water deficiency and gain a high net system benefit. Local economic targets, the decreasing of water permit and water trading ratio under varied levels of system-failure risk will be generated, which not only help decision makers optimally allocate water resources, but also gain insights into the tradeoffs between water trading and economic objective.

2. Modeling Formulation

In a water trading system, policymakers have allocated water permit target to each user in the beginning of this year according to the situations of water usage last year. However, the random water availability would result in water target being not met, which would be adjusted by policymakers. Since trading mechanism has introduced to improve the productivities of water from low value to high value, water permits can be allocated to by law of price. However, water is a quasi-public good, which should be allocated to satisfy the basic water demand of drink safety and food secure. Thus, water permit can be allocated to satisfy the basic demand of water user, then, the surplus water permit can be traded. Meanwhile, water bank is introduced into water trading, which can provide more water source participating into water market to allocate water more effectively and equally. However, multiple uncertainties may exist in water allocation and trading process, which affect the water resources management planning to be made. Thus, water trading model based on water bank can be formulated as follows:

$$\max f^{\pm} = \sum_{i=1}^{I}\sum_{j=1}^{J} B_{ij}^{\pm}W_{ij}^{\pm} - \sum_{i=1}^{I}\sum_{j=1}^{J} BC_{ij}^{\pm}BW_{ij}^{\pm} - \sum_{i=1}^{I}\sum_{j=1}^{J}\sum_{h=1}^{H} P_h C_{ij}^{\pm}Y_{ijh}^{\pm} + \sum_{i=1}^{I}\sum_{j=1}^{J}\sum_{h=1}^{H} P_h (TB_{ij}^{\pm} - TC_{ij}^{\pm})TY_{ijh}^{\pm} \tag{1}$$

Where $f^{\pm}$ is total system benefit; i is type of water user (1 is municipal; 2 is agricultural; 3 is industrial; 4 is ecological; j represent various district; h is water level (1 to 3 are high, medium and high level); $B_{ij}^{\pm}$ is the net benefit of one volume water being delivered to user i in district j; $W_{ij}^{\pm}$ is water target of user i in district j; $C_{ij}^{\pm}$ is loss of one volume water not being delivered to user i in district j; P_h is probability of water shortage; $Y_{ijh}^{\pm}$ is amount of water shortage; $BW_{ij}^{\pm}$ is initial water permit to satisfy the basic water demand in district j; $BC_{ij}^{\pm}$ is delivering cost; $TY_{ijh}^{\pm}$ is trading amount; $TB_{ij}^{\pm}$ is remedying cost from water trading; $TC_{ij}^{\pm}$ is trading cost.

$$\sum_{i=1}^{I}\sum_{j=1}^{J}(Q_{ijh}^{\pm} - BW_{ijh}^{\pm}) \le T_{ij}^{\pm}, \quad \forall i,j \tag{2a}$$

$$\sum_{i=1}^{I}\sum_{j=1}^{J} TY_{ijh}^{\pm} \le T_{ij}^{\pm}, \quad \forall i,j \tag{2b}$$

$$TC_{ij}^{\pm} + TS_{ij}^{\pm} \le C_{ij}^{\pm}, \quad \forall i,j \tag{2c}$$

$$\sum_{i=1}^{I}\sum_{j=1}^{J}(W_{ij}^{\pm} - Y_{ijh}^{\pm} + TY_{ijh}^{\pm}) \le Q_{ijh}^{\pm}, \quad \forall i,j,h \tag{2d}$$

$$\sum_{i=1}^{I}\sum_{j=1}^{J} TY_{ijh}^{\pm} \le (1-g) * \sum_{i=1}^{I}\sum_{j=1}^{J} S_{ijh}^{\pm} \le T_{ij}^{\pm}, \quad \forall i,j,h \tag{2e}$$

$$S^{\pm}_{ijh} \le (R^{\pm}_{ij} + V^{\pm}_{hij} - E^{\pm}_{ij}) - R^{\pm}_{ij\min}, \quad \forall i, j, h \tag{2f}$$

$$0 \le Y^{\pm}_{ij} \le Q^{\pm}_{ijh}, \quad \forall i, j \tag{2g}$$

$$W^{\pm}_{ij}, BW^{\pm}_{ij} \ge 0, \quad \forall i, j \tag{2h}$$

Where $T^{\pm}_{ij}$ is available water permit for trading; g is the reducing percent of water trading from water bank; $S^{\pm}_{ij}$ is available water from water bank under various water level; $R^{\pm}_{ij}$ is initial value of water in water bank; $V^{\pm}_{ijh}$ is water inflow into water bank; $E^{\pm}_{ij}$ is water loss in water bank; $R^{\pm}_{ijm\,in}$ is minimum water demand for water bank.

3. Case Study

Kaidu-kongque River Basin contains six counties with area of 62 × 10^3km^2, which has the population more than one million [8]. It is a typical arid region, where the water supply capacity of river is quite low due to extremely dry climate, low rainfall, and high evaporation, which can not satisfy water demands of 4 users in 6 counties easily. Water trading based on water bank can solve conflicts caused by water shortage, which not only improve the net system benefit. In study basin, under allocated water permit, since trading can release surplus water to join into water bank, water bank include three types of water (e.g., underground water, reservoir water and released surplus water). Competitive four users (e.g., municipal, agricultural, industrial, and environmental users) in six counties will sell released water or buy water from water bank to remedy water shortages and gain higher benefits. **Figure 1** present the relationship between water manager and users in water trading based on water bank.

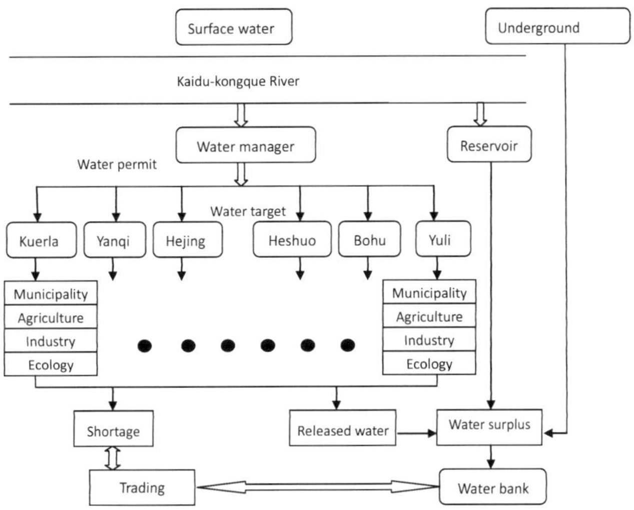

Figure 1. Framework of water trading based on water bank

However, a number of variations caused by factitious factors and natural factors exist in the trading system, which brings more complexities and uncertainties in water trading system. These complexities could become further compounded by not only interactions among the uncertain parameters but also their economic implications. The proposed ISFP model can be used for optimally allocating limited water to facilitate the regional sustainability. The parameters inputs for ISFP is based on calculation from field surveys, statistical data, and related research works previously. Table 1 shows water policy data and water target, in which is acquired by the water permit of water authority of Uygur Autonomous Region from 2005 to 2010 directly [8]. Table 2 shows economic data and trading costs. In order to appraise diverse net system benefits under different water permit and trading ratio in water trading system, numbers of scenarios with water permits (i.e., the value will change from 0% to 20%) and trading ratio (i.e., the value will change from 100% to 10%) will be assumed.

IOP Conf. Series: Earth and Environmental Science **178** (2018) 012005 doi:10.1088/1755-1315/178/1/012005

Table 1. Water targets and water permits

District		User			
		i = 1 Municipality	i = 2 Agriculture	i = 3 Industry	i = 4 Ecology
Water target (unit: 10^6 m^3)					
j = 1	Kuerle county	[8.80, 14.00]	[258.00, 275.00]	[53.30, 620.00]	[56.00, 76.00]
j = 2	Yanqi county	[6.00, 8.20]	[158.00, 165.00]	[28.00, 39.00]	[31.00, 47.00]
j = 3	Hejing county	[2.40, 4.30]	[81.00, 88.00]	[16.00, 20.10]	[14.70, 23.00]
j = 4	Heshuo county	[0.24, 0.50]	[9.70, 10.10]	[1.78, 2.25]	[1.28, 2.60]
j = 5	Bohu county	[2.20, 4.30]	[75.00, 85.00]	[15.60, 19.00]	[13.70, 23.00]
j = 6	Yuli county	[4.60, 6.00]	[110.00, 120.00]	[21.60, 27.00]	[24.00, 33.00]
Allocated allowable water permit (unit: 10^6 m^3)					
j = 1	Kuerle county	[9.72, 13.75]	[261.60, 275.08]	[55.44, 61.89]	[59.56, 75.65]
j = 2	Yanqi county	[[6.64, 8.14]	[158.33, 162.87]	[30.31, 36.65]	[32.26, 44.79]
j = 3	Hejing county	[2.89, 4.23]	[82.22, 84.50]	[15.70, 19.01]	[17.08, 23.24]
j = 4	Heshuo county	[0.36, 0.53]	[9.11, 9.38]	[1.89, 2.11]	[1.69, 2.70]
j = 5	Bohu county	[2.56, 4.09]	[78.89, 81.84]	[15.78, 18.41]	[15.58, 22.58]
j = 6	Yuli county	[4.94, 5.87]	[110.56, 117.41]	[22.33, 26.42]	[25.56, 32.29]

Table 2. Economic data

District	User			
	i = 1 Municipality	i = 2 Agriculture	i = 3 Industry	i = 4 Ecology
Trading fix cost (unit: US$/$10^3$ m^3)				
j = 1 to 6	[3050, 3150]	[550, 650]	[2400, 2600]	[280, 350]
Trading variable cost (unit: US$/$10^3$ m^3)				
j = 1 to 6	[1200, 1350]	[700, 800]	[150, 200]	[100, 150]

4. Result Analysis

In this study, five scenarios corresponding to different water permit levels and ten scenarios corresponding to different water trading ratio levels were examined by the ISFP model in Kaiduo-kongque River Basin. Different λ levels achieved different water availabilities, water targets, and varied water shortage. In general, actual water allocation would be target minus shortage, which can be impacted by random stream condition with an associated probability level. **Figure 2** shows that a higher λ level (higher credibility satisfaction) can lead lower water availability, which results in a lower allocation, leading a lower benefit, vice versa. Solutions for optimized net system benefits show that the sum of the first-stage benefit from the water allocation and the second-stage random losses of water deficiency.

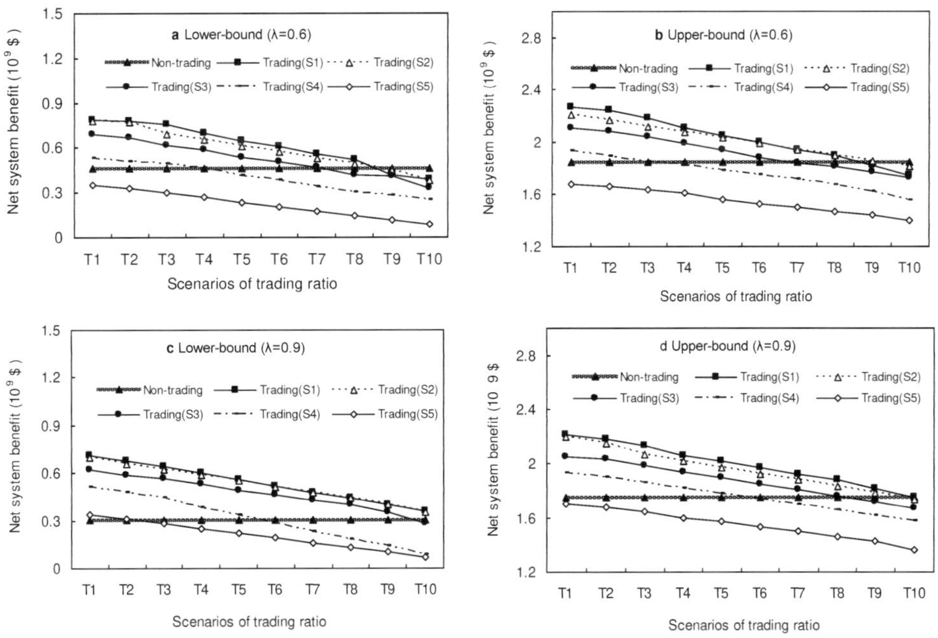

Figure 2. System benefits under various decreasing water permit levels / trading ratio (λ=0.6 and 0.9)

Figure 3 presents water shortages under various policy cases (S1 and S4). Due to water deficiency in study region, water permits have been allocated in advance to satisfy the basic water demand; then the surplus can be traded based on water bank, so as to saving more water or getting higher benefits. By the decreasing of water permits, less water would be allowanced but more water would be released to trade, which led much more shortage under lower water permit than that under higher water permit.

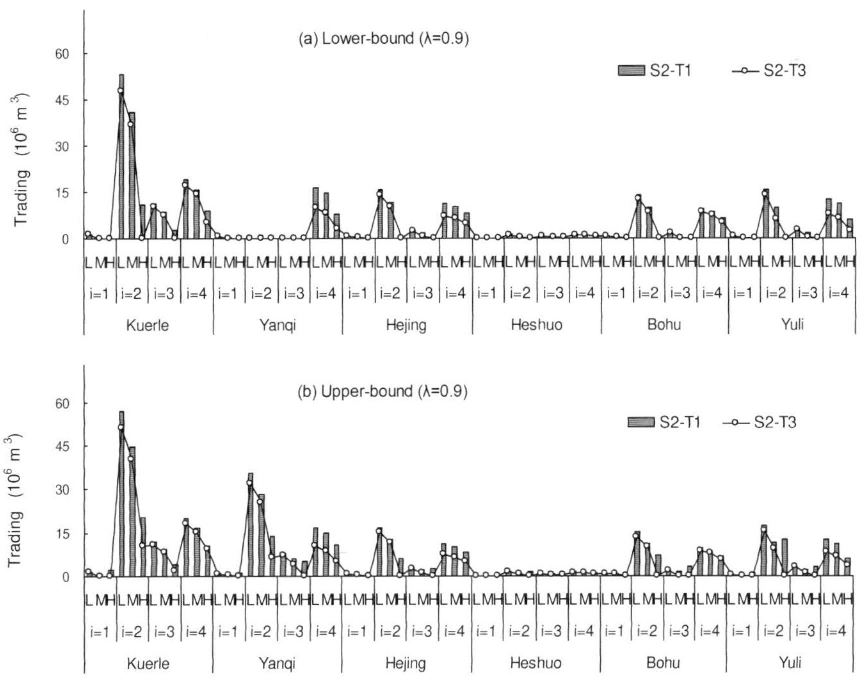

Figure 3. water trading from water bank under S1-T1 and S4-T1 (λ=0.6)

IOP Conf. Series: Earth and Environmental Science **178** (2018) 012005 doi:10.1088/1755-1315/178/1/012005

With S2-T1, amount of water trading of ecological users in Kuerle county (j = 1) would be [18.98, 19.97], [15.73, 16.77] and [8.92, 10.37] × $10^6 m^3$ at low, medium and high levels under S2-T1 (λ = 0.6), while it would be [17.07, 17.98], [14.16, 15.09] and [4.99, 9.34] × $10^6 m^3$ under S3-T2. The results indicated the more water surplus remedy water shortage, the less water trading from water bank. Meanwhile, the results demonstrate that net system benefits under trading were much higher than non-trading when decreasing permit levels were from 0 to 15%, but lower than non-trading when decreasing permit exceeded 15%. It implied that markets can provide incentives to adopt water-saving, since market prices make the opportunity cost of water explicit to users. Therefore, water trading was considered an effective way to not only reduce the shortages of water systems, but also gain a higher net system benefit in arid region.

5. Conclusions

In this study, an interval stochastic-fuzzy programming (ISFP) has been developed to quantify the effectiveness of water trading, which can deal with system uncertainties between allocation policy and violated risk. The developed method has been applied to Kaidu-kongque watershed for water trading under uncertainties of water supply and demand, which a number of scenarios were listed to compare net system benefits and water shortages. Because trading can reduce the "penalty", but increase "trading cost", system net benefit needed to be compared to get the optimal results based on the actual need such as saving water or maximizing system net benefit. The analysis shows that the effectiveness of a trading program is sensitive to water permit and trading ratio. Modeling results indicated that trading is more effective than non-trading under some designated situations. Although application of ISFP model to water trading is a new attempt, which could also be improved by introducing other advanced optimization methods to tackle more uncertainties and complexities in the future.

6. References

[1] Kondili, E., Kaldellis, J.K. & Papapostolou, C. 2010 A novel systemic approach to water resources optimisation in areas with limited water resources. Desalination, 250(1), 297-301.

[2] Calatrava, J, & Garrido, A. 2005 Spot water markets and risk in water supply. Agri. Econ. 33(2), 131-143.

[3] Luo, B., Maqsood, I. & Gong, Y. 2010 Modeling climate change impacts on water trading. Sci. Total. Environ. 408(9):2034-2041.

[4] Li, Y.P. & Huang, G.H. 2009 Fuzzy-stochastic-based violation analysis method for planning water resources management systems with uncertain information. Inform. Sciences. 179(24), 4261-4276.

[5] Huang, G.H. & Loucks, D.P. 2000 An inexact two-stage stochastic programming model for water resources management under uncertainty. Civ. Eng. Environ. Syst. 17(2), 95-118.

[6] Guo, P., Huang, G.H., Zhu, H., & Wang, X.L. 2010 Two-stage programming approach for water resources management under randomness and fuzziness. Environ. Model. Softw. 25(12), 1573-1581.

[7] Zadeh, L.A. 1978 Fuzzy sets as a basis for a theory of possibility. Fuzzy. Sets. Syst. 100(S1), 3–28.

[8] The statistical yearbook of Xinjiang Uygur Autonomous Region in Uygur Autonomous Region 2005-2010 (SYXU, 2005-2010). China.

Acknowledgments

This research was supported by the water rights confirmation project of Heilongjiang, Beijing social science fund project (Grant No. 17LJC010) and the start project of young teacher of CUEB.

IOP Conf. Series: Earth and Environmental Science **178** (2018) 012006 doi:10.1088/1755-1315/178/1/012006

Improvement on Entropy Weighting Model in Groundwater Quality Evaluation

Yan Feng[1, 2],Zhang Xinglei[1], Xiao Chengzhi[1], Tang Bei[1], Liu Chenglin[1,2]

1 School of Civil Engineering and Architecture, Nanchang University, Nanchang, 330031, China;
2 Key Laboratory of Poyang Lake Environment and Resource Utilization (Nanchang University), Ministry of Education, Nanchang, 330031, China
E-mail: yfmilan@163.com

Abstract. Entropy weighting model (EWM) is a widely used weighting method in groundwater quality assessment. EWM assigns weight according to the information amount principle. For the indicator with a higher dispersion degree, a larger weight parameter is assigned; and vice versa. However, through multiple practical applications, we find that the conventional EWM often neglects the indicator with high pollution degree, for it only takes the statistics characteristics of the observation data into consideration, but ignores their practical meaning in aquatic environment management. To solve this problem, this study makes an improvement on the conventional EWM through introducing a pollution degree principle. And then, the improved EWM is applied into the groundwater quality evaluation in Suihua city as an illustration. The results show that: (i) the conventional EWM theory neglects the indicator with high pollution degree, and makes the synthetic assessment result distorted and over-optimistic. (ii) Through introducing a pollution degree principle, the improved EWM highlights the heavily polluted indicators and excludes the "zero weight" phenomenon, which makes the final comprehensive groundwater quality assessment much stricter and more reasonable. (iii) The comprehensive groundwater quality of Suihua city is "Marginal", and its crucial pollutants are iron, manganese and ammonia nitrogen.

1. Introduction

Entropy weighting model (EWM) is a widely used weighting method in groundwater quality assessment [1]. EWM is designed by sociologists based on statistics and information theory [2]. For the indicator with a higher dispersion degree, which indicates that a larger amount of information is contained, a larger weight parameter is assigned; and vice versa [3].

EWM was firstly introduced into groundwater quality evaluation by Lu in 2010 [4]. And the results showed that EWM assigned weights according to the observation data, and none subjective factor was contained, which avoided the difference in weights parameters introduced by subjective experiences [4]. Since then, EWM has got a widely application in the groundwater management [5]-[8].

However, through multiple practical applications, we find that the weights parameters assigned by the conventional EWM are not always reasonable, for it often neglects the importance of the indicator with high pollution degree, which makes the synthetic assessment result over-optimistic.

To solve this problem, we make an improvement on the current EWM theory in this study. And to test the effectiveness of the improved EWM, it is applied into the groundwater quality evaluation of Suihua city as an illustration.

2. Materials and Methods

2.1. Study Area

As is illustrated in Fig. 1, Suihua city is an important industrial city and transportation hub in northeast China [9]. The land area of Suihua city is 35211 km^2, and the population is 5.2 million [9]. Groundwater is the major water resource for Suihua city [9]. To afford the industrial, commercial, and residential water demand, approximately 15×10^4 m^3 groundwater is pumped every month [9].

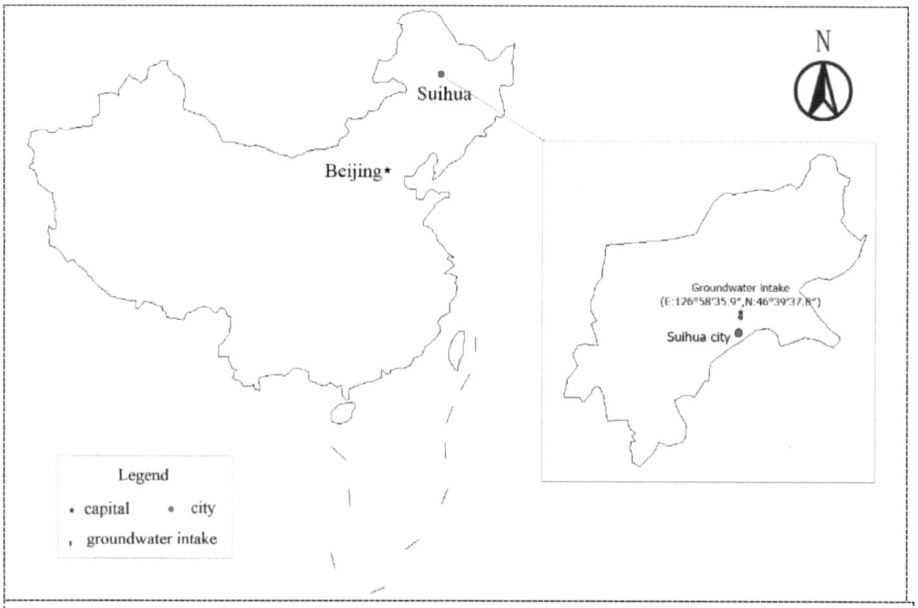

Figure 1. Figure with short caption (caption centred).

To protect the groundwater environment condition, the Ministry of Environmental Protection of China published the National Environmental Quality Standards of Groundwater in 1993. According to this standard, there are 22 indicators should be evaluated in the groundwater quality assessment, which are illustrated in Table 1[10].

Table 1. The assessment indicators and their acceptable domain

indicator	acceptable domain	indicator	acceptable domain
pH*	6.5-8.5	nitrate nitrogen	≤20
total hardness	≤450	nitrite nitrogen	≤0.02
sulfate	≤250	ammonia nitrogen	≤0.2
chloride	≤250	fluoride	≤1
manganese	≤0.1	total cyanide	≤0.05
iron	≤0.3	total mercury	≤0.001
copper	≤1	arsenic	≤0.05
zinc	≤1	selenium	≤0.01
volatile phenol	≤0.002	cadmium	≤0.01
anionic surfactant	≤0.3	chromium (VI)	≤0.05
permanganate index	≤3	lead	≤0.05

*pH is a dimensionless indicator; while the units of the other indicators are mg/L.

To investigate the groundwater quality condition, the Environmental Protection Bureau of Suihua City makes a detailed environmental monitoring at the groundwater intake every month, and publishes these data as water quality monitoring reports since 2017, May [11]. And in this study, all the data are cited from these monthly reports published by the Environmental Protection Bureau of Suihua City.

2.2. Synthetic Evaluation Model

In this study, we use the weighted Canadian Council of Ministers for the Environment Water Quality Index (weighted CCME WQI) model to evaluate the groundwater quality condition of Suihua City. Weighted CCME WQI is the improvement of the conventional CCME WQI model through introducing weights parameters [12]-[14]. Denote the number of indicators as m, the number of monitoring times as n, the threshold of the ith indicator as c_i , and the observation data of the ith indicator at the jth monitoring time as x_{ij}, respectively.

For the indicator the larger the better, if x_{ij} is less than c_i, it is determined to be failed, and the overproof degree e_{ij} is calculated by

$$e_{ij} = \frac{c_i}{x_{ij}} - 1. \tag{1}$$

For the indicator the smaller the better, if x_{ij} is larger than c_i, it is determined to be failed, and the overproof degree e_{ij} is calculated by

$$e_{ij} = \frac{x_{ij}}{c_i} - 1. \tag{2}$$

And then, weighted CCME WQI quantifies the synthetic water quality condition from three factors: scope (F_1), frequency (F_2) and amplitude (F_3), which represent the percentage of failed indicators, the percentage of failed tests, and the average overproof condition, respectively [12]-[14]. Use y_{ij} =1 to represent that x_{ij} is failed, and y_{ij} =0 represent that x_{ij} is failed, respectively. Scope (F_1) is generated by

$$F_1 = \sum_{i=1}^{m} \left(w_i \cdot \max_{j=1,\ldots,n} \left\{ y_{ij} \right\} \right) \times 100. \tag{3}$$

In Eq. (3), w_i is the weight of the ith pollutant.

Frequency (F_2) is calculated by

$$F_2 = \frac{\sum_{i=1}^{m}\sum_{j=1}^{n}\left(w_i \cdot y_{ij}\right)}{n} \times 100. \tag{4}$$

And amplitude (F_3) is generated by

$$F_3 = \frac{\sum_{i=1}^{m}\sum_{j=1}^{n}\left(w_i \cdot e_{ij} \cdot y_{ij}\right)}{\sum_{i=1}^{m}\sum_{j=1}^{n}\left(w_i \cdot e_{ij} \cdot y_{ij}\right)+n} \times 100. \tag{5}$$

All of scope (F_1), frequency (F_2) and amplitude (F_3) are the factors the smaller the better; and their values lie in the domain [0,100]. To make a comprehensive and intuitive representation of the water quality condition, CCME WQI uses the cube root principle to generate the final index [12]-[14]:

$$\text{CCME WQI} = 100 - \frac{\sqrt{F_1^2 + F_2^2 + F_3^2}}{1.732}. \tag{6}$$

According to the values of CCME WQI, the synthetic water quality condition is classified into five categories, which are illustrated in Table 2.

Table 2. Categories of CCME WQI

Category	CCME WQI	Description
Excellent	[95, 100]	The water quality is protected with a virtual absence of threat or impairment. And the aquatic environment condition is very close to pristine levels [12]-[14].
Good	[80, 95)	The water quality is protected with only a minor degree of threat or impairment. And the aquatic environment condition rarely departs from desirable levels [12]-[14].
Fair	[65, 80)	The water quality is usually protected but occasionally threatened or impaired. And the aquatic environment condition sometimes departs from desirable levels [12]-[14].

Marginal	[45, 65)	The water quality is frequently threatened or impaired. And the aquatic environment condition often departs from desirable levels [12]-[14].
Poor	[0, 45)	The water quality is almost always threatened or impaired. And the aquatic environment condition usually departs from desirable levels [12]-[14].

2.3. Entropy Weighting Model (EWM)

EWM is an application of information theory in management science [15], [16]. For the indicator with a higher dispersion degree, which indicates that a larger amount of information is contained, a larger weight is assigned; and vice versa [16]. In EWM, the entropy of the *i*th pollutant is defined as H_i:

$$H_i = -\frac{1}{\ln n} \cdot \sum_{j=1}^{n} \left(\frac{x_{ij}}{\sum_{j=1}^{n} x_{ij}} \ln \frac{x_{ij}}{\sum_{j=1}^{n} x_{ij}} \right). \tag{7}$$

H_i is a measurement of pollutant's information amount, its domain is [0,1]. The higher H_i is, the less its information amount is, and a smaller weight should be assigned. On the other hand, the less H_i is, the higher its information amount is, and a larger weight should be assigned [15], [16]. Based on this principle, the weights parameters are generated according to

$$w_i^* = \frac{1 - H_i}{m - \sum_{i=1}^{m} H_i}. \tag{8}$$

The current EWM is established in social sciences. It only takes the statistical characteristics of the observation data into consideration, but neglects its practical meaning. For example, assume that all the monitoring data of chromium (VI) in each month are 1mg/L. Obviously, chromium (VI) is significantly overproof and needs special attention. However, based on the current EWM, its weight is 0, for none discrimination is contained in its observation data. As a result, the indicator with high pollution degree is ignored, which makes the synthetic assessment result over-optimistic.

2.4. Improvement of Entropy Weighting Model

As is introduced in Section 2.3, the current EMW only takes the amount of the pollutant's information into consideration, but neglects its practical significance in environment management. This principle makes the weight parameter cannot represent the pollutant's importance correctly. To solve this problem, this study improves the current EWM theory through introducing a pollution degree weighting principle. Firstly, calculate the pollution degree of each observation data based on Eq. (9)

$$p_{ij} = \begin{cases} (x_{ij} / c_i)^{1/2} & \text{For the indicator the smaller the better} \\ (c_i / x_{ij})^{1/2} & \text{For the indicator the larger the better} \end{cases}. \tag{9}$$

In Eq. (4), p_{ij} is the pollution degree of x_{ij}. $p_{ij}>1$ indicates that x_{ij} is failed; while $p_{ij}\leq 1$ indicates that x_{ij} is acceptable. The larger p_{ij} is, the higher its pollution degree is; and vice versa. As a result, the pollution degree weighting principle can be defined as

$$w_i^{**} = \frac{\sum_{j=1}^{n} p_{ij}}{\sum_{i=1}^{m} \sum_{j=1}^{n} p_{ij}}. \tag{10}$$

In most cases, the weights parameters generated by Eq. (8) and Eq. (10) are different. Therefore we take their average value as the comprehensive weight result.

$$w_i = \frac{w_i^* + w_i^{**}}{2}. \tag{11}$$

Obviously, the improved EWM model assigns weight based on not only the indicator's discriminative degree, but also the indicator's pollution condition. For the indicator with a high discriminative degree and a high pollution degree, a larger weight is assigned; and vice versa.

3. Result and Discussion

3.1. Observation Data

According to the monthly groundwater quality report published by the Environmental Protection Bureau of Suihua City [11], all the observation data are illustrated in Table 3.

Table 3. The observation data of each indicator

Indicator	May	June	July	August	Indicator	May	June	July	August
pH*	7.09	6.91	7.10	7.10	nitrate nitrogen	0.018	0.018	0.018	0.016
total hardness	183	176	186	189	nitrite nitrogen	0.009	0.009	0.009	0.016
sulfate	4.46	4.56	6.40	4.73	ammonia nitrogen	0.466	0.491	0.487	0.473
chloride	3.79	3.86	1.83	5.50	fluoride	0.190	0.245	0.229	0.424
manganese	1.40	1.41	1.34	1.57	total cyanide	0.004	0.004	0.004	0.004
iron	1.34	1.32	1.39	1.46	total mercury	0.00004	0.00004	0.00004	0.00004
copper	0.001	0.001	0.001	0.001	arsenic	0.0039	0.0054	0.0046	0.0053
zinc	0.001	0.001	0.001	0.001	selenium	0.0004	0.0004	0.0003	0.0004
volatile phenol	0.0003	0.0003	0.0003	0.0003	cadmium	0.0001	0.0001	0.0001	0.0001
anionic surfactant	0.05	0.05	0.05	0.05	chromium (VI)	0.004	0.004	0.004	0.004
permanganate index	1.31	1.68	1.56	1.62	lead	0.001	0.001	0.001	0.001

*pH is a dimensionless indicator; while the units of the other indicators are mg/L.

As is illustrated in Table 3, there are 3 failed indicators: iron, manganese, and ammonia nitrogen. All of the other indicators are controlled within the acceptable domain.

In these failed pollutants, manganese is the worst indicator. Its observation data vary from 1.32mg/L to 1.46 mg/L, which is about 13 times larger than its acceptable threshold. Iron is the indicator with the second highest pollution condition, its observation data lie in the domain [1.34mg/L, 1.57mg/L], which is about 4 times larger than its acceptable threshold. Ammonia nitrogen is better than iron, its observation data vary from 0.466mg/L to 0.491 mg/L, which is about 2 times larger than its acceptable threshold.

The high pollutions of manganese and iron are caused by the hydro geochemical environment of the study area. The groundwater of Suihua city is pumped from the Quaternary gravel and sand porous phreatic aquifer [9]. The primary minerals in this aquifer contain large amount iron and manganese [9]. However, the pollution of ammonia nitrogen is introduced by human factors. According to the research of Li, most of the livestock farms, poultry farms, and the toilets in countryside are lack of seepage control measures [17]. As a result, the large amount of the residents' and animals' excreta seeps into underground reservoirs in rainy season, which brings high concentration of ammonia nitrogen into the groundwater.

3.2. Study Area

According to the conventional EWM, the weights parameters are generated and listed in Table 4.

Table 4. The weighting parameters generated by the conventional EWM theory

Indicator	Entropy	Weight	Indicator	Entropy	Weight
pH	1.0000	0.0004	nitrate nitrogen	0.9991	0.0067
total hardness	0.9997	0.0019	nitrite nitrogen	0.9737	0.1949
sulfate	0.9915	0.0630	ammonia nitrogen	0.9998	0.0012
chloride	0.9535	0.3451	fluoride	0.9635	0.2709
manganese	0.9987	0.0094	total cyanide	1.0000	0.0000

iron	0.9994	0.0041	total mercury	1.0000	0.0000
copper	1.0000	0.0000	arsenic	0.9943	0.0422
zinc	1.0000	0.0000	selenium	0.9949	0.0375
volatile phenol	1.0000	0.0000	cadmium	1.0000	0.0000
anionic surfactant	1.0000	0.0000	chromium (VI)	1.0000	0.0000
permanganate index	0.9969	0.0229	lead	1.0000	0.0000

As is illustrated in Table 4, according to the conventional EWM, the weights of chloride, fluoride and nitrite nitrogen are much larger than the other 19 indicators, for their dispersion degrees in observation data are significantly higher than the other indicators. The sum weight value of chloride, fluoride and nitrite nitrogen is 0.8109, suggesting that these three indicators have absolute advantages in the comprehensive groundwater quality assessment.

On the other hand, as is introduced in Section 3.1, iron, manganese and ammonia nitrogen are the indicators with the highest pollution conditions. However, their weights parameters are 0.0094, 0.0041 and 0.0012, respectively. In other words, the failed indicators only account for about 1.5% proportions in the synthetic evaluation, which is almost negligible.

Furthermore, all the weights parameters of copper, zinc, volatile phenol, anionic surfactant, total cyanide, total mercury, cadmium, chromium (VI) and lead are zero, for they have none dispersion degree in observation data at all. In other words, these nine pollutants are completely ignored in the comprehensive groundwater quality assessment.

Based on the improved EWM established in Section 2.4, the weights parameters are generated and listed in Table 5.

Table 5. The weighting parameters generated by the improved EWM

Indicator	**Sub-weight 1***	**Sub-weight 2***	**Comprehensive Weight**	**Indicator**	**Sub-weight 1***	**Sub-weight 2***	**Comprehensive Weight**
pH	0.0004	0.0500	0.0252	nitrate nitrogen	0.0067	0.0022	0.0045
total hardness	0.0019	0.0478	0.0248	nitrite nitrogen	0.1949	0.0544	0.1247
sulfate	0.0630	0.0106	0.0368	ammonia nitrogen	0.0012	0.1158	0.0585
chloride	0.3451	0.0090	0.1770	fluoride	0.2709	0.0385	0.1547
manganese	0.0094	0.1633	0.0863	total cyanide	0.0000	0.0212	0.0106
iron	0.0041	0.2776	0.1409	total mercury	0.0000	0.0150	0.0075
copper	0.0000	0.0024	0.0012	arsenic	0.0422	0.0231	0.0326
zinc	0.0000	0.0024	0.0012	selenium	0.0375	0.0145	0.0260
volatile phenol	0.0000	0.0290	0.0145	cadmium	0.0000	0.0075	0.0037
anionic surfactant	0.0000	0.0305	0.0153	chromium (VI)	0.0000	0.0212	0.0106
permanganate index	0.0229	0.0536	0.0383	lead	0.0000	0.0106	0.0053

*Sub-weight 1 is generated by the information amount principle; while sub-weight 2 is generated by the pollution degree principle.

As is illustrated in Table 5, according to the improved EWM, the sum weight value of chloride, fluoride and nitrite nitrogen is 0.4564, which is much smaller than that generated by the conventional

EWM. On the other hand, the sum weight value of iron, manganese and ammonia nitrogen is 0.2857, which is about 20 times larger than that generated by the conventional EWM.

The reason for this phenomenon is that the improved EWM takes the indicator's pollution condition into consideration. As a result, the importance of clean indicators is decreased, while the significance of heavily polluted indicators is highlighted.

Furthermore, through introducing the pollution degree principle, the "zero weight" phenomenon is excluded, too. As is shown in Table 5, the sum weight value of copper, zinc, volatile phenol, anionic surfactant, total cyanide, total mercury, cadmium, chromium (VI) and lead is 0.0698, which indicates that these pollutants cannot be ignored in the comprehensive groundwater quality assessment, though they have none dispersion degree in observation data at all.

Obviously, the major difference between the conventional EWM and the improved EWM is that, the conventional EWM only takes the information amount principle into consideration, while the improved EWM assigns weight based on not only the information amount principle, but also the pollution degree principle.

From the perspective of water resources management, the weight generated by the improved EWM is more reasonable than that generated by the conventional EWM.

(i) The failed indicators, especially the heavily polluted indicators, are the items deserve special attentions in the water resources management. These pollutants are neglected in the conventional EWM, for their dispersion degrees in observation data are small. However, through introducing an auxiliary pollution degree principle, the improved EWM highlights the importance of these indicators.

(ii) The improved EWM solves the "zero weight" phenomenon in the conventional EWM theory. As is discussed above, based on the conventional EWM theory, all the weights parameters of copper, zinc, volatile phenol, anionic surfactant, total cyanide, total mercury, cadmium, chromium (VI) and lead are zero, for they have none dispersion degree in observation data at all. However, these indicators are important toxicological indices in the water resources management; especially total cyanide, cadmium and chromium (VI) have been proved to be carcinogenic. Therefore, these indicators should not be completely neglected in evaluation, though they have none dispersion degree in observation data.

3.3. Comprehensive Evaluation Results

Based on the weight CCME WQI model introduced in Section 2.2, the comprehensive groundwater quality of Suihua city is evaluated and illustrated in Table 6. To make the expression more concise, we use CCME WQI (CEWM) and CCME WQI (IEWM) to represent evaluation results with the weights parameters generated by conventional EWM and improved EWM, respectively.

Table 6. The comprehensive groundwater quality of Suihua city

Evaluation Model	F_1	F_2	F_3	CCME WQI	Category
CCME WQI (CEWM)	1.46	1.46	8.19	95.12	Excellent
CCME WQI (IEWM)	28.57	28.57	68.81	53.93	Marginal

As is illustrated in Table 6, there is significantly difference between the evaluation results of CCME WQI (CEWM) and CCME WQI (IEWM). According to CCME WQI (CEWM), the comprehensive groundwater quality is "Excellent", indicating that water quality is protected with a virtual absence of threat or impairment However, based on CCME WQI (IEWM), the comprehensive groundwater quality is "Marginal", indicating that the aquatic environment is frequently threatened or impaired.

Obviously, the evaluation result of CCME WQI (CEWM) is over-optimistic. As is illustrated in Table 3, the concentrations of manganese are about 13 times larger than its acceptable threshold; the concentrations of iron are approximately 4 times as large as its acceptable threshold; and the concentrations of ammonia nitrogen are significantly overproof, too. These phenomena prove that the groundwater of Suihua city suffers from frequently and significantly pollutions of manganese, iron and ammonia nitrogen; and its environment often departs from desirable levels. Therefore, compared with CCME WQI (CEWM), the evaluation result of CCME WQI (IEWM) is much more reasonable.

The fundamental cause for the distortion in CCME WQI (CEWM) is that the conventional EWM only takes the statistics dispersion degree of the observation data into consideration, but neglects their

practical meaning. Therefore, the indicators whose monitoring values concentrate in the high pollution domains are easily to be ignored. As is shown in Table 6, for the importance of iron, manganese and ammonia nitrogen is neglected, both of the scope (F_1) and frequency (F_2) in CCME WQI (CEWM) are 1.46, indicating that percentages of failed indicators are failed tests are almost negligible.
However, through introducing an auxiliary pollution degree principle, the improved EWM highlights the importance of heavily polluted indicators. Therefore, both of the scope (F_1) and frequency (F_2) in CCME WQI (IEWM) are 28.57, which are about 20 times larger than these in CCME WQI (CEWM). And as a result, the comprehensive groundwater quality assessment conclusion of CCME WQI (IEWM) is much stricter and more reasonable than CCME WQI (CEWM).

4. Conclusions

(i) The conventional EMW only takes the statistics dispersion degree of the indicator's observation data into consideration, but neglects its practical significance in aquatic environment management. Therefore, the "zero weight" phenomenon is prone to existing, and the indicators with high pollution degrees are easily to be ignored. As a result, the conventional EMW often makes the synthetic groundwater quality evaluation result distorted and over-optimistic.
(ii) The improved EWM assigns weights according to both of the information amount principle and the pollution degree principle. In the improved EWM, the "zero weight" phenomenon is excluded and the importance of heavily polluted indicators is highlighted, which make the final comprehensive groundwater quality assessment much stricter and more reasonable
(iii) The comprehensive groundwater quality of Suihua city is "Marginal", indicating that the aquatic environment is frequently threatened or impaired. The crucial pollutants in the groundwater of Suihua city are iron, manganese and ammonia nitrogen. To improve the groundwater environment, the water resources managers should take more methods to decrease the natural pollution of iron and manganese; and strengthen the seepage control of livestock farms, poultry farms and the toilets in countryside.

5. References

[1] Amiri V, Rezaei M and Sohrabi N 2014 *Environmental Earth Sciences* 72(9): 3479-90
[2] Amiri V, Sohrabi N and Dadgar M 2015 *Environmental Earth Sciences* 74(7): 6163-76
[3] Li P, Wu J and Qian H 2016 *Arabian Journal of Geosciences* 9(1): 1-17
[4] Lu X, Li Y, Lei K, Wang L, Zhai Y and Zhai M 2010 *Environmental Earth Sciences* 60(8): 1693-99
[5] Li P Y, Qian H and Wu J H 2010 *Journal of Chemistry* 2010, 7(S1), 209-16
[6] Wu J, Li P and Qian H 2014 *Journal of Chemistry* 8(2), 787-93
[7] Wu J, Li P and Qian H 2013 *Asian Journal of Chemistry* 25(17): 9795-99
[8] Li P, Wu J and Qian H 2014 *Environmental Geochemistry & Health* 36(4): 693-712
[9] Tang R, Wen ZH, Shu LC, Liu B and Li S 2013 *Water Resources Protection* 29(4): 19-25
[10] Ministry of Environmental Protection of China (http://kjs.mep.gov.cn)
[11] Environmental Protection Bureau of Suihua City (http://www.shhbj.gov.cn).
[12] Yan F, Qiao DY, Qian B, Ma L, Xing XG, Zhang Y and Wang XG 2016 *Journal of Hydrology* 543: 316-23
[13] Abtahi M, Golchinpour N, Yaghmaeian K, Rafiee M, Jahangiri-Rad M and Keyani A 2015 *Ecological Indicators* 53: 283-91
[14] Hurley T, Sadiq R and Mazumder A 2012 *Water Research* 46:3544-52
[15] Huang S, Chang J, Leng G and Huang Q 2015 *Journal of Hydrology* 527: 608-18
[16] Chang J, Li Y, Wang Y and Yuan M 2016 *Journal of Hydrology* 540: 824-34
[17] Li GD 2013 *Heilongjiang Science and Technology of Water Conservancy* 6(41): 15-18

Acknowledgments
This work is supported by the National Natural Science Foundation of China under the contract No. 51709142.

IOP Conf. Series: Earth and Environmental Science **178** (2018) 012007 doi:10.1088/1755-1315/178/1/012007

Environmental Risk Assessment of a Water Diversion Project for Improving the Water Quality of Chaohu Lake: A Case Study

Xiaodong Liu[1], Lingqi Li[1], Zulin Hua[1], Siwei Peng[1], Dingyi Shen[1] and Ting Yang[1]

1 College of Environment Hohai University(HHU) Xikang Road 1 Nanjing 210098 P.R. CHINA
E-mail: xdliu@hhu.edu.cn
Key Laboratory of Integrated Regulation and Resource Development on Shallow Lake of Ministry of Education, College of Environment, Hohai University, Nanjing 210098, P.R. China

Abstract. New method for assessing the environmental risk of water diversion projects is proposed. Fault tree analysis was used to identify risk sources. The project route was divided into four regions: water source region, conveyance region, receiving water region, and drainage region. According to the different characteristics of the four regions, a risk evaluation index system and a multi-layer fuzzy comprehensive evaluation model were established. The results indicated that the environmental risk grade in the water source region was low and those of the other regions were medium-low. This work contributes to the environmental feasibility analysis of water diversion projects.

1. Introduction

Chaohu Lake is one of the five largest freshwater lakes in China and plays a huge role in water storage, flood control, water supply, agricultural irrigation, aquatic products, travel, and etc.[1]With the economy development and population increase, the gradual accumulation of pollution load has exceeded the carrying capacity of the lake, leading to serious eutrophication and loss of ecological function. Water diversion project has been considered as an important and helpful measure to improve water quality in the area. To improve the water quality of Chaohu Lake, a water diversion project was proposed by The Water Resources Department of Anhui Province. However, the water diversion project is complicated and can be influenced by many uncertain factors. So its environmental must be assessed before implementation.

The importance of environmental impact assessment of hydraulic engineerings has been widely recognized. A number of studies on water diversion risk assessment have been performed. Chang et al.[2]presented a comprehensive evaluation index system as a basis for evaluating the impact of inter-basin water diversion on ecological environment. The comprehensive evaluation index system included a target layer, a rule layer, and an index layer. Dou et al.[3]stated that the middle route of the South-to-North Water Transfer Project may produce adverse impacts on the ecological environment of Hanjiang. Yuan[4]established the comprehensive evaluation index system and a multilevel linear weighting and fuzzy mathematical comprehensive evaluation model to assess the environmental impacts of the Three Gorges Project. Xu et al.[5]analysed the "Two Rivers and Two Stations" project for Chaohu, discussed projects that may produce impacts on the social, economic, and ecological environment, and defined an index system. They used field investigation, analysis by analogy, expert consultation, and an analytic hierarchy to determine the main impact factors, which they conceptualized as a factor set to carry out a fuzzy comprehensive evaluation. Wang et al.[6]established an evaluation index system and used multi-level fuzzy comprehensive evaluation methods to analyse the social impacts of a water transfer project in Heihe River basin. Jin et al.[7]proposed a risk assessment system for water transfer projects that was structured according to the characteristics of water transfer project accidents. Chen et al.[8]analysed the risk factors of long-distance water transfer project system, which related to hydrology, buildings, economy, policy, the environment, and society. Extremum statistical methods, probability risk analysis methods, gray stochastic risk analysis, maximum entropy risk analysis, and many other methods have also been used[9]-[12].

Most previous studies involving environmental risk assessment of water diversion projects have focused on water quality assessment or safety evaluation of hydraulic structures. Only a small body of published research exists on comprehensive evaluation of environmental risk for a water diversion project. Therefore, in this paper, an environmental risk assessment model has been constructed and applied to the Chaohu water diversion project, which can provide a reference for feasibility analysis of water diversion projects.

2. Materials and Methods

2.1. Study Area

The route of the Chaohu Lake ecological water diversion project is shown in Figure 1.

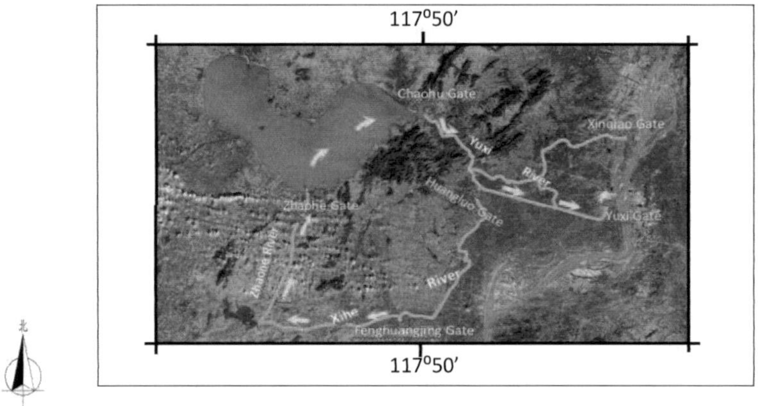

Figure 1. Route of the water diversion project.

According to the natural water diversion process, the route was divided into four regions: the water source region, the conveyance region, the receiving water region, and the drainage region:

①Water source region: the reach of the Yangtze River from a point 3km upstream of the FenghuangjingGate to a point 2km downstream;

②Conveyance region: the Xihe and Zhaohe Rivers;

③Receiving water region: East Chaohu Lake.

④Drainage region: the Yuxi River and the reach of the Yangtze River from 3km upstream of the YuxiGate to 7km downstream.

2.2. Environmental Risk Assessment Method

To control the negative influences and environmental risks which may occur during ecological water transfer, an environment risk assessment was undertaken. The procedure is shown in Figure 2.

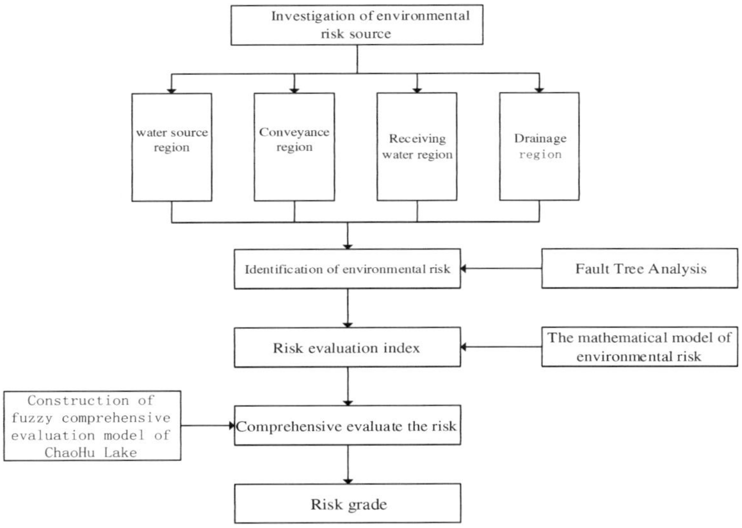

Figure 2. The flowchart of environment risk assessment

2.3. Source Identification of Environmental Risk
Figure 2 showsthat the first step is identifying the sources of risk. Here, fault tree analysis as an environmental risk identification method has been applied to the Chaohu Lake ecological water diversion project. This approach has been discussed in detail in the literature[13].

2.4. Construction of the Risk Evaluation Index System
According to the source identification of environmental risk and the characteristics of the four regions, a risk evaluation index system was introduced and is shown in Table 1. This system includes 50 indices in five layers: the object layer, the sub-object layer, the criterion layer, the sub-criterion layer, and the index layer.

2.5. Construction of the Fuzzy Comprehensive Evaluation Model
The fuzzy comprehensive evaluation method was used to evaluate the risk of the Chaohu Lake ecological water diversion project. This method was used because of its advantages in solving problems that are vague and hard to quantify. The basic steps are as follows[15]:

(1) According to environmental risk characteristics, the evaluation set is divided into five levels: $V=\{V_1,V_2,V_3,V_4,V_5\}$={low,medium–low, medium, medium–high, high}.

(2) Determine the classification standard that determines the membership degree of each index. Membership degrees lie between 0 and 1.

(3) Determine the weight matrix. The weight of an index depends on its importance.

(4) Determine a fuzzy evaluation model and obtain results. The calculation formula is:

$$D_i=w_{ij}R_{ij}$$

Where D_i is the evaluation result for index *i*, w_{ij} is the weight of index *i*, and R_{ij} is the membership degree of index *i*. Based on the formula, the risk value can be calculated layerbylayer. The last result is the overall evaluation result provided by the model.

Assign an environmental risk grade according to the following classification: low risk is between 0 and 1, low-medium risk is between 1 and 2, medium risk is between 2 and 3, medium-high risk is between 3 and 4, and high risk is between 4 and 5.

3. Results and Discussion
The environmental risk assessment method was used in the current study, with the results shown in Table 2. It is apparent that the risk grade of the water source region is low, whereas those of the other regions are medium-low.

Table 1. Risk evaluation index system of the Chaohu Lake ecological water diversion project

Object layer	Sub-object layer	Criterion layer	Sub-criterion layer	Index layer
Environmental risk assessment of water diversion project	Risk evaluation of water source region	Water quantity risk	Reliability of water quantity	Available amount of water in the Yangtze
			Reliability of the hydraulic facilities	Reliability of water transfer facilities
				Operating skills and level of management
		Water quality risk	Upstream water quality	Water quality assessment
				Comparison with receiving waterquality
			Influence of the drainage outlet	Discharge capacity
				Distance from water inlet
			Shipping pollution index	Shipping sewage volume
				Petroleum emissions
				Crew operating level
				Risks of oil spill
		Ecological risk	Oncomelania diffusion risk	Appearance rate of living oncomelanias contained in box
				Average density of living oncomelanias
				Infection rates of oncomelanias
	Risk evaluation of conveyance region	Water quality risk	Water quality in the water transport channel	Exceed the water quality standard about Xihe
				Exceed the water quality standard about Zhaohe
				Water quality compared to the conveyance region
			Analysis of the drainage outlet	Scope of normal discharge
				Scope of accidental discharge
			Vessel pollution index	Oily wastewater emissions
				Petroleum emissions
				Operating status of environmental protection facilities
			Influence of aquaticbreeding	Proportional area of aquatic breeding
		Ecological risk	Possibility of oncomelania diffusion by floating	Number of oncomelanias outside the gate
				Number of oncomelanias inside the gate
			Possibility of oncomelania diffusion with vessels	Proportion of vessels carrying oncomelanias
	Risk evaluation of receiving water region	Water quality risk	Water quality of the area	The condition of exceed the water quality standard
				Risks regarding the distribution of water quality differences

		Influence of the river on the lake	Area of influence
			Impact on intake (concentration increment)
		Eutrophication of water bodies	Ratio of nitrogen to phosphorus
			Change in water temperature
		Vessel pollution index	Oily wastewater emissions
			Petroleum emissions
			Operating status of environmental protection facilities
	Ecological risk	Possibility of oncomelanias breeding and multiplying in Chaohu	Survival rates of oncomelanias in the wild
			Development status of the offspring of oncomelanias
	Other Risks	Flood risk	Flood risk
			Drainage risk
			Risk of sediment deposition
		Risks of water resource utilization	Risk of irrigation water
			Shipping risk
Risk evaluation of drainage region	Risk of the Yuxi River	Influence on water quality in theYuxi River	Water quality compared to water source region
			Water quality compared to conveyance region
		Impact on intake	Increment of COD concentration
	Risk of the Yangtze River	Water quality of the Yangtze in the outlet section	Water quality compared to water source region
			Water quality compared to conveyance region
		Influence on Yangtze water quality	Length of the influence
			Width of the influence
		Impact on intake	Concentration increment

Table 2. Comprehensive evaluation results for each region.

Region	Water source region	Conveyance region	Receiving water region	Drainage region
Evaluation results	0.192	0.352	0.384	0.345
Risk standard values	0.975	1.76	1.92	1.72
Risk grade	low	medium–low	medium–low	medium–low

4. Conclusions

In this paper, a new method for assessing the environmental risk of water diversion projectshas been proposed. Fault tree analysis was used to identify risk sources during water diversion. The project route was divided into four regions: the water source region, the conveyance region, the receiving water region, and the drainage region.To reflect the different characteristics of the four regions, a risk evaluation index system was introduced, including 50 indices in five layers: the object layer, the sub-object layer, the criterion layer, the sub-criterion layer, and the index layer. A fuzzy comprehensive evaluation model was used to evaluate the risk of the Chaohu Lake ecological water diversion project. Using this method, project risk wassuccessfully assessed. The results indicated that the environmental risk forthe water source region was in the low range and for the other regions were in the medium-low range. This work contributes to feasibility analysis of water diversion projects.

5. References

[1] HongbingTang, QianTan. The diversion scheme comparison of improving the water quality of drinking water in Chaohu, *Water Resources and Power*, 2011, 29(10) pp 60-62, 214 (in Chinese)

[2] Yumiao Chang, Min Zhao. The comprehensive evaluation index system research about inter-basin water transfer impacts ecological environment, *Water conservancy economy*, 2007, 25(2) pp 6-7 (in Chinese)

[3] Ming Dou, Qiting Zuo and Cainan Hu. The ecological environment impact assessment of the south-north water diversion project, *Journal of Zhengzhou University* (Engineering Science), 2005(2) pp 63-66 (in Chinese)

[4] YunxiangYuan. The comprehensive evaluation method about large hydropower projects produce impacts on environmental and ecological, *Acta Scientiae Circumstantiae*, 1989(2) pp 76-86 (in Chinese)

[5] FuliuXu, Xiaoyan Lu and Jiagui Zhou, etc. The evaluation index system and fuzzy comprehensive evaluation of large-scale water conservancy projects produce impacts on environmental, *Bulletin of Soil and Water Conservation*, 2001(4) pp 10-14 (in Chinese)

[6] Ruina Wang and Deshan Tang. The multi-level fuzzy comprehensive evaluation of water control projects in Heihe, *Yangtze River*, 2006(7) pp 27-28, 86 (in Chinese)

[7] Jian Jin and Jianyong Hong. The risk evaluation system research of water diversion Project, *Water Conservancy Science and Technology and Economy*, 2007, 13(4) pp 260-262 (in Chinese)

[8] Jin Chen and Wei Huang. The risk and countermeasures of the cross-basin long distance water transfer engineering system, *Water Resources and Hydropower Engineering*, 2004, 35(5) pp 95-97, 103 (in Chinese)

[9] Ping Feng. The water resources risk management while water supply system of drought, *Journal of Natural Resources*, 1998, 4(13) pp 139-143 (in Chinese)

[10] Tao Tao, Xiang Fu and Changming Ji. Application research on risk analysis of regional water resources supply and demand, *Engineering Journal Wuhan University*, 2002, 35(3) pp 9-12 (in Chinese)

[11] Qiting Zuo, Zening Wu and Wei Zhao. The uncertainty in water resources system and risk analysis method, *Arid Land Geography*, 2003, 26(2) pp 116-121 (in Chinese)

[12] Nianhua Xue ,Changming Ji. The research status about risk analysis of water resources system, *Water Resources and Electric Power*, 1993, 20(2) pp 33-36 (in Chinese)

[13] Chunyan Shi, Gongyu Hu, Zulin Hua, Xiaodong Liu, Li Gu and Yu Chen. Environmental Risk Identification of Water Quality in Ecological Water Transfer Project Based On Fault Tree Analysis, *Environmental Science & Technology*, 2007, 30(6) pp 52-54 (in Chinese)

Acknowledgements

This research was supported by the National Nature Science Foundation of China (Grant No. 51479064, 51379058, 51379060), a Project Funded by the Priority Academic Program Development of Jiangsu Higher Education Institutions（PAPD）and Qing Lan Project.

IOP Conf. Series: Earth and Environmental Science **178** (2018) 012008 doi:10.1088/1755-1315/178/1/012008

Safety Assessment of Urban Water Metabolism Based on PSR Framework—— Taking Tianjin City as an Example

Jin Xu[1, 2], Binke Li[2, *], and Zhihao Yu[1]

1 School of Environmental Science and Engineering, Tianjin University, Tianjin 300072, China
2 School of Management, Tianjin University of Technology, Tianjin 300384, China
Email: 15737129420@163.com

Abstract. The complex urban water circulation system is developing in a more severe and complicated direction, and the phenomenon of urban water metabolism imbalance and water metabolism crisis become more and more obvious. The PSR framework model is introduced to construct the urban water metabolism safety evaluation index, and the entropy weight method is used to determine the weight of the corresponding evaluation index and the calculation of the safety index, at the same time, taking Tianjin city as an example. Research results show that, the evolution process of water metabolism in Tianjin showed a trend of development during the evaluation period, that is, obviously improved in the early stage, stable in the medium term, and decreased in the later period, the empirical data show that the evaluation results are reasonable.

1. Introduction

Compared with the orderly and self-organized natural water circulation system, the urban complex water circulation system is moving in a more intense and complicated direction by the comprehensive role and profound impact of the scale of population and the pressure of economic and social development. With the continuous expansion of the impact of social and economic activities, the competition between water consumption in different industries is fierce, environmental capacity and emissions of contradictory sharp, urban water shortage, urban water pollution and urban water ecological degradation is particularly prominent, the resulting urban complex water cycle metabolic imbalance and metabolic security crisis is more significant. Therefore, based on this related to promote the metabolic healthy operation of urban water cycle system and the realization of urban water resources security and efficient use of research will have an urgent practical significance.

With urban water metabolism safety connotation analysis and conceptual system development process continues to deepen, the domestic and foreign related research is gradually turning to explore the evaluation method based on the deep understanding of urban water metabolism safety connotation, and the evaluation index system of urban water metabolism safety was established by using different methods and research paths such as fuzzy sets ([1]; [2]), entropy weight method ([3]), cluster analysis ([4]), multiple regression structure ([5]), water footprint([6]; [7]) etc., and the evaluation method of urban water metabolism safety was enriched. However, some scholars believe that, because the concept of water metabolism safety itself is still expanding, so far there is no uniform water metabolism safety evaluation standards and methods ([8]). More importantly, due to the limitations of the relevant evaluation tools and methods, the study of urban water metabolism safety assessment has not been able to play an ideal interpretation and prediction of water metabolism imbalance. In view of this, this paper explores the problem of urban water metabolism safety evaluation from the perspective of complex system, using PSR framework model and entropy weight method to take the metabolic PSR mechanism of urban complex water circulation system as a fundamental angle of view, in order to establish scientific and reasonable urban water metabolism safety evaluation index, and to a certain extent, promote the development of water metabolism safety research field evaluation method.

2. Method

2.1. Construction of Water Metabolism Safety Evaluation Index

The determination of the evaluation index is the basic condition and important prerequisite for the comprehensive evaluation and quantitative analysis of the degree of urban water metabolism safety. The PSR framework model is widely used in the development of human society to produce the pressure on the ecosystem caused a series of ecological security issues of evaluation and early warning research for its "natural - social" binary coupling system between the elements of the causal relationship between the scientific interpretation and logical advantages, and with its simple optimization, structured and logical clarity and other characteristics in the environment, ecology, land use and other research areas to obtain good application and feedback. The urban water metabolism process as a typical "natural-social" binary complex coupling system characterization phenomenon, fully based on the PSR framework analysis tools within the scope of systematic evaluation. In this paper, the PSR framework model is introduced to construct a reasonable and scientific evaluation index of urban water metabolism safety, at the same time to Haihe River Basin in Tianjin as an empirical object for the collection of indicators of data, At the same time to Haihe River Basin in Tianjin as an empirical object for the collection of indicators of data, the data collected in this paper are mainly based on the "China Statistical Yearbook", "Tianjin Statistical Yearbook" and "Tianjin Water Resources Bulletin" in the evaluation period (2006-2015).

Table 1. Urban water metabolism safety PSR evaluation index

Target layer	System layer	Numbering	Index layer	Dimension	Naturc
Urban Water Metabolic Safety Assessment (PSR)	Urban water metabolic pressure Pressure	P1	Per capita GDP	yuan	negative
		P2	Water resources development intensity	%	negative
		P3	COD emission intensity	Million tons	negative
		P4	Fertilizer application strength	Million tons	negative
		P1	Pesticide application intensity	Million tons	negative
	Urban water metabolic state State	S1	Per capita water resources	m³/per	positive
		S2	Water quality condition index	—	positive
		S3	Water use efficiency	%	positive
		S4	Ecological environment	%	negative
		S5	Groundwater overrun rate	%	negative
	Urban water metabolic response Response	R1	Environmental management investment rate	%	positive
		R2	Industrial water repetition rate	%	positive
		R3	Domestic sewage treatment rate	%	positive
		R4	Water quality comprehensive compliance rate	%	positive
		R5	Sewage discharge compliance rate	%	positive

2.2. Determination of Index Weight and Safety Index

In view of the accuracy of the subjective evaluation method, too much depends on the level of experience of the evaluators, the inevitable result bias often exists and the evaluation does not conform to the facts, so it is possible to control distractors and ensure the rigorous and credibility degree of the system evaluation, the use of the system to measure the degree of disorder degree of entropy method for urban water metabolism safety evaluation index weight and safety index to determine the work. The evaluation principle is to analyze the relative merit of the object to be evaluated, and to calculate and compare the close distance between the optimal solution of the analysis unit, evaluation standard is that when the object is close to the optimal value and away from the worst when more ideal. The specific operation flow is as follows:

2.2.1. Construction of evaluation matrix. Constructs the initial evaluation matrix X, and contains the judgment object m and the judgment index n, the matrix form is as follows:

$$X = \left(x_{ij}\right)_{m\times n}, i = 1,2,...,n; j = 1,2,...,m$$

2.2.2. Standardized matrix processing. In view of the existence of certain differences in the indicators measurement units, and have a greater impact on the follow-up index entropy weight determination and comparison work. In addition, the different indicators on the urban water metabolism safety state of the role of the nature of the difference, leading to its direction of play (positive and negative effects) are also different, as shown in table 1. Therefore, the initial data of the original price matrix X is standardized by dimensional standardization, and the evaluation indexes of different properties are processed as follows:
For the evaluation of the positive nature of the indicators, the increase in such indicators will be conducive to the improvement of urban water metabolism safety status, the standardization formula is as follows:

$$x_{ij}^{'} = \frac{x_{max} - x_{ij}}{x_{max} - x_{min}}$$

For the evaluation of the negative nature of the indicators, the increase in such indicators will lead to urban water metabolism safety state out of control, the standardization formula is as follows:

$$x_{ij}^{'} = \frac{x_{ij} - x_{min}}{x_{max} - x_{min}}$$

Among them, x_{ij} indicates the initial value of the object to be evaluated, $x_{ij}^{'}$ is the index value after dimension normalization, x_{max} and x_{min} represent the best value of the corresponding evaluation index respectively.

2.2.3. Determination of entropy weight of index. The entropy weight method is used to determine the index weight of each evaluation index, this method is used as an objective weighting method in information theory to represent information entropy, can fully tap and reflect the implicit information and make the system within the evaluation of the resolution of the meaning of a more clear, the specific calculation process is as follows:

$$\begin{cases} y_{ij} = \dfrac{x_{ij}^{'}}{\sum_{i=1}^{m} x_{ij}^{'}}, 0 < y_{ij} < 1 \\ h_j = -\dfrac{1}{\ln m}\sum_{i=1}^{m} y_{ij} \ln y_{ij} \\ W_j = \dfrac{1 - h_j}{\sum_{i=1}^{n}(1 - h_j)} \end{cases}$$

Among them, the information entropy of each evaluation index is calculated as h_j, the entropy weight of each evaluation index is calculated as W_j.

2.2.4. Determination of relative merits. After the entropy weight is determined, the evaluation matrix Y_{ij} after normalization is weighted to establish the weighted decision matrix, that is $V = \left|V_{ij}\right|_{m\times n} = W_j \times Y_{ij}$. Among them, the weighted solution matrix positive and negative ideal solution (with the corresponding V_j^+ and V_j^- respectively) can be expressed as the following form:

$$\begin{cases} V_j^+ = \{\max_{1\le i\le n} v_{ij} \left| i = 1,2,\cdots,n \right|\} = \{v_1^+, v_2^+, \cdots v_n^+\} \\ V_j^- = \{\min_{1\le i\le n} v_{ij} \left| i = 1,2,\cdots,n \right|\} = \{v_1^-, v_2^-, \cdots v_n^-\} \end{cases}$$

Finally, in order to determine the relative merit of each index, the distance between the evaluation index and the positive and negative ideal solution needs to be determined and can be expressed as the following form. At the same time, the lower the T value is, the smaller the distance between the evaluation index and the ideal solution is, and the better the urban water metabolism safety state. Similarly, the lower the Y value is, the smaller the distance from the negative ideal solution, and is reflected in the city water metabolism safety state out of control.

$$\begin{cases} D_i^+ = \sqrt{\sum_{j=1}^{n} (v_{ij} - v_j^+)^2}, (i = 1,2,\cdots,m) \\ D_i^- = \sqrt{\sum_{j=1}^{n} (v_{ij} - v_j^-)^2}, (i = 1,2,\cdots,m) \end{cases}$$

2.2.5. Calculation of safety index. The safety degree of urban water metabolism safety is expressed by the degree of closeness C_i between the evaluation index and the optimal scheme, where the value of the proximity degree C_i should be within the range of (0, 1), when the bigger to the degree of C_i, that is, the corresponding evaluation of the object and the optimal solution closer to the city water metabolism more healthy state of health.

$$C_i = \frac{D_i^-}{D_i^+ + D_i^-}, (1 \le i \le m)$$

At the same time, drawing on the relevant literature results ([9]; [10]; [11]) and the status of water metabolism in Tianjin, divided to characterize the degree of urban water metabolism safety and health of the safety index, as shown in Table 2.

Table 2. urban water metabolism safety and health level classification

Urban water metabolism safety index interval	Degree of urban water metabolism safety and health
>0.0~0.1	Severe safety
>0.1~0.3	Poor safety
>0.3~0.6	Critical state
>0.6~0.9	Good security
>0.9~1.0	Excellent security

3. Result and Discussion

3.1. Empirical Evaluation Results

First of all, after the above experimental process, the entropy of each index of urban water metabolism safety PSR is shown in table 3. According to the evaluation results, the water resources development strength (WP2=0.069) and COD emission intensity (WP3=0.046), water quality index (WS2=0.072) and the utilization rate of water resources (WS3=0.130) and environmental governance investment rate (WR1=0.109) and the sewage discharge standard rate (WR5=0.107) were the main influencing factors of urban water metabolism safety pressure, safety status and safety response system respectively within the experimental period.

At the same time, the effect of urban water metabolism safety response on the overall urban water metabolism safety evaluation (WR = 0.476) is the most significant, indicating that the health status of water metabolism in Tianjin depends largely on the social level of urban water metabolism response measures. In addition, it can be seen that the improvement of environmental management investment rate as a typical government water governance behavior measures have a strong positive effect on the overall urban water metabolism safety (WR1 = 0.109, the highest weight of the evaluation index), this point to a certain extent for the evaluation results provided a confirmation.

Table 3. Urban water metabolism safety PSR evaluation index entropy weight

PSR evaluation index entropy weight

P	P1	P2	P3	P4	P5
entropy	0.03	0.06	0.04	0.03	0.02
weight	2	9	6	8	6
S	S1	S2	S3	S4	S5
entropy	0.02	0.07	0.10	0.05	0.05
weight	6	2	6	0	9
R	R1	R2	R3	R4	R5
entropy	0.10	0.08	0.09	0.08	0.10
weight	9	0	4	6	7

Through the treatment of weighted decision matrix and the calculation of urban water metabolism safety indicators, the comprehensive evaluation results are shown in Fig 1. According to the comprehensive evaluation results, it is concluded that the safety status of water metabolism in Tianjin shows a trend of "obvious improvement in the early period, medium-term stable but late decline", among them, 2008 and 2012 for two important turning point, and accordingly divided into three cities metabolic safety trends change year interval. On the other hand, the urban water metabolism safety indicators of each year are above the critical state, and the system safety and health degree in 2008 - 2012 and 2014 are "safe", but in recent years the overall security trend has declined significantly, need to take measures to actively respond.

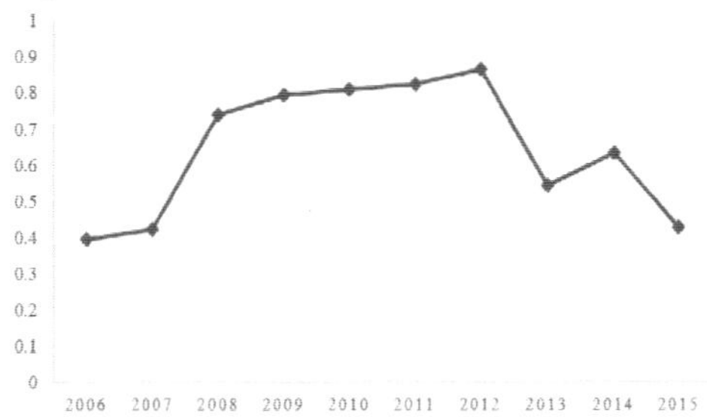

Figure 1. Urban water metabolism safety index evolution trend

3.2. Empirical Evaluation Analysis

The comprehensive evaluation conclusion shows that the water quality of Tianjin in 2006-2008 has been improved continuously, the safety status of water metabolism in 2008 - 2012 has maintained a good and healthy development trend and gradually stabilized, but the degree of security has dropped significantly and showing uncertain fluctuations in the state in 2012 ~ 2015. Overall, the urban water metabolic safety situation in Tianjin is not ideal, and according to the actual feedback from the relevant data (such as the public data of Tianjin Water Resources Bulletin over the years), this general trend can better explain the evolution of water metabolism in Tianjin. 2006 (and before) Tianjin has long been in urban water metabolic disorders caused the shortage of urban water resources, water supply tense relations and urban water pollution seriously and other serious situation, in order to cope with and alleviate the crisis of water metabolism in Tianjin, the relevant departments have actively carried out the response measures such as the Luan River and the Yellow River water into Tianjin and water source protection, and achieved remarkable results to achieve a significant improvement in the urban water metabolism safety situation in the 2007-2008 year. At the same time, with the relevant policies (such as water transfer projects, water restrictions, sewage treatment, etc.) landing and efficient use of water resources and awareness of conservation, making the mid-term (2008 - 2012) ushered in the Tianjin water metabolic security status of good maintenance and sustained stability. However, late (after 2012) urban water metabolism safety and health status is not ideal, or even a significant ups and downs and fall. Through the investigation of the historical data of water metabolism in Tianjin, it can be found that the main causes of the above-mentioned situation are as follows: 1. since 2013, the annual precipitation in Tianjin has been gradually reduced and continuously entered into the low water years (including 2015, the level of flat water), And with the spatial and temporal distribution of precipitation, the city's surface water resources and reservoir water storage capacity continued to decline, and with the increase in water demand intensity led to the city water supply and demand relationship disorder; 2. The late waste water discharge increased rapidly, this index for 2015 data for the 6.136 million tons, compared with 2012 increased by 7.88%, in the context of urban water recycling capacity has not been greatly improved, make urban water pollution further deteriorated. The two driving forces together contributed to the non-sustainable development of water metabolism in Tianjin.

4. Conclusion

In this paper, the urban water metabolism safety evaluation as the research object, the use of complex system research perspective, the introduction of PSR framework model to build urban water metabolism safety evaluation index. In order to avoid the deviation of the evaluation result caused by the subjective evaluation method, the entropy weight method is used to determine the weight of the evaluation index and the calculation of the safety index, and the data acquisition and conclusion analysis are made in Tianjin as the empirical case. According to the results of Tianjin water metabolism safety evaluation, it is found that the evolution process of water metabolism safety in the empirical area shows a development trend of "obvious improvement in the early stage, stable in the middle period and ups and downs in the later period". The research is beneficial to the development of the evaluation method of water metabolism safety research field, and it has certain reference value for Tianjin water metabolism safety evaluation and management policy.

5. References

[1] LU S, BAO H, PAN H. Urban water security evaluation based on similarity measure model of Vague sets [J]. International Journal of Hydrogen Energy, 2016, 41 (35): 15944-15950.

[2] LU S, WANG J, BAO H. Study on Urban Water Security Evaluation Based on the Vague Set Similarity Model [J]. Energy Procedia, 2016, 88: 309-312.

[3] HAN B, LIU H, WANG R. Urban ecological security assessment for cities in the Beijing–Tianjin–Hebei metropolitan region based on fuzzy and entropy methods [J]. Ecological Modelling, 2015, 318: 217-225.

[4] ESLAMIAN S A, LI S S, HAGHIGHAT F. A new multiple regression model for predictions of urban water use [J]. Sustainable Cities and Society, 2016, 27: 419-429.

[5] VEETTIL A V, MISHRA A K. Water security assessment using blue and green water footprint concepts [J]. Journal of Hydrology, 2016, 542: 589-602.

[6] Ma Jing, Peng Jian.Study on water footprint research [J]. Acta Ecologica Sinica, 2013, (18): 5458-5466

[7] Dong-guo Shao, Feng-shun Yang, Yu-long Liu, Xiao Chun. City Water Safety Index and Its Evaluation Criteria [J]. South-to-North Water Diversion and Water Conservancy Science & Technology, 2013, (01): 122-126.

[8] Chunyang Lu, Wen Feng, Qingyuan Yang, Linlin Chen, Huiming Zong. Performance Evaluation and Obstacle Factor Diagnosis of Urban Land Use Based on Improved TOPSIS Method - A Case Study of Chongqing City [J]. Resources Science, 2011, (03): 535-541.

[9] Huikun Hong, Heping Liao, Zhaofu Wei, Li Tao, Deti Xie. Study on the Health Evaluation of Land Use System in the Ecological Sensitive Area of the Three Gorges Reservoir Area Based on the Improved TOPSIS Method [J]. Acta Ecologica Sinica, 2015, (24): 8016-8027.

[10] Xun-ping Lei, Guang-hua Qi. Empirical Study on Evaluation of Regional Resources and Environment Carrying Capacity Based on Entropy TOPSIS Model [J]. Acta Scientiae Circumstances, 2016 (01): 314-323.

[11] NOIVA K, FERN NDEZ J E, WESCOAT JR J L. Cluster analysis of urban water supply and demand: Toward large-scale comparative sustainability planning [J]. Sustainable Cities and Society, 2016, 27: 484-496.

Preliminary Study on Dimensionless Expression of Bacterial Chemotaxis in Simulated Contaminated System

Xiaopu Wang[1 2], Beibei Gao[1 3], Shuaiwei Gu[4], Wei Zhong[1], Kenneth S. Kihaule[1], Roseanne M. Ford[3]

1. School of Petroleum Engineering, China University of Petroleum (East China), Shandong 266555, China
2. National Engineering Laboratory for Subsea Equipment Testing and Detection Technology, Shandong 266555, China
3. Department of Chemical Engineering, University of Virginia, Charlottesville, VA 22904, the United States of America
4. Shandong Provincial Key Laboratory of Oil & Gas Storage and Transportation Security, China University of Petroleum (East China), Shandong 266555, China
Email: wxp@upc.edu.cn, rmf3f@virginia.edu

Abstract. The use of chemotactic bacteria in bioremediation may improve the efficiency and decrease the cost of restoration of contaminated groundwater system. However, most previous studies focused at the laboratory scale and could hardly apply to the field scale. In this study, a dimensionless equation was formulated to solve this problem. First, the main influential factors were extracted from previous researches and then one set of dimensionless numbers was obtained based on Buckingham theory. After collecting parameter values and supplementing missing data, Chemotaxis number (Ch), the ratio of accumulated bacterial concentration to its initial value, was correlated with a combination of dimensionless numbers. For a Chemotaxis number greater than one, the bioremediation strategy is expected to be effective, and chemotactic bacteria are expected to accumulate around nonqueous phase liquid (NAPL) contaminant sources efficiently.

1. Introduction

Soil and groundwater polluted by refined petroleum products such as gasoline, kerosene, diesel, and lubricating oil has been frequently reported, like the oil spill in Gulf of Mexico, requiring appropriate measures to counteract their harmful effects on ecology and human health [1]. Conventional physical and chemical treatments and remediation of petroleum refinery waste have been shown inefficient in eliminating genotoxicity of these pollutants. The use of chemotactic bacteria in bioremediation is theoretically a potential efficient and low-cost technology in restoring contaminated soils and groundwater. They have perceptions of contaminant concentration gradients in water and can make response to it by preferentially swim towards regions of higher contaminant concentration [2].

Bacteria can swim through aqueous media at speeds comparable to typical groundwater flow rates. At particularly low flow rates, chemotaxis may be the main influence factor on bacterial transport through advection-dominated systems, and direct microbial populations towards contaminants retained in regions of low permeability [2]. In this study, we aim to acquire a Chemotaxis number, like Reynolds number, that can incorporate experimental results to evaluate the efficiency of chemotaxis and direct chemotactic biodegradation in the field application. We first extracted influential parameters in bioremediation [3]-[5], and then would combine them into a dimensionless number based on Buckingham theory. When Ch number greater than one, chemotactic bacteria show greater accumulation near contaminated regions and can effectively degrade contaminant chemicals into nontoxic byproducts.

2. Dimensionless Analysis

This section presents the dimensionless analysis for all the parameters that controls bacterial chemotaxis in contaminated porous media as shown in Table 1. Parameters are extracted from previous researches about bacterial chemotaxis [2]-[5], and their values can be directly obtained from experimental results except for contaminant distribution and concentration gradient.

Table 1. Influential parameters in bioremediation.

Factors	Parameters	Dimensional unit
Media	porosity, ε	[-]
	permeability, K	$[L^2]$
	tortuosity, τ	[-]
	groundwater velocity, v_f	$[LT^{-1}]$
Chemoattractant	concentration, a	$[Mol.\ L^{-3}]$
	concentration gradient, Δa/Δl	$[Mol.\ L^{-4}]$
	diffusion coefficient, D	$[L^2T^{-1}]$
Bacteria	concentration, b	$[Mol.\ L^{-3}]$
	motility coefficient, μ	$[L^2T^{-1}]$
	chemotactic sensitivity coefficient, χ_o	$[L^2T^{-1}]$
	chemotactic receptor constant, K_c	$[Mol.\ L^{-3}]$
	swimming speed, v	$[LT^{-1}]$

According to Buckingham theory, a function including all the parameters above can be written as,

$$F(\varepsilon, K, \tau, a, \Delta a/\Delta l, D, b, \mu, \chi_0, K_c, v, v_f)=0 \quad (1)$$

Then χ_0, K_c and $\Delta a/\Delta l$ are selected as recurring parameters by means of linear algebra, and all the other parameters can be expressed as dimensionless π numbers accordingly,

$$\pi_1=\frac{K_c^2}{K(\Delta a/\Delta l)^2},\ \pi_2=\frac{K_c}{a},\ \pi_3=\frac{\chi_0}{D},\ \pi_4=\frac{K_c}{b},\ \pi_5=\frac{\chi_0}{\mu_{eff}},\ \pi_6=\frac{vK_c}{\chi_0(\Delta a/\Delta l)},\ \pi_7=\frac{v_f K_c}{\chi_0(\Delta a/\Delta l)}$$

where tortuosity and porosity are incorporated into motility coefficient $\mu_{eff}=\frac{\epsilon}{\tau}\mu_0$ [2]. Combine these dimensionless groups, and we can get the main dimensionless numbers that have influence on bacterial transport[6] and may help us predict the bacterial distribution in contaminated layers.

$$P_1=\pi_1=\frac{K_c^2}{K(\Delta a/\Delta l)^2},\ P_2=\pi_3\times\pi_5=\frac{\chi_0^2}{\mu_{eff}D},\ P_3=\frac{\pi_7}{\pi_6}\frac{v_f}{v},\ P_4=\pi_2=\frac{K_c}{a}$$

especially $P_2=\frac{\text{Chemotaxis Bacterial Diffusive Force}}{\text{Normal Bacterial Diffusive Force}}$, $P_3=\frac{\text{Ground Flow Rate}}{\text{Swimming Velocity}}$.

P_2 can further adopted to explain the difference in diffusion between chemotactic and nonchemotactic bacteria. As for P_3, researchers have found that bacterial chemotaxis is significant with typical groundwater velocity [2], [4]. In P_4, K_c and a are normally considered in the same order of magnitudes. P_1 combines the factors of bacteria, porous media and chemoattractant, and requires further studies to explore its significance on directing bacterial towards sources of contaminants.

Here we define Chemtaxis number $Ch=b_{acc}/b_0$ as the ratio of accumulated bacterial concentration near contaminants to its initial value, and it is a combination of the influential parameters as the following [6],

$$Ch=f(P_1, P_2, P_3, P_4)=\xi P_1^{x_1}P_2^{x_2}P_3^{x_3}P_4^{x_4} \quad (2)$$

3. Results and Discussion

In this study, we selected three papers about bacterial chemotaxis out of data integrity, and Table 2 displays the parameters available in X.Wang 2016 as an example.

Table 2. All the parameters available in X.Wang 2016.

Bacteria	v_f (m/d)	ε_p	K(m^2)	a (Mol/m^3)	D(m^2/s)
P.putida F1	0.25	0.43	1×10^{-11}	5.65	9.5×10^{-10}
b (Mol/m^3)	μ (m^2/s)	v (m/s)	χ_0 (m^2/s)	K_c (mM)	τ
1	1.3×10^{-9}	4.4×10^{-5}	1.8×10^{-8}	1	$\sqrt{2}$

3.1. Chemoattractant Concentration Gradient Acquisition via Numerical Simulation

However, we still need to get chemoattractant concentration distribution before we can figure out the exponents x_1, x_2, x_3 and x_4 in equation (2). We rebuilt numerical modules of the experimental devices in references within COMSOL and solved the distribution of attractant concentration.

3.1.1. Heterogeneous Contaminated Porous Media

Figure 1(a) is the rebuilt numerical module of microfluidic device used in X.Wang 2016 and steady chemoattractant distribution under specific groundwater velocity. NAPL droplets were trapped within a fine pore network, and bacterial suspension was injected through a highly conductive adjacent macrochannel. We exported the distribution along series of cut lines parallel to black dash line in Figure 1(a) and got concentration distributions of attractant as shown in the Figure 1(b).

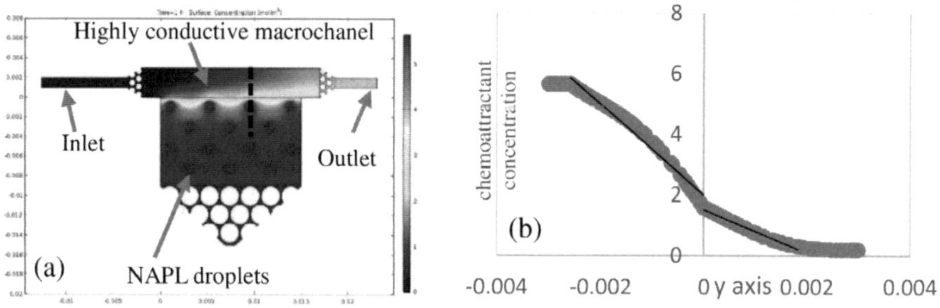

Figure 1. (a) Attractant distribution under 0.25 m/d; (b) Attractant distribution along black dash line.

It can be found the plotted data points around y=0 tend to be two lines with different slopes. y=0 represents the interface between porous media and groundwater flow path in the experiments, and we took the slope of line when y<0 as the concentration gradient Δa/Δl in pore network.

3.1.2. Pore-scale Chamber with NAPL Source

Wang used microfluidic device to study the dissolution of an organic-phase contaminant from a single pore into a larger macropore and migration of microorganisms that were carried along by groundwater flow. Figure 2(a) is a magnification of region near NAPL/water interface. We simulated the chemoattractant concentration distribution in the vicinity of NAPL/water interface shown in Figure 2(b). Similar to the calculation of Δa/Δl in X.Wang 2016, and we also obtained Δa/Δl under different groundwater flow rates as shown in Figure 2(c).

3.1.3. T-shaped Channel

Lanning used a T-shaped microfluidic device shown in Figure 3(a) to study chemotaxis transverse to advective flow. Two fluid streams, namely bacteria and attractant, were brought into contact by impinging flow, and then flowed adjacent to each other along a transparent channel. Figure 3(b) shows the attractant concentration distributes at specific flow rate, and (c) is chemoattractant concentration distribution vertical to the direction of stream flow.

IOP Conf. Series: Earth and Environmental Science **178** (2018) 012009 doi:10.1088/1755-1315/178/1/012009

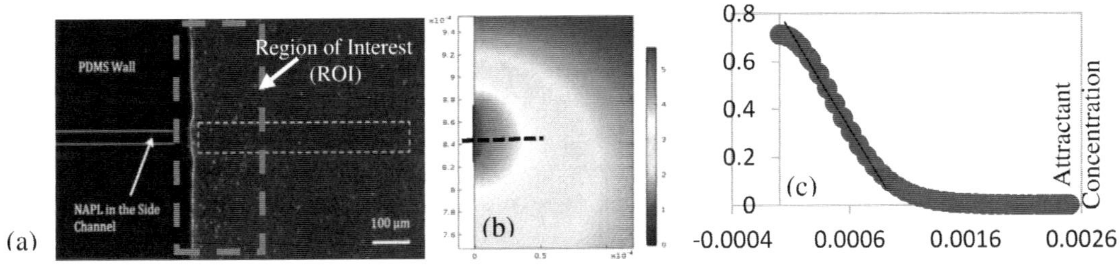

Figure 2. (a) NAPL/water interface; Attractant distribution in ROI (b) and along black dash line (c).

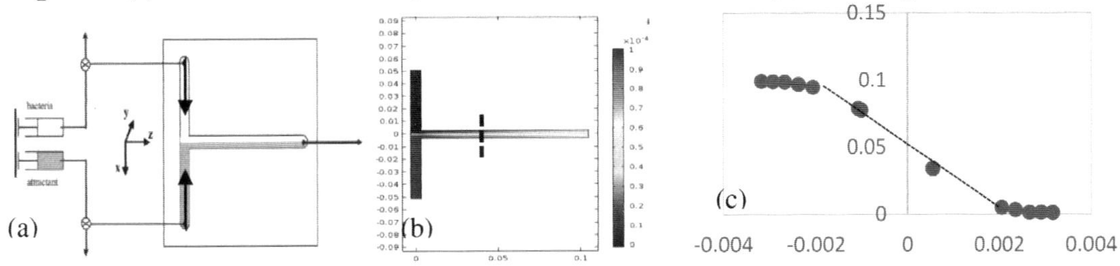

Figure 3. (a) T-shaped sensor; (b) T-shaped model; (c) Attractant distribution along black dash line.

Table 3 gives the concentration gradients of chemoattractant under different groundwater flow rates in discussed three papers.

Table 3. Concentration gradients under different conditions.

	v_f (m/s)	a_0 (mM)	$\Delta a/\Delta l$ (Mol/m^4)
X.Wang 2016	0	5.65	162.430
	5.79×10^{-6}		3.733×10^{3}
	5.79×10^{-5}		4.644×10^{3}
X.Wang 2012	5.79×10^{-6}	5.65	3.130×10^{4}
	1.16×10^{-5}		3.358×10^{4}
	5.79×10^{-5}		3.513×10^{4}
	1.16×10^{-4}		3.865×10^{4}
L.Lanning 2008	5.5×10^{-5}	0.1	20.041
	5.5×10^{-4}		21.566

3.2. Solution of Chemotaxis Number

The four dimensionless numbers P_1, P_2, P_3, P_4 are calculated accordingly and shown in Table 4, together with chemotaxis numbers directly obtained from published researches.

Table 4. Four dimensionless numbers and chemotaxis number in published articles.

	P1	P2	P3	P4	Ch
X.Wang 2016	3.79×10^{6}	0.177	0	826.83	1.15
	7.17×10^{3}		0.132		1.05
	4.64×10^{3}		1.315		1.02
X.Wang 2012	1.307×10^{-3}	0.177	0.132	0.822	1.24
	1.135×10^{-3}		0.263		1.2
	1.038×10^{-3}		1.315		1.12
	8.570×10^{-4}		2.630		1.08
L.Lanning 2008	2.543×10^{3}	1.25	2.412	2232.56	1.788
	2.196×10^{3}		24.123		1.7

The exponents and constant can be determiend by fitting experimental and simulated results in previous section, and the chemotaxis number is solved,

$$Ch = 0.266 P_1^{-0.2248} P_2^{0.0031} P_3^{-0.0588} P_4^{0.4852} \quad (3)$$

From P_3 and its exponent, we found that accumulation of chemotactic bacteria will decrease with the increasing of groundwater velocity, which agrees with experimental results shown in Figure 4. In Figure 4, the groundwater flow rate increases steadily from left to right in each reference, and Ch number, namely the accumulated bacteria concentration around contaminants, all shows a decline trend. P_1 is under the influence of groundwater, and its exponent also accords with the trend in Table 4.

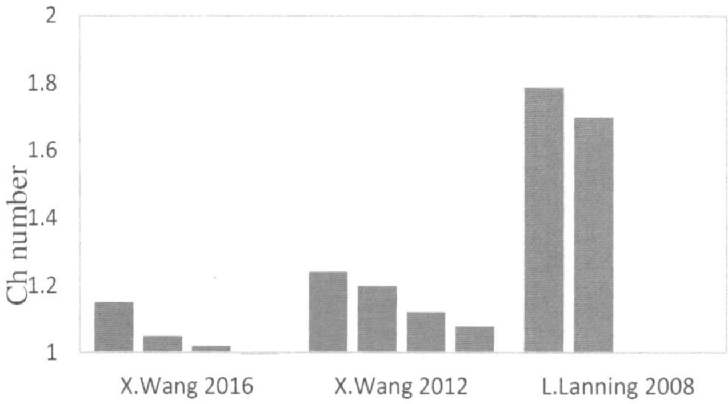

Figure 4. Ch number under different groundwater rates in references

However, we are still not sure about the effects of P_2 and P_4 on the bacterial chemotaxis. Available research results are difficult to be put together since they were obtained under different experimental conditions. At this moment, we have designed microfluidic devices with three parts of different geological properties, resembling different layers in realistic groundwater system. Next we will conduct experiments on the devices to verify the Chemotaxis number, and study systematically the influential factors on chemotactic bioremediation.

4. References

[1] Wang Z, Fingas M F. Development of oil hydrocarbon fingerprinting and identification techniques.[J]. Marine Pollution Bulletin, 2003, 47(9–12):423-452.

[2] Olson M S, Ford R M, Smith J A, et al. Quantification of bacterial chemotaxis in porous media using magnetic resonance imaging[J]. Environmental Science & Technology, 2004, 38(14):3864-70.

[3] Wang X, Lanning L M, Ford R M. Enhanced Retention of Chemotactic Bacteria in a Pore Network with Residual NAPL Contamination [J]. Environmental Science & Technology, 2016, 50(1):165.

[4] Wang X, Long T, Ford R M. Bacterial chemotaxis toward a NAPL source within a pore - scale microfluidic chamber[J]. Biotechnology & Bioengineering, 2012, 109(7):1622-1628.

[5] Lanning L M, Ford R M, Tao L. Bacterial chemotaxis transverse to axial flow in a microfluidic channel.[J]. Biotechnology & Bioengineering, 2008, 100(4):653-63.

[6] Abdelfatah E, Pournik M, Shiau B J B, et al. Mathematical modeling and simulation of nanoparticles transport in heterogeneous porous media[J]. Journal of Natural Gas Science & Engineering, 2017, 40:1-16.

Acknowledgements

This material is based upon work supports from National Basic Research Program of China [grant number 2015CB251200], National Natural Science Foundation of China [grant number 51509260], National Sciences and Technology Major Project [grant number 2016ZX05044003], Program for Changjiang Scholars and Innovative Research Team in University [IRT_14R58].

Enhanced Coagulation for the Different Mixture Ratio Raw Water of Yellow River Water and South-to-North Water Diversion Water

Jianhong Lu[1], Jin Zhang[1] and Feng Yin[2]

1 North China University of Water Resources and Electric Power, Zhengzhou, 450003, China.
2 Jiangxi University of Technology, Nanchang,330098, China.
E-mail: zhangjinhs@126.com

Abstract. This study analysed the water quality of the mixed raw water of Yellow River Water(YW) and South-to-North Water Diversion Water (SNW) in the proportion of 2:1, 1:1, 1:2. According to the actual operation condition of most water treatment plant in Zheng Zhou, PAC was used as coagulant and activated silica acid as coagulant aid for enhanced coagulation. Finally, the removal efficiency of turbidity, UV_{254}, NH_3-N for different raw water was gave. The optimum coagulant and coagulant aid (activated silica acid) dose of enhanced coagulation were determined. The results of this study would provide guidelines to the optimization operation of the local water treatment plants which used the mixed raw water of YW and SNW.

1. Introduction

After the operation of the middle route of the South-to-North Water Diversion Project, the water quality of SNW is better than that of the YW in Zhengzhou. However, it is difficult to completely switch the water source from YW to SNW in a short term. As a result, the water treatment plants in Zhengzhou would adopt dual water source supply mode which combined use YW and SNW as raw water for a long time. The change of water source mode poses new challenge to the original water treatment processes of the local drinking water treatment plant. The designs and operating parameters of the original treatment process may not be the most effective for the treatment of the mixed raw water of YW and SNW.

Enhanced coagulation refers to dosing coagulant and coagulant aid to the raw water or control the pH value, thereby enhancing the removal of natural organic matter in conventional treatment. This study aimed to investigate optimum coagulant and coagulant aid and its dosage based on the result of pilot test for the raw water that YW and SNW mixed in different proportion.

2. Experimental Procedure

The pilot test processes were composed by coagulation, inclined tube sedimentation, sand filtration, and disinfection. The flow chart was shown in Figure 1.

During the experiment, the flow rate of each system was controlled at 1 m^3/h. The coagulation process was carried out using the following conditions, rapid mixing (t=30s, G=500 s^{-1}) and three-stage flocculation (t1=5 min, G1=70 s^{-1}, t2=5 min, G2=40 s^{-1}, t3=8 min, G3=20 s^{-1}). The sedimentation pond was equipped with a 60° inclined pipe and it is 1 m in length. The depth of the sand filtration bed was 0.95 m, and sand with diameter range 0.8-1.2 mm was used as filter media. The value of turbidity,

UV_{254} and NH_3-N of filtration effluent were measured and the efficiencies were used to evaluate the treatment effect of enhanced coagulation.

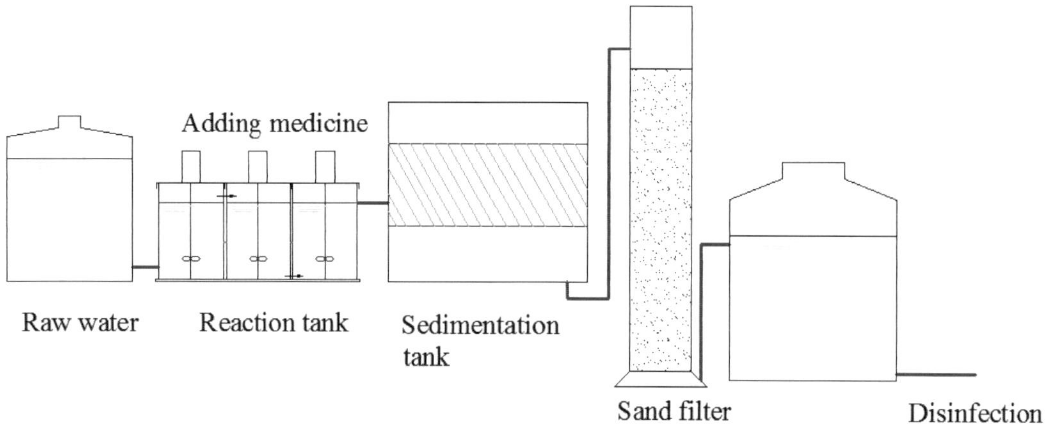

Figure 1. Flow chart of pilot treatment process.

3. Results and Discussion

The PAC dose of 15, 25 and 35 mg/L were used for coagulation, activated silica acid was dosed in the PAC and activated silica acid proportion of 5:1, and the raw water were the mixed water of YW and SNW in the ratio of 2:1, 1:1 and 1:2, respectively. The performance of PAC enhanced coagulation was investigated for three kinds of raw water.

3.1. Turbidity removal for Three Kinds *of* Raw Water

It was indicated that for different kinds of raw water, the optimum coagulant and coagulant aid dose varied greatly. It can be seen from Figure 2 that the turbidity removal efficiency for three kinds of source water all reached to 90% or even more regardless of the coagulant dosage. A highest turbidity removal efficiency of 96.5% could be achieved at the PAC dosage of 35 mg/L and activated silica acid dosage of 7 mg/L for the source water YW and SNW mixed in the ratio of 2:1. To get the 96.5% turbidity removal ratio, 25 mg/L PAC and 5 mg/L activated silica acid was needed for the source water YW and SNW mixed in the ratio of 1:1. Approximately 96.16% turbidity could be removed by PAC coagulation at the PAC and activated silica acid dosage of only 15 mg/L and 3 mg/L respectively for the source water YW and SNW mixed in the ratio of 1:2. The coagulant dosage needed to achieve the optimum turbidity removal became less with the increase of the proportion that SNW accounted for in raw water. The more of the SNW portion in raw water, the lower of raw water turbidity was.

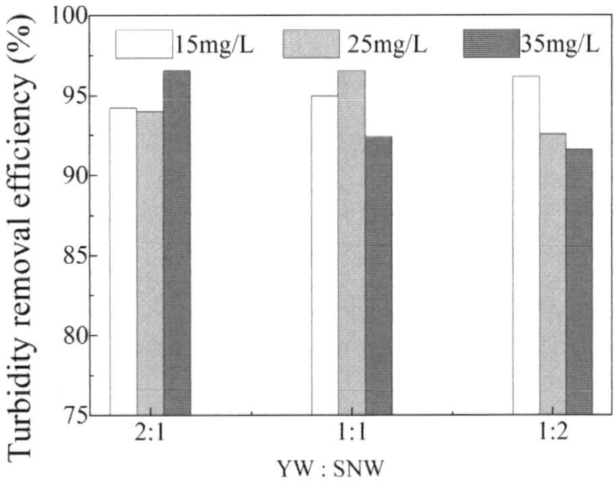

Figure 2. The turbidity removal by PAC enhanced coagulation for different kinds of raw water.

3.2. UV_{254} Removal for Three Kinds of Raw Water

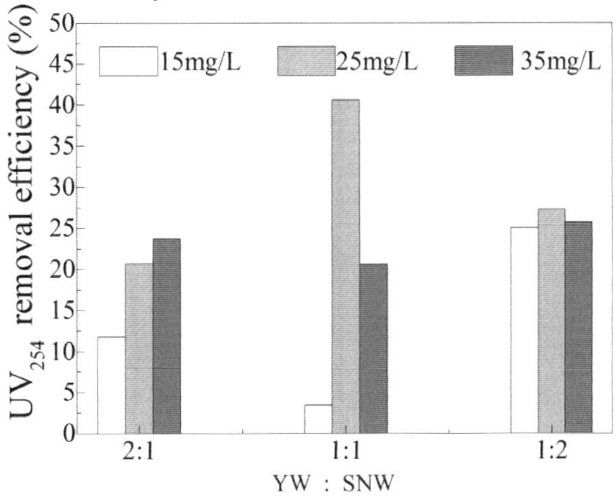

Figure 3. The UV_{254} removal by PAC enhanced coagulation for different kinds of raw water.

As shown in Figure 3, the UV_{254} removal behavior by PAC coagulation for three kinds of raw water did not follow the turbidity removal case. UV_{254} removal efficiency was relatively low in the PAC dose range of 15-35 mg/L. For the raw water YW and SNW mixed in the ratio of 2:1, UV_{254} removal efficiency increased with the increase of PAC dosage. The highest removal efficiency was achieved at PAC dosage of 35 mg/L and activated silica acid dosage of 7 mg/L. The removal of UV_{254} is positively correlated to the coagulant dosage. However, the highest UV_{254} removal efficiency was only 23.68%, which was extremely low relative to the high turbidity removal efficiency 96.55% at this PAC and activated silica acid dosage. The mechanisms involved in colloidal particle and Natural organic matter (NOM) removal could be significantly different, colloidal particle could be removed by charge neutralization and sweep flocculation, but formation of insoluble complexes between NOM and coagulant species as well as the adsorption of NOM onto freshly formed hydroxide precipitate could be the removal mechanisms of NOM [1]. Coagulation of humic acid might not follow the same charge neutralization rules associated with coagulation of mineral colloids. Due to the differences between coagulation of mineral colloids and NOM, the designs and operating procedures that are effective for removing turbidity may not be the most effective for removing NOM. Preformed polymeric Al coagulants are efficient in removing mineral colloids, but not in removing dissolved organic matter [2]. For the raw water YW and SNW mixed in the ratio of 1:1, it was observed that the

UV_{254} removal sharply increased with the increase of PAC dosage from 15 to 25 mg/L but decreased when PAC dosage increased from 25 to 35 mg/L. 40% UV_{254} could be removed by 25 mg/L PAC and 5 mg/L activated silica acid. The PAC and activated silica acid dosage for optimum turbidity removal could also remove UV_{254} efficiently, so the recommended PAC and activated silica acid dosage were 25 and 5 mg/L. For the raw water YW and SNW mixed in the ratio of 1:2, UV_{254} removal efficiency varied little with the increase of PAC dosage. The maximum UV_{254} removal efficiency was approximate 25%, and the recommended PAC and activated silica acid dosage were 15 and 3 mg/L, respectively.

3.3. NH_3-N Removal for Three Kinds of Raw Water

The NH_3-N removal performance of PAC coagulation for three kinds of raw water is shown in Figure 4.

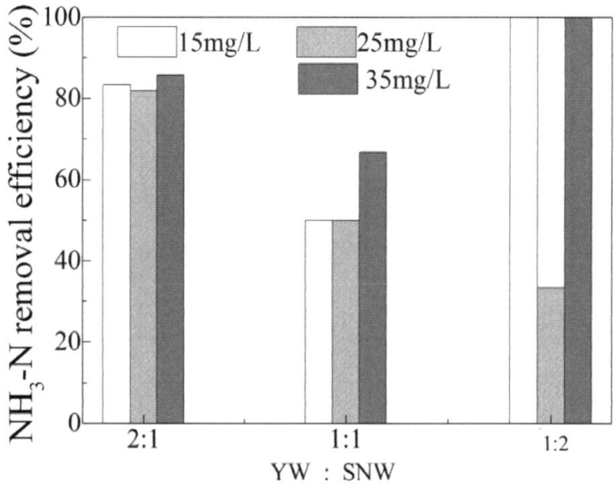

Figure 4. The NH3-N removal by PAC enhanced coagulation for different kinds of raw water.

For the raw water YW and SNW mixed in the ratio of 2:1, PAC dosage of 35 mg/L and activated silica acid dosage of 7 mg/L, which gave the maximum NH_3-N removal efficiency of 85.71%, was the optimum conditions for turbidity removal, PAC dosage had little influence in NH_3-N removal, it might because that the initial NH_3-N was relatively low (0.18 mg N /L), it was difficult to improve the NH_3-N removal in the coagulant dosage range applied in this study. NH_3-N removal efficiency was only used as a reference indicator for the optimization of PAC and activated silica acid dosage. For the raw water YW and SNW mixed in the ratio of 1:1, 50% NH_3-N could be removed by 25 mg/L PAC and 5 mg/L activated silica acid. It was 20% lower than the NH_3-N removal at the PAC and activated silica acid dosage of 35 and 7 mg/L. However, given the maximum turbidity removal achieved by 25 mg/L PAC and 5 mg/L activated silica acid, the recommended PAC and activated silica acid dosage for the raw water YW and SNW mixed in the ratio of 1:1 was 25 mg/L and 5 mg/L, respectively. For the raw water YW and SNW mixed in the ratio of 1:2, when PAC and activated silica acid were dosed at 15 mg/L and 3 mg/L respectively, NH_3-N could be almost 100% removed, moreover, the turbidity removal efficiency was high up to 96.16%. 15 mg/L PAC and 3 mg/L activated silica acid were recommended to be the optimum coagulation conditions for the treatment of the raw water YW and SNW mixed in the ratio of 1:2.

4. Conclusions

In this study, the optimum enhanced coagulation conditions were investigated by pilot test for the raw water that Yellow River water and South-to-North Diversion Water mixed in the proportion of 2:1, 1:1, 1:2, respectively. The main conclusions drawn are as follows:

(1) The turbidity removal efficiency of PAC enhanced coagulation all reached to 90% and more for three kinds of raw water. For the raw water YW and SNW mixed in the proportion of 2:1, the recommended PAC and activated silica acid dose was 35 and 7 mg/L, respectively. For the raw water

YW and SNW mixed in the proportion of 1:1, the recommended PAC and activated silica acid dose was 25 and 5 mg/L, respectively. For the raw water YW and SNW mixed in the proportion of 1:2, the recommended PAC and activated silica acid dose was 15 and 3 mg/L, respectively. The optimum PAC and activated silica acid dosage decreased with the increase of the SNW portion of raw water. Replacing YW with SNW partially would be significant to cost saving for the local water treatment plants.

(2) The UV_{254} removal efficiency of PAC enhanced coagulation was all less than 40% for three kinds of raw water in the PAC dose range of 15-35 mg/L, and UV_{254} removal varied significantly for different raw water. The NH_3-N removal efficiency was during the range of 80-100%, the PAC and activated silica acid dose had little influence on NH_3-N removal.

5. References

[1] Lu X, Chen Z and Yang X 1999 Spectroscopic study of aluminum speciation in removing humic substances by Al coagulation *J. Water Research.* 33 3271-80.

[2] Shi B, Wei Q and Wang D 2007 Coagulation of humic acid: the performance of preformed and non-preformed Al species *J. Colloids and Surfaces A: Physicochemical and Engineering Aspects.* 296 141-8.

Acknowledgments

The authors wish to express their thanks for Henan Key Laboratory of Water Environment Simulation and Treatment and the financial support from "Research and demonstration on the key techniques of technic adaptability and water quality upgrading and reconstruction of typical urban waterworks in Henan water-receiving area"(2012ZX07404-004).

Water Quality Control Technology of Sewage Treatment in Nanpu Oilfield

Guangbo Ma[1,*], Guohua Luan[2], Haoyang Li[3] and Mingjie Shi[2]

1 China National Petroleum Corporation Jidong Oilfield Branch, Tangshan City, Hebei Province, Caofeidian District 116.
2 CNPC Research Institute Of Safety&Environment Technology, Beijing City, Block A12, West Petroleum Technology Park, Xishantaxi, Shahe Town, Changping District.
3 China University of Petroleum, Beijing City, Changping District House Xue Road on the 18th.
E-mail: shpeng@cc.ctu.edu.tw

Abstract. An oilfield sewage treatment process during use often cannot continue to have good adaptability. (An oilfield sewage treatment process couldn't possess a good adaptability when putting into use.)The new equipment and new process should not be introduced fragmentarily to solve the existed problems since the sewage treatment equipment has been put into operation. Starting from the whole, many aspects should be treated such as the management standard, the operation mode, technological transformation and optimization of reagent, etc. in order to realize overall upgrade of sewage treatment system.

1. Introduction

Nanpu Oilfield which belongs to the marine oil using land drilling oilfield is located in the Bohai Sea[1]. (Nanpu Oilfield is located in the Bohai Sea, which belongs to the marine oil using land drilling.) There are 5 water treatment stations located in NP1-1D, NP2-3LP, NP3-2LP, NP403X1LP and NP4-2D, including 1 sewage treatment station and 4 water treatment stations. The summary of each water treatment station is shown in Table 1.

Table 1. Statistics of water treatment stations in Nanpu's operating area.

Serial number	name	Process	Water quality	Processing capacity	Remarks
1	NP1-1D Sewage Treatment Station	Flotation degreasing + 3-level filtration	B1	9600 square / day	flotation degreasing + 2-level filtration before 2014
2	NP2-3LP water treatment station	Single-stage dual filter filtration	B1	1700 square / day	
3	NP3-2LP water treatment station	Single-stage dual filter filtration	B1	400 square / day	
4	NP403X1 water treatment station	Fine membrane filtration	A1	800 square / day	
5	NP4-2D water treatment station	Fine membrane filtration	A1	800 square / day	

The sewage treatment station in Nanpu Oilfield undertakes the water treatment and supply of the No. 1 construction. Its daily processing capacity is much higher than the other four water treatment stations[2]. Compared with the water treatment station, NP1-1D sewage treatment station has the following three characteristics: First, The qualification rate of water quality in sewage treatment system is much lower than the qualification rate in water treatment system; the second is the suspended solid particles and qualified rate of median diameter in the output water are lower than the qualification rate of oil content; Third, the water quality in sewage treatment system has obvious volatility[3].

NP1-1D sewage treatment system is mainly responsible for water supply task of the No. 1 construction. The current formation pressure maintains less than 0.8. Compared with the water treatment, sewage treatment is trapped in the low qualified rate of the suspended solid particles and median diameter, the large fluctuations of the water quality in sewage treatment system and the situation that water quality is difficult to be guaranteed during water injection. Therefore, it has important practical significance to study and find new ways to improve the effect of sewage treatment[4].

2. The Main Process of NP1-1D Sewage Treatment

NP1-1D sewage treatment system adopts flotation degreasing and two-level filtration in the early production, which includes the main process and auxiliary process. The main process achieves the coagulating sedimentation of sewage, flotation degreasing and filtration. The auxiliary process achieves drug injection, system backwash and pollutant recovery. After the oil-gas-water mixture is separated by the three-phase separator, the oil-contaminated water enters the receiving tank and reaches the buffer tank by flotation degreasing and coagulation settling. Finally the water reaches the treated water tank for external output after being filtered through two stages. The flow chart of sewage treatment system is shown in Figure 1.

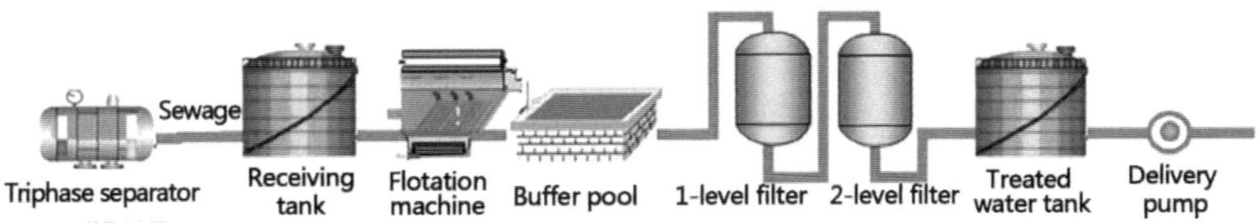

Figure 1. NP1-1D The flow chart of sewage treatment system.

With the continuous development of Nanpu Oilfield, the source of sewage in the system has been increased and the water quality has changed obviously. [5] After four years of operation, the circulating volume of internal sludge continues to increase and the operating efficiency of each equipment decreases. The original process structure, operating parameters and pharmaceutical selection have been unable to meet the requirements of the current water quality.

3. Improvement Measures and Application of Sewage Treatment System

Focused on the problems of NP1-1D sewage treatment system, we have made treatments through the aspects of the management standards, management models and process optimization in order to improve the efficiency of the equipment and achieve the system's overall upgrade. [6]

3.1. Standardization Management

According to the company's water-quality management principle of "taking control of the source, controlling by stages, strengthening the monitoring and ensuring compliance", the segmented assessment system of water quality has been established. The water treatment system is divided into seven sections and the water quality standards for each section are formulated. In the meantime, nine improvement measures are analyzed and formulated to achieve the goal of controlling water quality, take measures on the management of water quality, and strive to form a long-term mechanism. [7]

3.2. Implement the Model of Note Management

Based on functional division, the model of node management is implemented[1]. According to the independence and synergy of sewage treatment facilities, the system is divided into crude oil dehydration zone, sewage sedimentation zone, sewage filtration zone and output zone. The four-node water quality standard in Table 2-1 is formulated. By improving each module's qualified rate, this model of node management maximizes the efficiency of the sewage treatment system. System node is divided as shown in Figure 2.

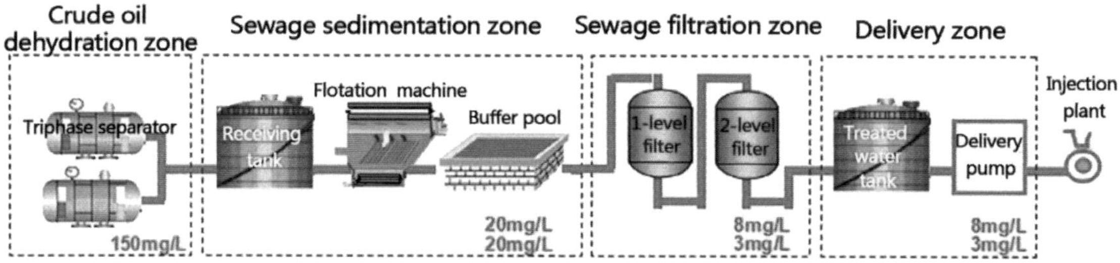

Figure 2. Node Partition map of sewage treatment system

Table 2. Node water quality standards of sewage treatment system.

	crude oil dehydration zone	sewage sedimentation zone	sewage filtration zone	outp ut zone
Oil content（mg/L）	150	20	8	8
Suspended particle content（mg/L）	/	20	3	3
Diameter of suspended particle（μm）	/	/	2	2

3.3. Carry Out Optimization and Transformation of Process

Following the principle of node control, the main process facilities and auxiliary process facilities of the node module are transformed to improve the qualified rate of node water quality.

3.3.1. Crude oil dehydration zone. The function of crude oil dehydration zone is liquid separation dehydration, including three-phase separator dehydration process and de-emulsifier injection process, in which fluid quality and oil-water separation effect are the core elements. In order to improve fluid quality and oil-water separation effect, this node takes four measures: the first is to implement the policy of liquid approval system and control flushing wells, rainwater and industrial sewage into the collection and transportation system; the second is to control the three-phase separator's operating parameters, adjust the oil-water interface and water level to improve the separation Quality; the third is to install stirring device of de-emulsifier to achieve uniform concentration of drug; the fourth is to optimize the amount of de-emulsifier by controlling the oil content of effluent and adjusting the injection concentration.

Through the optimization, the dosage of de-emulsifier is gradually adjusted from 500kg / d to 250kg / d. The oil content of effluent is controlled within 50mg / L which is far below the index of 150mg / L.

3.3.2. Sewage sedimentation zone. The function of the sewage sedimentation zone is to complete sewage receiving, flotation degreasing and coagulation settlement, including receiving tank, flotation machines, buffer tanks and medicament injection systems. The degreasing effect, sedimentation effect and mud-removing effect are the core elements. To enhance the degreasing effect, sedimentation effect and mud-removing effect, the node carries out the measures of process infrastructure maintenance and pharmaceutical compatibility test: The first is to replace flushing wheel in buffer tank, optimize automation control parameters of flotation machine, implement artificial silt-removing and mud valve replacement of receiving tank; the second is orthogonal test designed by using water purification agent, flocculant and coagulant. The best treatment effect is the use of 50mg / L water purification agent, 60mg / L flocculant and 50mg / L coagulant resulted from sampling three-phase water and analyzing turbidity indicators.

Through facilities maintenance and pharmaceutical research, it greatly reduces the amount of circulating sludge in the system and lowered the turbidity of the effluent from the sedimentation zone. The oil content and suspended solids in the effluent are strictly controlled within 20mg / L.

3.3.3. Sewage filtration zone. Sewage filtration zone includes 1-level dual filter, 2-level fiber bundle filtration. Due to filtration system is the end of sewage treatment equipment, water quality must meet the B1 standard. However, there are a series of problems during the operation, which are the poor filter effect, the serious filter leakage, the incomplete backwash, the filter board knot and short regeneration period of the filter. In order to improve water quality, system maintenance and optimization have been implemented.

Through the equipment maintenance and filter replacement of 1-level dual filter, 2-level fiber bundle filtration, the suspended solids content and the median diameter in the output water have been significantly reduced. The qualified rate of water quality has increased from 80.9% to 92% with 6 months of validity.

Through the research, it is found that 1-level filter is poor because of small cloth area and fast water flow caused by irrational fabric structure, leading to poor filtration and backwashing. The 2-level filter is poor because the traditional fiber bundle filter has been unable to meet the demand, which needs to be replaced by a new filter. For the existing problems, the following aspects have been restructured: the water distribution mechanism of the filtration system, the type selection of the filter material and the overall layout. The first method is to replace the screen tube with a radial-shaped wire tube to increase the cross-sectional area of water distribution and improve the effects of filtering and backwashing; the second method is to use modified fiber ball filter instead of fiber bundle filter to improve the filtration effect; the third method is to change the four 1-level filtration into a 3-level filtration to improve filtration accuracy and system stability. Comparison chart of dual filter before and after the transformation is shown in Figure 3.

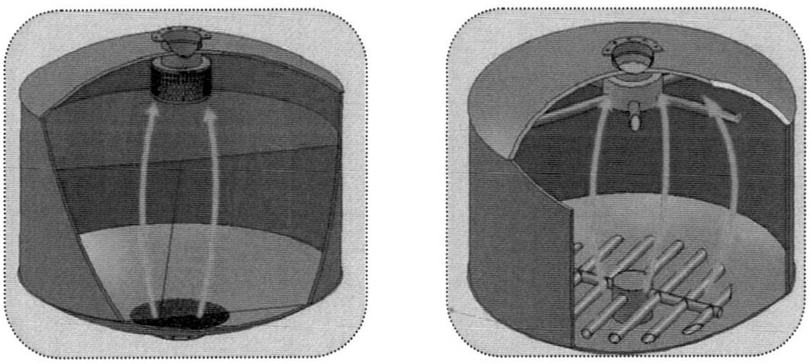

Figure 3. The comparison chart of dual filter before and after the transformation.

Through the optimization and transformation of filtration system, the problems of the filter board knot, short regeneration period of the filter, poor effluent effect and unstable water quality are effectively solved. The suspended solids content and the median diameter of the output water have been obviously decreased.

3.4. Treatment Effect

By standardizing management standards, implementing node management mode and carrying out technological transformation measures, the qualified rate of water quality of output water B1 in the sewage treatment system has been raised from 80.9% to 100%. There are 11 consecutive months which have achieved the standard stably and 7 of which reached standard A2. The comparison chart of water quality in NP1-1D sewage treatment system before and after treatment is shown in Figure 4:

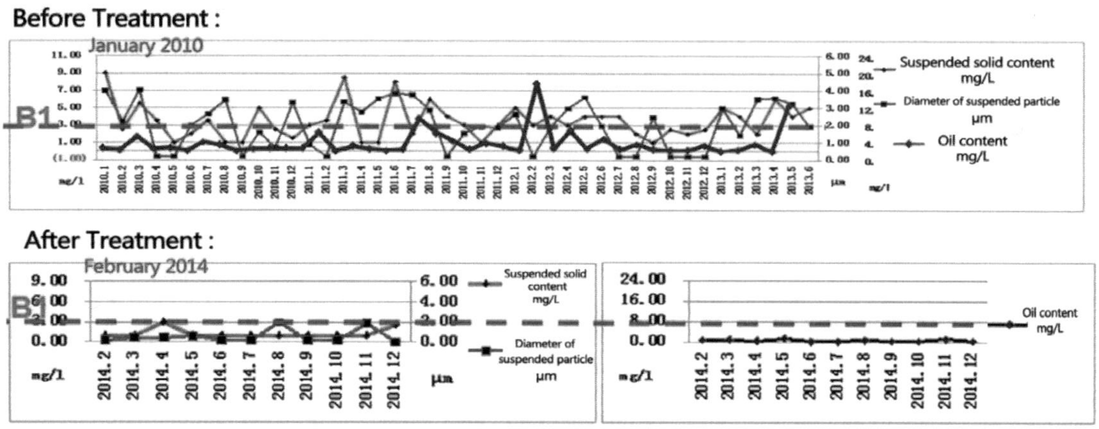

Figure 4. The comparison chart of water quality in NP1-1D sewage treatment system before and after treatment

4. Conclusion

Firstly, the management system, operating mode and process optimization should be proceeded at the same time to upgrade system in the face of the problem of poor output water quality in sewage treatment system.

Secondly, the node management mode should be innovated based on functional division and the maximum efficiency of the sewage treatment system could be achieved by improving the qualified rate of each module.

Thirdly, the transformation of the system must rely on the current facilities. At the beginning, the process improvement within the system should be set out in order to achieve the desired treatment effect without adding new equipment.

Finally, the management of sewage treatment system must pay attention to every step. The filter replacement, tank dredging, equipment maintenance and pharmaceutical optimization should be carried out periodically.

5. References

[1] Hua Wei. *The improvement of sewage treatment filtration process in Sabei Oilfield* [J]. Oil-Gasfield Surface Engineering, 2012.10.

[2] Wu Baoguo, Yu Yilian, Wang Hongxia. *Transformation of sewage treatment process and improvement of water-injection quality* [J]. Inner Mongolia Petrochemical Industry, 2008, (15).

[3] Yonggui Wang, Bernard A Engel, Panpan Huang, Hong Peng, Xiao Zhang, Meiling Cheng, Wanshun Zhang. Accurately early warning to water quality pollutant risk by mobile model system with optimization technology[J].Journal of Environmental Management, 2017.2, Pages 122–133.

[4] DongliangLu, KeqiangLi, ShengkangLiang, GuohongLin, XiulinWang. *A coastal three-dimensional water quality model of nitrogen in Jiaozhou Bay linking field experiments with modelling* [J]. Marine Pollution Bulletin, 2017.1, (114).

[5] Peng Li, Dexi Zhao, Yunge Zhang, Linquan Sun, Huimin Zhang, Meirong Lian, Baoan Li. *Oil-field wastewater treatment by hybrid membrane-aerated biofilm reactor(MABR) system* [J]. Chemical Engineering Journal, 2015.5, (264).

[6] Qing Xin, Xingwang Zhang, Zhongjian Li, Lecheng Lei. *Sterilization of oil-field re-injection water using combination treatment of pulsed electric field and ultrasound.* [J]. Ultrasonics Sonochemistry,, 2009.1, (16).

[7] Jian Jiang, Zongkuan Liu, Yanling He. *Development Trend:Rotary fibre disk filtration in the treatment of oilfield wastewater* [J]. Filtration + Separation, 2014, (51).

Applications of Biological Magnetic Effects in Water Treatment

Li Tianhang[1,a], Jia Xinqiang[2], Wang Xiaoyu[1], Yu Cong[1] and Wang Jiabin[1,b]

1 School of Architecture and Civil Engineering, University of Jinan, Jinan, China
2 IOP Publishing SAES Environmental Science and Science and Technology Co.,Ltd., Jinan, China
E-mail: a 474386519@qq.com ; corresponding author:b cea_wangjb@ujn.edu.cn

Abstract. In order to enhance the efficiency of wastewater treatment, magnetic field is usually used to promote microbial activity, and it has become a novel water treatment technology. In this paper, the biological magnetic effect in water treatment are systematically and logically reviewed with respect of mechanisms and applications. The major problems existing in the current magnetic field strengthening technology are pointed out, and the promising application of the biological magnetic effect in water treatment is prospected.

1. Introduction

Recently, serious pollutions of water environment attract more attentions. One of the general water treatment technology is aerobic biological treatment, but with disadvantages of low volume load and unstable treatment efficiency. Aiming at these disadvantages, magnetic field was introduced and it could affect the growth and metabolic activity of microbes, which is called the biological magnetic effect[1]-[3] and is expected to improve the volume load of traditional biological treatment technology and improve its ability to resist impact load[4] [5]. The issues in the applications of the biological magnetic effect have been paid much attention in the scientific fields of water treatment. This novel water treatment technology based on the biological magnetic effect have a good application potential.

2. Current Research in the Biological Magnetic Effect

There are various forms of biological magnetic effects in water treatment. They often reflected through the rate of biological metabolism, the growth cycle, the treatment efficiency of the characteristic pollutants, and the form of the biological enrichment rate.

2.1. *Roles of Magnetic Field in Microorganism Sensitivity*

Xing Yicun[6] studied the growth and metabolism of microbes under the condition of low frequency magnetic field. The study found that the sensitivity of different microbes to the magnetic field was different. The growth and reproduction of microorganism can be promoted or suppressed under the same magnetic field. By comparing the effects of different intensity magnetic fields on the growth of microorganisms, Li Jie and others[7] found that: 100 mT field strength showed a positive effect on actinomycetes and fungi, while 100 mT field strength showed a inhibitory effect on the growth of bacteria. However, the number of soil microbes of different groups also increased in varying degrees under the effect of 300mT field strength.

2.2. *Effects of the Biological Magnetic Effect on Metabolic Rate*

Justo[8] in the extremely low frequency magnetic field under the condition of cultivation of Escherichia coli 1 ~ 12 hours, found the activity of Escherichia coli is 100 times of the group without magnetic field under the magnetic field strength of 0.1 T for 6.5 hours. And under the magnetic field intensity of 0.055T for 12 h, the maximum dry weight of Escherichia coli appeared, which indicated the growth rate of the fastest.

2.3. Effects of the Biological Magnetic Effect on Pollutant Removal Efficiency
ShiBo[9] investigated the start-up characteristics of the membrane of the reactor under the magnetic field intensity between 0-400Gs. The study found: 100Gs magnetic field can inhibit the growth of microorganisms and reduce the removal effect of organics; 200Gs, 260Gs and 320Gs magnetic field can improve the removal effect of organics. Among them, the 260Gs magnetic field has the greatest increase in the removal rate of organic matter, and the time required for the biofilm to achieve steady state operation is the shortest. Tu Ningyu[10] found that the magnetic field promoted the growth of the strain and the promotion became stable as the time of magnetization prolonged by studying the effect of magnetic field on a strain of copper bacteria. Xu Yanbin[11] [12] studied the colony size, bacterial density, the number of mixed bacteria and the effect of chromium removal of two anaerobic chromium removal bacteria (A, B) under different magnetic induction intensity (0 mT, 2.4 mT, 6 mT, 10 mT, 17.4 mT). The results showed that the magnetic field has a cumulative effect on the growth of microbes, and the biological magnetic effect is lagging, and the sensitivity of different strains to magnetic field intensity is also different. Under the magnetic induction intensity of 10 mT, two strains of anaerobic chromium removal showed inhibition of growth. Under the magnetic induction intensity of 6 mT, the colony of B bacteria reached the maximum, and the A bacteria showed the strongest adaptability. And A and B mixed bacteria (ratio 1 to 1) had the fastest propagation speed, and bacterial density increased by 65%. The removal rate of chromium is 36.17%, which is 8.84% higher than that of the unadded magnetic field.

2.4. Mechanisms in the Biological Magnetic Effect
Wang Hongshuo[13] et al investigated the effect of the extremely alternating magnetic field on the relative absorbance of the leaks in the cell and the DNA of the thalliEscherichia coli, staphylococcus aureus and yeast as experimental strains. The results showed that the two polar alternating magnetic fields could damage the cell wall of microorganism and change the structure of DNA
Gu Jiguang[14] studied the changes of catalase, pulse enzyme and polyphenol oxidase activity in soil microorganism under different magnetic field strength. The results showed that the effect of magnetic field intensity on soil microbial enzyme activity is also different. Han Qingxiang[15] and others studyed the effect of magnetic field effect on water treatment. It was found that under medium magnetic field (below 50mT), the activity of microbial enzymes increased, and the processing efficiency increased by 11%. Coneracka[16] and others have found that if the proteins and enzymes are immobilized on the surface of the magnetic particles, the activity can be kept above 90% of the original activity.Wang Xiangsan[17] determined the effect of magnetic treatment on dehydrogenase activity by measuring the decolorization time of methyl blue. It was found that the time needed for decolorization of methyl blue reduced from 29 hours to 24 hours with the increase of magnetic field intensity, indicating that magnetic treatment enhanced the ability of microorganism to oxidize pollutants. The main reason for improving the efficiency of water treatment by the biological magnetic effect is that the existence of magnetic field improves the activity of microbial enzyme

3. Applications of the Biological Magnetic Effect in Water Treatment

3.1. Applications of Biological Magnetic Effect in Activated Sludge Process
Hulya Yawz and other [18] set the magnetic field outside the reactor of the activated sludge process to treat the simulated wastewater. It was found that the removal rate of organic pollutants was 44% higher than that in the non magnetic field. Geng Shuying [19] found that adding a magnetic field to SBR can improve the removal efficiency to a certain extent. When the magnetic field intensity is 700Gs, the total nitrogen removal rate is increased by 30.89%, and the abundance of the biological

community is also the highest. Lu Guangli[20] and other add the right amount of magnetic powder to the reflux sludge. The results showed that the removal rate of COD can be increased by about 15%, the structure of the sludge flocs and settling performance can be improved compared with the conventional activated sludge process. Sun Shuiyu and other [21] think that the effect of magnetic particle enhanced activated sludge process is better than the traditional activated sludge process. The removal rate of COD and turbidity is greatly improved, and the floc structure and sedimentation performance of sludge are improved obviously. Sui Weiyan[22] studied the effect of nitrogen and phosphorus removal under the conditions of magnetic field intensity of 1500Gs and 800Gs, respectively. The results showed that: COD, ammonia nitrogen and total phosphorus removal rate has improved to a certain extent in the two kinds of magnetic field intensity, and activated sludge under the action of 1500Gs magnetic field can rapidly improve the removal capacity of organic matter in the early stage of the experiment, and activated sludge under the action of 800Gs magnetic field can improve the removal ability of organic matter and phosphate
Zhao Xiang [23] used magnetic flyash fillers in MBBR reactor for hanging film experiment. The results showed that magnetic fly ash fillers were better than non magnetic fly ash fillers in film-forming performance and pollutant removal efficiency of sewage, and the time of hanging film was shorter. Zhu Ling Li [24] quickly activated anammox reactor by adding magnetic particles. It was found that the activity of anammox bacteria was the strongest and the total nitrogen removal rate was the highest under the 0.78mT magnetic field intensity. Zhang Dong[25] put the new high-strength magnetic particles prepared into the self nutrient denitrification system, and studied the effect of magnetic field intensity on the enrichment rule and pollutant removal efficiency of anammox bacteria. The results showed that the start-up time of the reactor adding magnetic particles is shorter, and the removal rate of total nitrogen and ammonia is higher, and the abundance of the biological community is higher.

3.2. Application of the Biological Magnetic Effect in Biofilm Method

Pan Yongzhang[26] applied new magnetic carrier to biofilm reactor for experiment of wastewater treatment. It was found that the removal rate of COD and NH3-N in magnetic carrier biofilm reactor is 5% and 20% higher than that of non magnetic carrier. The amount of biofilm and the amount of suspended sludge in the magnetic carrier reactor is only 70% of the non magnetic carrier reactor, which indicated that the presence of magnetic field is more beneficial to sludge reduction. Wang Weisheng[27] compared the magnetic biological carbon filler and ordinary biochar biofilm performance and pollutant removal in BAF. It was found that the biofilm formation time of magnetic biochar BAF increased by 13.3% compared with the conventional biochar BAF. The removal rate of ammonia nitrogen and total phosphorus of magnetic biochar BAF was significantly higher than that of conventional biochar BAF, and the impact load was stronger.Li Chihao[28] developed an integrated magnetic filter based on A/O technology, and hung the membrane into two kinds of fillers, ordinary activated carbon filler and magnetic activated carbon filler in turn. It was found that the magnetic packing was higher than the ordinary filler at the speed of film hanging and the removal rate of COD and ammonia nitrogen. Jiao Fuqiang[29] prepared magnetic filter media and started the hanging film experiment under 6mT magnetic field strength. The removal efficiency of simulated sewage was studied under four conditions of 12mT, 20mT, 30mT and 40mT at room temperature and low temperature respectively. It was found that the magnetic field of 12mT has an inhibitory effect on the removal efficiency of the filter, and the removal rate of COD and NH3-N was reduced. Under the magnetic field intensity of 20mT, 30mT and 40mT, the magnetic field played a positive role in the removal effect of the filter. The average removal rate of COD increased by 4.23%, 8.41% and 3.87% respectively, and the average removal rate of ammonia nitrogen increased by 4.28%, 14.38% and 6.72%, respectively. The removal of phosphorus can be promoted under the magnetic field intensity of 12mT and 20mT and the removal of phosphorus was inhibited under the magnetic field intensity of 30mT and 40mT. Cheng Yangyang[30] used magnetic anaerobic biofilter to treat acid polymetallic smelter wastewater. It was found that when the magnetic carrier dosage was 6g/L, the removal rates of COD, F- and Cd2+ in wastewater could reach 60%, 70% and 90%, respectively. Compared with anaerobic biological system without magnetic carriers, the removal rate of COD, F- and Cd2+

increased by 7.84%, 7.37% and 1.32%, respectively. Jin Xiao Bei[31] immobilized the microorganism after acidity of acidic copper plating wastewater under the presence of magnetic field on the surface of magnetic carrier to treat acid copper plating wastewater. The results showed that the magnetic field had a significant effect on the immobilization of microbes and the removal of copper ions, but the magnetic field intensity needed was different. When the center magnetic field of the carrier is 260Gs, the fixed amount of microorganism is the largest. While the center of the carrier is 140Gs, the removal rate of copper ion is the highest

4. Conclusion

In summary, the water treatment process based on the biological magnetic effect are promising of water treatment with the advantages of high efficiency, without secondary pollution and low energy consumption. However, the mechanisms and process of the biological magnetic effect is still confused. Their key influencing factors should be determined and then optimized on account of biological growth. The behaviors and mechanisms in the biological magnetic effect promoting the degradation of characteristic pollutants should be further investigated.

5. References

[1] Xu Xilin, Guo Siyuan, Li Lin. Effect of Dynamic Magnetic Field on Microorganisms[J]. Journal of South China University of Technology.2006, (12):47-50.
[2] In vitro biological effects of magnetic nanoparticles[J]. Chinese Science Bulletin,2012,57(31):3972-3978.
[3] Chathurika, D.A., Peter, M.F., Malka, N.H.. Analysis of biological effects and limits of exposure to weak magnetic fields[P]. Information and Automation for Sustainability (ICIAFs), 2010 5th International Conference on,2010.
[4] Yang Fang, Yao Cuang, Luo Huan. The effect of magnetic microorganism and its application in waste water treatment[J]. Pearl River. 2014,35(05):76-79.
[5] Bingrui Maac;Sen Wangb;Zhiwei Lia;Mengchun Gaoac .Magnetic Fe3O4 nanoparticles induced effects on performance and microbial community of activated sludge from a sequencing batch reactor under long-term exposure[J].Bioresource Technology,2017Vol.225： 377-385
[6] Xing Yicun,Zhou Yifan, Effeets of low frequeney magnetic fields on microorganisms[J].Journal of Hainan Normal University. 2001,(03):34-39.
[7] Li Jie, Yi Yanli, Jiao Ying. Effect of Magnetic Field on Soil Microbes and Soil EnzyineActivities in Brown Earth[J]. Chinese Journal of Soil Science. 2007,(05):957-961.
[8] Justo O R,Perez V H,Alvarez D C,et al.Growth of Escherichia coli under extremely low-frequency electromagnetic fields[J].AppliedBiochemistry and Biotechnology,2006,134:155~163.
[9] Shi Bo Study on the Effects of Low-intensity Steady Magnetic Field onWastewater Treatment with Biofilm Reactor[D]. Jinan University.2010
[10] Tu Ningyu. Study on the effect of magnetic powder on the treatment of heavy metal wastewater by biological method[D]. Guangdong University of Technology. 2004.
[11] Xu Yanbin, LI Xun, Sun Shuiyu. Effect of magnetic field on the growth of two strains ofanaerobic Cr(VI)-removing bacteria[J]. Electroplating&Finishing,2005,(07):46-49.
[12] Xu Yanbin, LI Xun, Sun Shuiyu. Effect of magnetic field on removal efficiency of Cr6+ in an ASBR[J]. Water Resources protection,2005,(04):26-29+46.
[13] Wang Hongshuo, Ye Shengying, Song Xianliang, et al. Effect of two-pole alternating magnetic fields on microbes at the micro-level[J].Transactions of the CSAE, 2007,23(8): 253-256.
[14] Gu Jiguang, Zhou Qixing. Effects of Soil Magnetization on DefensiveEnzymes in Rape Plants[J]. Journal of basic science and Engineering,2002,10(3):263-267.
[15] Han Qingxiang, Shao Fengqin. The Increasing Effect of Magnetic Field on WastewaterTreatment by Activated Sludge Process[J]. Journal of Fushun Petroleum Institute ,2002,(03):8-10

[16] M. Koneraka, Kopcansky P, Antalik M et al. I Immobilization of proteins and enzymes to finemagnetic particles. Joural of Magnetism and Magnetic Materials. 1999, 201(1): 427430

[17] Wang Xiangsan, Wang Ping. Biological effects of magnetization in sewage treatment[J]. Environmental Science and technology,2000,(02):33-36

[18] Hulya Yavuz, Serdar Swastewater treatment withcelebi. A typical application of magnetic field influidized bed biofilm reactor. Chen Eng Commun,2003, 190: 5 — 8

[19] Geng Shuying，Fu Weizhang，Wang Jing. Treatment Efficiency and Microbial Community Diversity in a Magnetic Field Enhanced Sequencing Batch Reactor (SBR) [J]. Environmental Science,2017,(11):1-12

[20] Lu GuangLi, Zhao Qingxiang. Investigation on the Ferromagnetic Powder-SupplementedActivated Sludge Process Technology[J]. Urban environment and urban ecology. 1998,11(2): 10-12.

[21] Shuiyu, Liu Hong, Xie Guangyan, et al. A study on the treatment of restaurant wastewater by magnetic powder-intensified activated sludge process[J]. Environmental pollution and Prevention,2003,25(3):170-172.

[22] Sui Wei-yan, Song peng, Han Tiantian. Contrastive Studies on Experiment of Activated Sludge Method Enhanced by Magnetic Field for Treating Urban Wastewater[J].Journal of Anhui Agricultural Sciences,2011,(33):20469-20471+20540.

[23] Zhao Xiang, Membrane Experimental Study of Magnetic Fly Ash Ceramsite[J]. Guangdong chemical industry,2017,(09):36-38+63.

[24] Zhu LingLi. Magnetic Particles Induce Anammox bacteriaEnrichment and the Autotrophic Nitrogen RemovalSystem quickly Start[D]. University of Jinan.2016

[25] Zhang Dong. Study on the Start up and Operating Characteristics of Autotrophic Nitrogen Removal Process Enhanced by NdFeB Magnetic Particle[D]. University of Jinan.2017

[26] Pan Yongzhang, Zhang Zhiyong, Shi Bo. Domestic sewage treatment using a magnetic carrier bioreactor[J]. China Environmental Science,2011,31(05):734-739.

[27] Wang Weisheng. Experimental investigation on municipal wastewater treatment by BAF packed with magnetic biochar[D]. Jiangsu University.2016

[28] Li Chihao. A Dissertation Presented to Kunming University of Science and Technology for Master Degree of Engineering[D]. Kunming University of Science and Technology.2015

[29] Jiao Fuqiang. The Magnetic Strengthen Biological Filter Building Based on The Microbial Magnetic and Its Application Research[D]. University of Jinan.2016

[30] Cheng Yangyang. The characteristic analysis on disposal of acidmulti-metal and organic smelting compositewastewater through the magnetism-anaerobicbiofilter[D]. Xiangtan University.2014

[31] Jin Xiaobei, Chao Yunlong, Li Hongjun. Treatment of acid copper plating wastewater by magnetic combinedimmobilized microorganism technique[J]. Chinese Journal of Environmental Engineering2016,10(03):1041-1047.

IOP Conf. Series: Earth and Environmental Science **178** (2018) 012013 doi:10.1088/1755-1315/178/1/012013

Biodegradation and Decolorization of Dye Wastewater: A Review

Yingying Gao [2], Bo Yang [1, 2], Qing Wang [2]

1 State Environmental Protection Engineering Center for Pollution Treatment and Control in Textile Industry, Shanghai 201620, PR China.
2 College of Environmental Science and Engineering, Donghua University, Shanghai 201620, PR China.
E-mail: yangbo@dhu.edu.cn

Abstract. The total production of all kinds of dyes in China has reached 900,000 tons. Dyes are used in various industries, such as textiles, foods, paper-making and plastic industries. In these industries, a large quantity of dye wastewater is produced and dye contamination changes the water color, enhances the eutrophication, depletes oxygen and ultimately disturbs aquatic organisms to great extent. Dye wastewater has become one of the key pollution sources. In recent years, dye biodegradation has been extensively explored. The paper reviews biodegradation of dyes under aerobic, anaerobic, anaerobic-aerobic conditions and the degradation of dyes by microorganisms. Moreover, we summarized the possible degradation mechanisms of dyes to provide the basis for biological treatment of dye wastewater.

1. Introduction

In recent years, the production of dye wastewater is increasing rapidly. It is estimated that more than 100,000 tons of dyes are produced each year and about 10% of the dyes are discharged into the environment, thus causing serious harm to aquatic organisms and human health [1]-[3]. In the dye production process, benzene, naphthalene, anthraquinone and other compounds are used as raw materials and usually chelate with metals or salts to generate dye wastewater containing acids, alkali, salts, halogen, hydrocarbons, nitro, amine, dyes, corresponding intermediates and other substances. Moreover, some dyes may be decomposed to produce carcinogenic aromatic amines or dyes containing pyridine, cyanide, phenol and heavy metals such as mercury, cadmium, and chromium. The wastewater components are complex and toxic [4]-[6]. It is necessary to develop effective ways to degrade dyes and other organic contaminants from wastewater. Many researchers at home and abroad are exploring new methods for the treatment of dyestuff wastewater. China has initially developed some treatment methods through experimental studies, including physical, chemical, biochemical and several combined treatment methods. Physical and chemical methods have some limitations including high operating cost, interferences from other components in the wastewater and large amounts of sludge, whereas biological methods have low operation costs and stable effects. Therefore, biological methods have been widely used [7]-[10]. Biological treatment methods of azo dye have some degradation effects and the biodegradation mechanisms and ways of the dyes have been explored. The isolation and identification of dye intermediates is important in the exploration of the biodegradation mechanism and degradation pathway of dyes [11]. Different dye structures involve different biodegradation pathways. The biodegradation pathways of dyes mainly depend on the molecular structures of dyes, microorganisms, related enzymes and other factors. However, in general, the biodegradation process of dyes is generally divided into two phases. In the first phase, under anaerobic conditions, decolorization is realized through reduction and pyrolysis of dyes, but harmful aromatic amines are generally produced. In the second phase, under aerobic conditions, the intermediate aromatic amines are further utilized by aerobic bacteria in aerobic reactors and decomposed into small inorganic molecules such as CO_2 and NH_3, which are utilized by microorganisms to synthesize protoplasms and ultimately realize dye degradation [12], [13]. Reductive degradation of dyes may be ascribed to azo-reductases, redox mediators, and chemical reducing agents such as sulfides or compositions. Another degradation mechanism of dyes may involve cytosolic flavin-dependent reductions, transfer the electrons to the azo dye via soluble flavin, and finally decolorize dyes. Therefore, the biodegradation of dyes has gradually become one of the hotspots in the environmental field. The dye biodegradation methods include aerobic treatment, anaerobic treatment, combined aerobic-anaerobic treatment, and microbial treatment. The four methods of decolorization and degradation of dyes are discussed below.

2. Decolorization and Degradation of Dye Under Aerobic Conditions

Dye degradation under aerobic conditions is mainly catalyzed by specific enzymes and most of them are azo reductase. Some aerobic microorganisms in the presence of azo reductase and other factors can reduce azo dyes and produce intermediate metabolites such as aromatic amines, which can be further degraded by microorganisms under aerobic conditions so as to achieve the complete mineralization of azo dyes. Liang et al. [14] studied the degradation of azo dye alizarin yellow R (AY) in a combined process of iron-carbon micro-electrolysis and aerobic bio-contact oxidation. Under the optimum conditions (the hydraulic retention time of 6 h and the reflux ratios of 1 and 2), AY degradation efficiency in the final effluent was more than 96.5% and the total organic carbon (TOC) removal rates were 69.86% and 79.44%, respectively. Iron carbon micro-electrolysis significantly increased the removal rate of AY and the generated small-molecule acids and aldehydes could be further degraded in aerobic reactors. The biological method combined with physical and chemical methods provides a new method and choice for the treatment of dye wastewater.

3. Decolorization and Degradation of Dye under Anaerobic Conditions

Anaerobic digestion of dye wastewater is a promising technology. The decolorization and degradation of azo dyes under anaerobic conditions is a relatively simple and nonspecific reduction process involving the azo reduction mechanism relying on redox intermediates. In direct enzymatic catalysis, non-specific azo reductase in microorganisms directly catalyzes the reduction of azo dye. However, the existence of this enzyme has not been directly verified. The reduction reaction relying on cofactors and redox mediators may be the main mechanism for the degradation of azo dyes under anaerobic conditions. At present, the main redox intermediates are FAD and NAD(P)H, etc. and they can receive electrons from an azo-reductase and then transfer the electrons to the azo dye. Then azo-bonds (-N=N-) are destroyed to produce aromatic amines, which decolorize the azo dye. The spontaneous non-specific reduction reaction between the azo dye and redox intermediates mainly depends on the redox potential. According to related reports, under the anaerobic conditions, the redox potential of 50 mV can lead to effective decolorization of azo dyes. However, the cleavage of the chromophore groups of dyes results in a colorless, odorless, and toxic intermediate metabolites of aromatic amines, which are commonly mineralized under aerobic conditions. Furthermore, it has been reported that the presence of oxygen usually suppresses the reduction activity of the azo bond because aerobic respiration may dominate the utilization of NADH and impede the transfer of electrons from NADH to the azo bond. It has also been reported that aromatic amines containing functional hydroxyls and carboxyl groups can be mineralized under anaerobic conditions. Some aromatic amines which are prone to undergo self-oxidation can be degraded under anaerobic conditions.

Cao et al. [15] used the electro-assisted microbial system to degrade Reactive Brilliant Red X-3B (RBRX-3B) dye under anaerobic conditions and compared the degradation of RBRX-3B by an electro-assisted microbial system (EAMS), a microbial system (MS) and an electrochemical system (ECS). The comparison results showed that the degradation efficiency of RBRX-3B in EAMS was 10% higher than the total degradation efficiency of MS (61.9%) and ECS (27.1%). The EAMS in which the anaerobic microbial system was combined with the electrochemical system could accelerate RBRX-3B degradation. RBRX-3B was transformed to nontoxic hydrocarbons and toxic compounds in

EAMS so as to achieve the complete mineralization of the dye. This technology provided a new option for the treatment of dye wastewater.

4. Decolorization and Degradation of Dyes under Anaerobic-Aerobic Conditions

In general, some azo dyes with simple structures can be completely biodegraded, but the aromatic amines of azo dyes under anaerobic conditions often cannot be further degraded under anaerobic conditions, thus resulting in the accumulation of toxic substances. These aromatic amines are biodegradable under aerobic conditions, so the combined aerobic-anaerobic process for treating azo dye wastewater is a good way to completely degrade the dye.

Melgoza R M and Cruz A [16] studied the degradation of dyes in textile wastewater by anaerobic/aerobic reactors and observed that the colorant disperse blue 79 (DB79) was converted into aromatic amines in the anaerobic phase to decolorize wastewater by microbes. The generated aromatic amines were subsequently mineralized in the aerobic section, thus eliminating the dye toxicity. Mohan S V and Babu P S [17] found that the degradation efficiency of dyes under anaerobic conditions was higher than that under aerobic and anaerobic conditions, so they believed that anaerobic conditions had the dual advantages of aerobic and anaerobic degradation of dyes. Ong et al. [18] compared the biodegradability of four heavy azo dyes in anaerobic/aerobic (SBR1) reactors and aerobic (SBR2) reactors and found that anaerobic/aerobic (SBR1) and aerobic (SBR2) reactors realized 95% COD removal, whereas the anaerobic/aerobic (SBR1) and aerobic (SBR2) reactors realized the decolorization rates of 93% and 50%. The decolorization rate of anaerobic/aerobic (SBR1) reactor was higher than that of aerobic (SBR2) reactor.

5. Microbial Decolorization and Degradation of Dyes

In recent years, the previous studies on the microbial degradation of wastewater were mainly focused on the breeding of a variety of excellent decolorization strains for degradation and adsorption of dyes in wastewater, efficient engineering bacteria and other enhanced technologies. The azo dyes can be decolorized by a wide variety of microorganisms, including bacteria, fungi, actinomycetes, algae and so on. Differences in the growth environment and metabolism mechanisms among different microorganisms affect the decolorization efficiency of azo dyes. Therefore, it is of great significance to understand the mechanisms of decolorization and degradation of different microorganisms for creating a suitable living environment for microorganisms, which contributes to the decolorization and degradation of azo dyes. Li et al. [19] studied the decolorization effect of the electroactive bacterium *Shewanella oneidensis MR-1* on the azo dye cationic red X-GRL. Under optimal anaerobic conditions (pH 5.5-8.0, 30-40 °C, and reaction time of 12 h), *Shewanella oneidensis MR-1* showed a higher decolorizing capacity for X-GRL. Liu et al. [20] studied the decolorization of methyl orange by *Bacillus circulans BWL1061,* a newly isolated salt tolerant strain. The results showed that the newly isolated salt tolerant strain azo reductase, NADH-DCIP reductase and laccase were involved in the degradation of methyl orange by *BWL1061*. In addition, the strain *BWL1061* could also degrade aromatic amines, thereby reducing the toxicity of the dye. Song et al. [21] isolated a salt-tolerant yeast, G1, which could decolorize various azo dyes, and studied the decolorization performance of the strain against acid red B (ARB). Under optimal conditions, the decolorization rate of 50 mg/L of ARB within 16 h was up to 98% and the intermediate aromatic amines could be further mineralized. NADH-DCIP reductase was identified as key decolorization reductase. Peroxidase, manganese peroxidase, and laccases were important oxidoreductases for further degradation of decolorization intermediates. Therefore, it is necessary to screen the dominant strains of efficient degradation of dyes for the decolorization of dye wastewater.

6. Conclusion

Among the treatment methods of dye wastewater, the dye biodegradation is important. The physical and chemical treatment processes of dye wastewater are inefficient and costly and produce refractory pollutants, thus resulting in secondary pollution. The biological methods are eco-friendly and efficient with low cost and widely concerned. However, to date, the mechanisms of biological decolorization of dyes are still not clear, and it is difficult to screen efficient decolorizing bacteria. In the future, it is necessary to explore the biological decolorization and degradation of dyes, screen efficient decolorization bacteria, and provide the theoretical basis for the studies on the decolorization of dye wastewater. However, laboratory studies of dye wastewater treatment are far from treatment practices.

Therefore, the future studies should focus on the treatment practices of dye wastewater so as to provide a feasible process for the treatment of dye wastewater.

7. References

[1] Castro E, Avellaneda A, Marco P. Combination of Advanced Oxidation Processes and Biological Treatment for the Removal of Benzidine-Derived Dyes [J]. *Environ Prog Sustain Energy*. 2014, 33(3): 873-85.

[2] Shafqat M, Khalid A, Mahmood T, et al. Evaluation of bacteria isolated from textile wastewater and rhizosphere to simultaneously degrade azo dyes and promote plant growth [J]. *Journal of Chemical Technology and Biotechnology*. 2017, 92(10): 2760-8.

[3] Almeida F P, Botelho M B S, Doerenkamp C, et al. Mesoporous aluminosilicate glasses: Potential materials for dye removal from wastewater effluents [J]. *Journal of Solid State Chemistry*. 2017, 253(406-13).

[4] Pan F, Yu Y, Xu A, et al. Application of magnetic OMS-2 in sequencing batch reactor for treating dye wastewater as a modulator of microbial community [J]. *Journal of hazardous materials*. 2017, 340(36-46).

[5] Zeng Q F, Fu J, Zhou Y, et al. Photooxidation Degradation of Reactive Brilliant Red K-2BP in Aqueous Solution by Ultraviolet Radiation/Sodium Hypochlorite [J]. *Clean-Soil Air Water*. 2009, 37(7): 574-80.

[6] Lumbaque E C, Gomes M F, Carvalho V D, et al. Degradation and ecotoxicity of dye Reactive Black 5 after reductive-oxidative process [J]. *Environmental Science and Pollution Research*. 2017, 24(7): 6126-34.

[7] Wijetunga S, Li X-F, Jian C. Effect of organic load on decolourization of textile wastewater containing acid dyes in upflow anaerobic sludge blanket reactor [J]. *Journal of Hazardous Materials*. 2010, 177(1-3): 792-8.

[8] Nouren S, Bhatti H N, Iqbal M, et al. By-product identification and phytotoxicity of biodegraded Direct Yellow 4 dye [J]. *Chemosphere*. 2017, 169(474-84).

[9] Bilal M, Asgher M. Sandal reactive dyes decolorization and cytotoxicity reduction using manganese peroxidase immobilized onto polyvinyl alcohol-alginate beads [J]. *Chemistry Central Journal*. 2015, 9

[10] Holkar C R, Jadhav A J, Pinjari D V, et al. A critical review on textile wastewater treatments: Possible approaches [J]. *Journal of Environmental Management*. 2016, 182(351-66).

[11] Khalid A, Arshad M, Crowley D, et al. Biodegradation of Azo Dyes [M]. *Springer Berlin Heidelberg*. 2010.

[12] Bonakdarpour B, Vyrides I, Stuckey D C. Comparison of the performance of one stage and two stage sequential anaerobic-aerobic biological processes for the treatment of reactive-azo-dye-containing synthetic wastewaters [J]. *International Biodeterioration & Biodegradation*. 2011, 65(4): 591-9.

[13] Ong S-A, Ho L-N, Wong Y-S, et al. Performance and Kinetic Study on Bioremediation of Diazo Dye (Reactive Black 5) in Wastewater Using Spent GAC-Biofilm Sequencing Batch Reactor [J]. *Water Air and Soil Pollution*. 2012, 223(4): 1615-23.

[14] Liang B, Yao Q, Cheng H Y, et al. Enhanced degradation of azo dye alizarin yellow R in a combined process of iron-carbon microelectrolysis and aerobic bio-contact oxidation [J]. *Environmental Science and Pollution Research*. 2012, 19(5): 1385-91.

[15] Cao Z, Zhang J, Zhang J, et al. Degradation pathway and mechanism of Reactive Brilliant Red X-3B in electro-assisted microbial system under anaerobic condition [J]. *Journal of Hazardous Materials*. 2017, 329(159-65).

[16] Melgoza R M, Cruz A, Buitron G. Anaerobic/aerobic treatment of colorants present in textile effluents [J]. *Water Science and Technology*. 2004, 50(2): 149-55.

[17] Mohan S V, Babu P S, Srikanth S. Azo dye remediation in periodic discontinuous batch mode operation: Evaluation of metabolic shifts of the biocatalyst under aerobic, anaerobic and anoxic conditions [J]. *Separation and Purification Technology*. 2013, 118(196-208).

[18] Ong S A, Ho L N, Wong Y S, et al. Comparative study on the biodegradation of mixed remazol dyes wastewater between integrated anaerobic/aerobic and aerobic sequencing batch reactors [J]. *Rendiconti Lincei-Scienze Fisiche E Naturali*. 2017, 28(3): 497-501.
[19] Li Q, Feng X-L, Li T-T, et al. Anaerobic Decolorization and Detoxification of Cationic Red X-GRL by Shewanella oneidensis MR-1 [J]. *Environmental technology*. 2017, 1-29.
[20] Liu W, Liu C, Liu L, et al. Simultaneous decolorization of sulfonated azo dyes and reduction of hexavalent chromium under high salt condition by a newly isolated salt tolerant strain Bacillus circulans BWL1061 [J]. *Ecotoxicology and Environmental Safety*. 2017, 141(9-16).
[21] Song L, Shao Y, Ning S, et al. Performance of a newly isolated salt-tolerant yeast strain Pichia occidentalis G1 for degrading and detoxifying azo dyes [J]. *Bioresour Technol*. 2017, 233(21-9).

Acknowledgements
The authors are grateful to college of Environmental Science and Engineering, Donghua University, Shanghai, Project of the Shanghai Science and Technology Committee (Grant No. 17DZ1202204), and National Key Technology Support (Grant No. 2015BAB07B09).

IOP Conf. Series: Earth and Environmental Science **178** (2018) 012014 doi:10.1088/1755-1315/178/1/012014

Process Parameters Optimization of Multilevel Bio-Contact Oxidation System Treating Automobile Painting Wastewater

Y F Zhu[1], T Zhu[1], M Groetzbach[2], Y G Ma[3] and B Ma[4]

1 School of Mechanical Engineering & Automation, Northeastern University, Shenyang, China
2 BMW Brilliance Automotive Ltd., Shenyang, China
3 Liaoning Provincial Machinery Research Institute Co., Ltd., Shenyang, China
4 SNF China FLOERGER, Taixing, China
E-mail: tongzhu@mail.neu.edu.cn

Abstract. A laboratory experiment of biological treatment has been performed with multilevel bio-contact oxidation system to treat the actual automobile painting wastewater pre-treated by physicochemical processes. The influence of three important process operating parameters (namely, hydraulic retention time, dissolved oxygen concentration, organic loading) on treating efficiency of multilevel bio-contact oxidation system has been investigated, and the optimum process operating parameters have been determined.

1. Introduction

With rapid development of automobile industry, automobile painting wastewater has become a difficulty and key point in the process of industrial pollution treatment. Painting wastewater is a kind of typical degradation-resistant industrial wastewater and has such features as great variation in discharge, high concentration of organic pollutants, complex compositions, and low biodegradability[1]. For automobile painting wastewater with pollutants of complex compositions and high concentration, different pretreatment in physicochemical method is commonly required for various kinds of painting wastewater, and biological method is then applied for further treatment after balanced water quality and quantity[2]-[3]. Currently, physicochemical and biological method are becoming main method for treatment of automobile painting wastewater by replacing pure physicochemical method, which not only delivers good treatment effect, saves operation cost, but also ensures stability of the entire process operation[4].

At present, there are few researches on Multilevel Bio-contact Oxidation (MBCO) system treating automobile painting wastewater at home and abroad, and the technical data is not complete[5]-[6]. The MBCO technique is based on the traditional contact oxidation method, and the reactor is designed into multilevel form, so that the microorganisms in original unipolar reactor could create natural differentiation of dominant bacteria in the multilevel reactor, and it could realize selection and allocation in space of microorganisms at different growing stages and with different functions spatially, promote biophase separation, enhance the synergistic effect in the same microbial population, and overcome antagonism among different microbial populations, thus treatment efficiency of the entire reactor is improved[7]-[8].

A joint venture automobile factory in Liaoning Province requires that the final effluent of painting wastewater receiving independent treatment in paint shop should reach more rigorous internal control standard; existing painting wastewater treatment process in paint shop uses a series of physicochemical methods, and organic matter concentration of effluent is still high and fails to meet internal control standard. This research adopts the MBCO system to perform biological treatment on actual automobile painting wastewater receiving pre-treatment of physicochemical process and optimize process operating parameters.

2. Methods

2.1. Experiment Device

The experiment device system mainly includes MBCO reactor, biological carrier, adjust tank, discharge tank, peristaltic pump and aerating system, as shown in Figure 1.

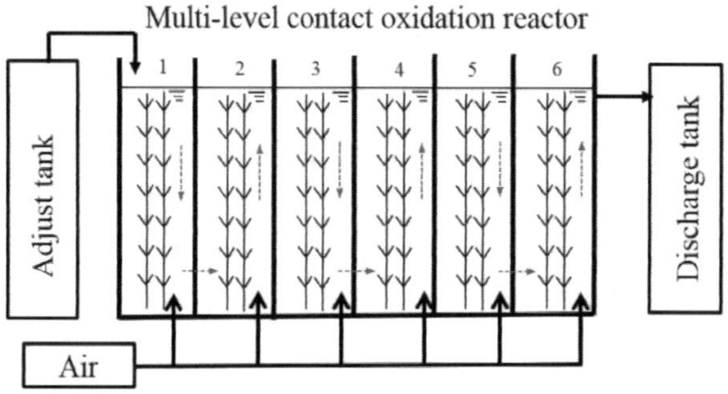

Figure 1. Experiment system of multilevel contact oxidation

The MBCO reactor is the core apparatus of experiment system, materials of which are transparent organic glass, with length, width and height of 90cm, 25cm and 45cm respectively, and total effective volume of 80L. The whole biological reactor is separated internally into 6 tanks with the same size spatially, and left upper corner or right bottom corner of each tank partition is reserved with an overflowing port in 8cm × 8cm, so that wastewater can flow in the upper and at bottom to pass through the whole biological reactor and fully contact and react with microorganism on the biological carrier. Biological carrier is made of polypropylene fiber, and each piece of carrier is fixed at carrier holder at certain interval to form a carrier unit, which can directly be put in each segment of biological tank, with carrier amount of 2.5 kg/m^3. Such biological carrier has super large specific surface area of 1200 m^2/m^3, and with density of 25 kg/m^3 and porosity of 97%; Each segment of biological tank is configured with aeration pipeline and aeration disk; air enters each segment of biological tank through aeration pump to provide sufficient oxygen to microorganism. The air pressure flow of aeration pump is 120 L/min. Aeration intensity can be adjusted via the flow valve.

2.2. Seeding Sludge and Experimental Wastewater

The return sludge from secondary sedimentation tank of the automobile factory wastewater treatment station is obtained as seeding sludge, which appears to be yellowish-brown, has abundant microorganism and great activity. The MLSS of seeding sludge added initially is about 3600 mg/L. The quality of experimental wastewater has been shown in Table 1.

Table 1. Quality of wastewater used for experiment

Pollutants	PH	COD (mg/L)	BOD (mg/L)	TP (mg/L)	TN (mg/L)	NH_4^+-N (mg/L)	SS (mg/L)
Actual value	7±0.5	1200±50	180±10	0.5±0.1	30±5	10±2	150±10
Standards	6.5-8.5	500	250	5.0	50	40	300

2.3. Start-up and Acclimation

Starting process takes place under condition of room temperature of 20-25°C; DO concentration is controlled at 2.0-4.0 mg/L or so, and pH at 6.5-8.5. After two days of aeration only, the suspended

MLSS in biological tank is 200 mg/L or so; the majority of activated sludge in biological tank has covered on the surface of the entire biological carrier evenly, generating a layer of sludge-yellow biological film, and sludge floc has a good growing condition. After successful biofilm formation, start inject water to the biological reactor and gradually cultivate and domesticate activated sludge slowly, which enables the microorganism which can effectively treat painting wastewater to further breed and grow through survival of the fittest, and to adapt to the biological environment gradually.

3. Results and Discussion

3.1. Effect of Hydraulic Retention Time

Set the hydraulic retention time (HRT) to be 24h, 16h, 12h and 8h respectively, and the system operates constantly for 10 days under conditions of each HRT. As Figure 2 shows, COD value of influent has a great fluctuation range, with maximum value of 1480 mg/L, minimum value of 869 mg/L, and average value of 1214 mg/L. After biological treatment with MBCO system, COD fluctuation range of final effluent of painting wastewater is 187-490 mg/L, which reaches internal control discharge standard of COD<500mg/L stably. After shortening the HRT each time, the variation tendency of COD removal rate in each segment is basically the same, namely reducing firstly, rising subsequently, and tending to stability at last. When HRT is 24h, 16h and 12h, total removal rate of biological system rapidly rises and keeps stable state. However, when HRT is reduced to 8h, the system requires longer period to adapt to the influent load, and total removal rate rises slowly and needs a longer period to achieve stable state.

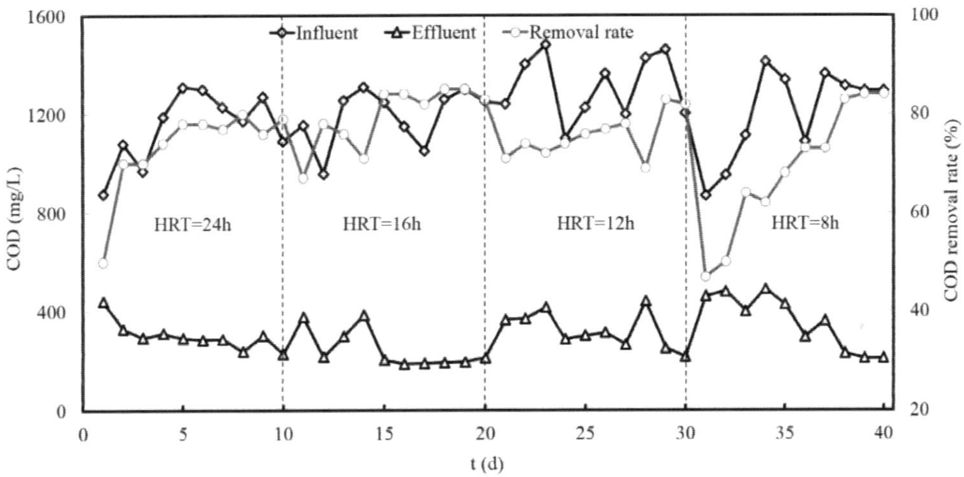

Figure 2. Remove efficiency of COD in MBCO system on condition of different HRT

In Figure 3, the fluctuation range of influent total nitrogen (TN) value is 25.7-35.8mg/L, and as HRT decreases gradually, average TN value of final effluent rises little by little. Extension of HRT is conducive to the removal of TN in wastewater, which may because that the longer HRT is, and the less influent is, the slower water flow is, resulting in less washing force on biological film on carrier and lower oxygen transmission efficiency. This is also beneficial for the formation of anaerobic environment in the internal three-dimensional space of carrier or inside biological film, which could facilitate the denitrifying bacteria performing denitrifying reaction [9]-[11].

Under different HRT conditions, the removal situation of ammonia nitrogen (NH_4^+-N) by the biological system is similar with variation tendency of TN. Small fluctuation occurs to both TN and NH_4^+-N concentration of final effluent and can be kept in a range of low concentration stably; average of TN and NH_4^+-N concentration of final effluent is 7.9 mg/L and 0.7 mg/L respectively, which is much less than discharge standard of internal requirement. When biological system reaches stable state under conditions of HRT being 24h, 16h, 12h and 8h respectively, average removal rate of TN is 77%,

76%, 76% and 65% separately; average removal rate of NH_4^+-N is 97%, 96%, 94% and 86% separately.

When HRT is 12h, the MBCO system has a high removal rate on COD, TN and NH_4^+-N, and the values of COD, TN and NH_4^+-N of final effluent all could satisfy the discharge standards of internal requirement. If HRT is extended continuously, carbon and nitrogen removal efficiency of biological system increases slowly. Given operation cost, occupying area and removal effect comprehensively, the optimum HRT is determined to be 12h.

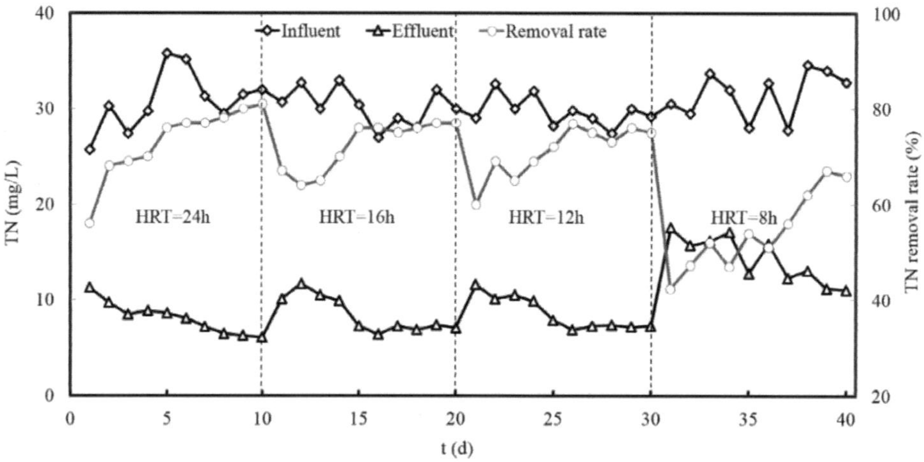

Figure 3. Remove efficiency of total nitrogen in multilevel contact oxidation system on condition of different HRT

3.2. Effect of Dissolved Oxygen Concentration

To explore efficient process treatment efficiency and further energy saving and consumption reduction, orthogonal experiment is employed to analyze the dissolved oxygen (DO) concentration in each segment of the MBCO system, as shown in Table 2. The optimum DO concentration in each segment is determined via evaluating the influence of DO concentration in each segment of biological tank on carbon and nitrogen removal.

Table 2. Factor and level table of orthogonal experiment

Factor / Level	A DO_F[a] (mg/L)	B DO_M[b] (mg/L)	C DO_L[c] (mg/L)
1	0.5~1	1~2	1~2
2	1~1.5	2~3	2~3
3	1.5~2	3~4	3~4

[a] DO concentration in the front segments.

[b] DO concentration in the middle segments.

[c] DO concentration in the back segments.

Choose comprehensive removal rate as the evaluation index of orthogonal experiment; comprehensive removal rate is the sum of COD removal rate and TN removal rate, and a higher comprehensive removal rate indicates that the corresponding DO concentration condition is more in favor of efficient operation of biological system. It is available to clearly judge primary and secondary sequence of influence of three experiment factors on treatment efficiency of biological system with visual analysis and calculation table of orthogonal experiment, and to analyze optimum experiment level corresponding to each experiment factor as well, thus optimum level combination of three experiment factors is confirmed. The orthogonal experiment results is shown in Table 3.

Wherein, T_{xy} refers to total of three experiment results corresponding to factor in column x with y level (x referring to factor, i.e. A, B, C respectively; y referring to level, i.e. 1, 2, 3 respectively ;) K_{xy} refers to average of three experiment results corresponding to factor in column x with y level; R refers to range, namely difference between maximum value and minimum value of K_{xy}. A greater K_{xy} means a better treatment effect. A less range value (R) means factor corresponding to range has less influence on experiment result. It can be indicated via comparison range corresponding to various factors that DO concentration of each segment has the greatest influence on middle segment of decarbonization and denitrification effect of multilevel contact oxidation system, then back segment, and front segment with the least.

Table 3. Intuitive analysis and calculation of orthogonal experiment results

Factor / **Level**	A DO_F (mg/L)	B DO_M (mg/L)	C DO_L (mg/L)	COD removal rate (%)	TN removal rate (%)	K General removal rate (%)
1	0.5~1	1~2	1~2	76.4	58.6	135.0
2	0.5~1	2~3	2~3	89.3	69.5	158.8
3	0.5~1	3~4	3~4	91.7	58.6	150.3
4	1~1.5	1~2	2~3	78.1	60.9	139.0
5	1~1.5	2~3	3~4	84.6	58.2	142.0
6	1~1.5	3~4	1~2	81.7	62.6	144.3
7	1.5~2	1~2	3~4	80.1	57.5	137.6
8	1.5~2	2~3	1~2	83.8	61.8	145.6
9	1.5~2	3~4	2~3	88.2	62.1	150.3
T_{x1}	444.099	411.600	424.899			
T_{x2}	426.099	447.201	448.101			
T_{x3}	433.500	444.900	430.701			
K_{x1}	148.033	137.200	141.633			
K_{x2}	142.033	149.067	149.367			
K_{x3}	144.500	148.300	143.567			
R	6.000	11.867	7.734			

As shown in Figure 4, when DO concentration is in range of 0.5-1 mg/L, COD and TN have the greatest average comprehensive removal rate, so the optimum DO concentration in front segment should be controlled in 0.5-1 mg/L. When DO concentration of middle segment and back segment is in range of 2-3 mg/L, the greatest average comprehensive removal rate occurs. Given full consideration of energy consumption and carbon and nitrogen removal efficiency, it can be determined that the optimum combination of DO concentration in each segment of MBCO system is: DO concentration in the front segments is 0.5-1 mg/L, with that in the middle segments of 2-3 mg/L, and that in the back segments of 2-3 mg/L. Under the optimum DO concentration, the average removal rate on COD and TN of MBCO system is 89.3% and 69.5% respectively.

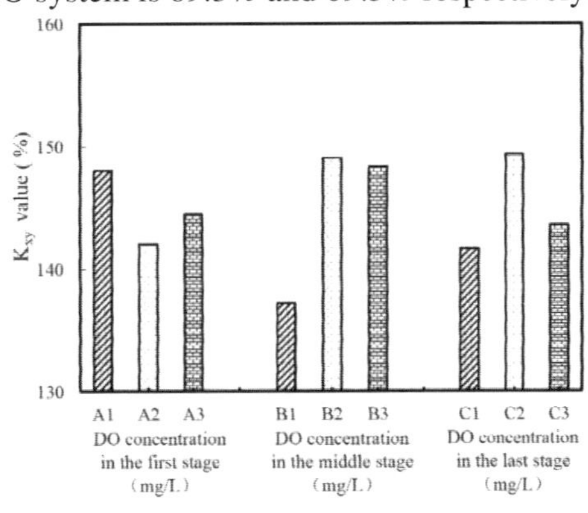

Figure 4. Efficiency varation of comprehensive removal rate

3.3. Effect of influent organic loading

To explore organic loading of influent influence on treatment efficiency of MBCO system, adjust the organic loading of influent with other process parameters unchanged to analyze the variation tendency of COD, NH_4^+-N and TN when COD valve of influent is 1000 mg/L, 2000 mg/L, 3000 mg/L, 4000 mg/L and 5000 mg/L.

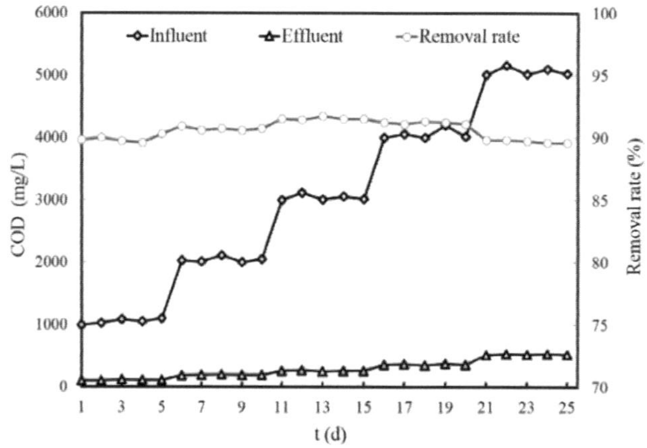

Figure 5. COD removal efficiency under different organic loading conditions

Figure 5 shows that as organic loading of influent increases, the variation tendency of total COD removal rate in biological system is not obvious and maintains at 90% or so basically. COD value of final effluent rises little by little as the COD value of influent increases; when the COD value of influent is 5000mg/L, the COD value of final effluent exceeds 500mg/L and fails to reach internal control discharge standard of the enterprise. It shows that the increase of influent organic loading has little influence on the COD removal efficiency of biological system, and the biological system has a strong ability to resist the impact of high organic matter concentration. To better meet the discharge standard, the influent COD value of biological system had better remain 5000 mg/L below.

Figure 6 shows that the total removal rate of ammonia nitrogen has a small variation range under five different organic loading conditions, and maintains in range of 86.4%-92.5%, and the ammonia nitrogen of final effluent is low and almost maintains in range of 0.7-1.3 mg/L. It is clear that with the impact of influent organic loading variation, the nitrification effect of nitrifier in biological system still could occur normally and is subject to little influence of organic loading variation, and the ammonia nitrogen concentration of effluent is stable and completely could meet the discharge standard.

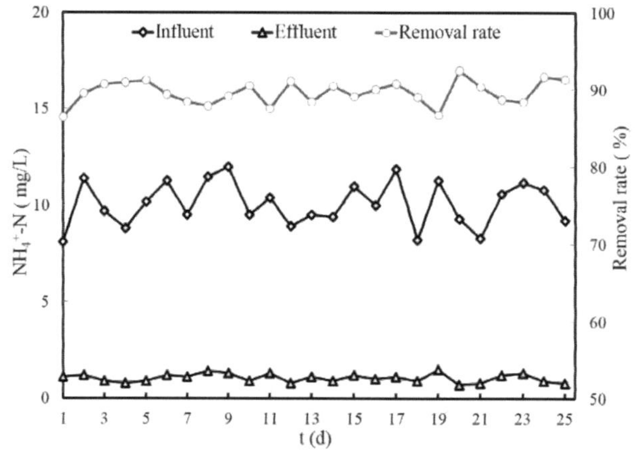

Figure 6. Removal efficiency of ammonia nitrogen under different organic loading conditions

Figure 7 shows that it is identified that as organic loading of influent increases, the TN removal rate of biological system presents rising trend, and the average TN removal rate is 70.1%, 71.8%, 75.7%, 80.7% and 85.6% successively, and the average TN value of final effluent is 9.4 mg/L, 9.0 mg/L, 7.4 mg/L, 5.9 mg/L and 4.7 mg/L successively. It results from that the nitration in biological system could remain in normal process, the increase of influent organic loading could provide more carbon source for denitrification, which could make for denitrification of denitrifying bacteria and further improve the denitrification effect of biological system.

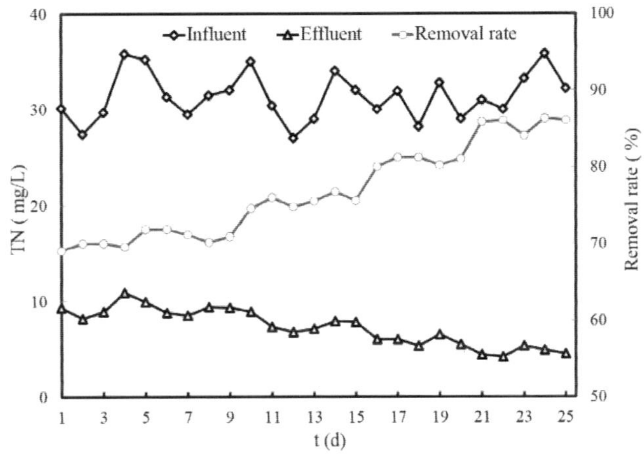

Figure 7. Removal efficiency of total nitrogen under different organic loading conditions

4. Conclusions

Three influence factors are investigated in the process of biological treatment of automobile painting wastewater by multilevel bio-contact oxidation system via a laboratory experiment and the optimum process operating parameters are determined. When HRT is 12 h, DO concentration in the front segments, the middle segments and the back segment of biological system is in range of 0.5-1 mg/L, 2-3 mg/L and 2-3 mg/L respectively; when the organic matter concentration of influent remains under 5,000 mg/L, the total COD removal rate in the biological system is 90% or so; the total removal rate of NH_4^+-N maintains in range of 86.4%-92.5%, and the total removal rate of TN maintains in range of 70.1%-85.6%. Relevant results and analysis may provide important theoretical basis and technical reference for the promotion and application of multilevel bio-contact oxidation system in future.

5. References

[1] El-Gohary F A, Abo-ElEla S I and Ali H I 1989 *Water Sci. Technol.* 21 255-63
[2] Wahaab R A 2001 *Bull. Environ. Contam. Toxicol* 66 770-6
[3] Aboulhassan M A, Souabi S, Yaacoubi A and Baudu M 2005 *Environ. Technol.* 26 705-11
[4] Krithika D and Philip L 2016 *Int. Biodeter. Biodegr.* 107 31-41
[5] Li P, Zheng T, Wang Q, Yang S, Liu S, Li L and Huang P 2015 *Environ. Prog. Sustain.* 34 339-45
[6] Zheng T L, Li P, Wang Q H, Li X S, Ai H Y, Gan K M and Sharavsambuu A 2015 *Desalin.Water Treat.* 55 1142-51
[7] Yang Q X, Wang J and Wang H T 2012 *Bioresource Technol.* 117 155-163
[8] Zheng T, Li P and Wu C *Int. J. Environ. Technol. Manage.* 18 3-12
[9] Hu Z, Zhang J, Xie H, Liang S and Li S 2013 *J. Biosci. Bioeng.* 115 272
[10] Zhang Y, Wang X, Hu M and Li P 2015 *Appl. Microbiol. Biot.* 99 1977-87
[11] Kargi F and Uygur A 2004 *Enzyme Microb. Tech.* 35 167-172

Experiments on Microalgae Mud Adsorbing Heavy Metal Pb(II) Ions and Study on its Thermodynamics

Binhui Jiang, Guangtai Zheng, Jiaqi Zhang, Shuqi Yang

College of Resources and Civil Engineering, Northeastern University, Shenyang, Liaoning, China
E-mail: jiangbinhui@mail.neu.edu.cn

Abstract. By using microalgae mud (MAM) taken from a high-efficiency algae pond as raw material to treat domestic sewage, this paper investigated the effects of such factors as the amount of microalgae added, shaking time, pH, temperature, the concentration of Pb ions and the rotating speed of the shaking table on the performance of MAM in the adsorption of the heavy metal Pb (II) by use of the single factor control variable method. Research results show that when the initial concentration of Pb(II) is 30mg/L, the amount of MAM added is 30g/L, the pH is 4, and the rotating speed of the shaking table is 180rpm, the maximum adsorption rate of Pb(II) may reach 85.51% after the solution is shaken and reacts at 25℃ in a thermo shaker for 30min. By further studying the adsorption rate of Pb(II) and fitting the adsorption isothermal equations, we found that the best effect can be achieved when the pH was 4 and the adsorption time was kept for 30min or above, with a maximum adsorption capability of 22.52mg/g in theory, and that the adsorption process was in line with the Langmuir and Freundlich adsorption isothermal models, and the correlation coefficients R2 of these two models were both greater than 95%, which demonstrates MAM was easy to adsorb Pb(II) ions, and the adsorption process was of monomolecular physical adsorption.

1. Introduction

With the development of industry in recent years, emissions of industrial waste water containing heavy metals, particularly those from such industries as chemical engineering, electroplating, mining, smelting, electronics and leader tanning, are increasing year by year[1], causing serious water body pollution. While among these types of pollution, industrial waste water containing heavy metals brings the most serious environmental pollution. Industrial waste water contains a large amount of metal ions, such as Cd(II), Cu(II), Zn(II) Cr(II), Hg(II), Ni(II) and Pb(II), and these metal ions have a high stability in water bodies, so they may do harm to aquatic plants and animals when accumulated to a certain level, and further have a direct or indirect effect on the health of human beings themselves through food chains, therefore, heavy metal pollution has become one of the most serious environmental issues across the world[2]. It is reported that among the 21 cities along China's Yangtse River, Chongqing, Yichang, Panzhihua, Wuhan, Nanjing and Shanghai are facing a severe situation, with over 65% of their water bodies having been polluted by heavy metals[3]. Furthermore, some of other major water systems, such as the Pearl River, the Tai Lake, the Huai River, the Yellow River, the Hai River, the Songhua River and the Liao River, also face quite serious challenge from heavy metal pollution[4]. In such a circumstance, it has become crucial to control and treat heavy metal pollution in water bodies.

The treatment of heavy metals in water by physical or chemical method tends to have its limitation and bring secondary pollution, while bio-adsorption technology[5] is to adsorb, degrade and enrich pollutants in the environment by use of plants, animals and/or micro-organism co-existing with them so as to lower the concentrations of pollutants in water bodies. Compared with the chemical precipitation method, physical remediation method and membrane separation method commonly used by physical or chemical method, the bio-adsorption method features its low cost, wide resources, small secondary pollution, simple operation and reusability of raw materials, and is expected to materialize the resource utilization of solid waste[6]-[8]. In recent years, bacteria, microzyme, mold, algae and macromolecular organisms such as chitosan, starch, plant fiber and cellulose were commonly studied to treat heavy metals as bio-adsorbents.

Compared with the traditional physical and chemical adsorption methods, the bio-adsorption method is an innovation, which will have a broad prospect in the treatment of heavy metal waste water due to its high efficiency, energy saving, easy operation and desirable environmental protection.

This paper studied the adsorption process of MAM on Pb(II) ions in water, as well as its mechanism and affecting factors, so as to lay a foundation for the development of high-efficient and low-cost heavy metal adsorbents, the exploration of new approaches and the industrial application of the treatment of waste water containing heavy metals.

2. Materials and Methods

2.1. Experimental Materials and Instruments

MAM: taken from Liaozhong District Biological Sewage Treatment Plant in the city of Shenyang as MAM precipitation collected from a high-efficiency algae pond used for treating domestic sewage through filtration and drying and stored in 4 refrigerators for future use. The MAM contains green microalgae, diatoms, cyanobacteria and other microorganism cells.

Main experimental reagents: ethanol, phosphate buffer, glutaraldehyde, isoamyl acetate, NaOH, $Pb(NO_3)_2$, HCl, etc., which are all analytical pure reagents and purchased from Sinopharm Group Chemical Reagent Co., Ltd.; water used in the experiments is deionized water.

2.2. Experimental Method

The main affecting factors for MAM to adsorb Pb(II) were investigated by the single factor experimental method. In the experiments, the amount of MAM added, shaking time, pH, temperature, the concentration of Pb(II), the rotating speed of the shaking table were selected as the factors that have remarkable effects on the adsorption rate of the heavy metal Pb(II) in the water. Three parallel samples and a control group (with no MAM added) were set in the experiments.

2.3. Determination Method

Determination of the concentration of Pb(II): Pb ions were tested using a full-spectrum direct-reading plasma emission spectrometer (PES), with the analytical spectral lines of Pb(II) being 220.353r and 217.000r respectively. The concentrations of heavy metal ions were calculated by the standard curve method as below: under the given analytical conditions, the standard series solutions of not less than three different concentrations of to-be-determined elements were determined (the medium used in the standard solutions and their acidities should be consistent with those of the test solutions), so as to plot standard curves and calculate regression equations by taking the response value of the analysis line as ordinate and concentration as abscissa, with relevant regression coefficients higher than 0.99. Determine the response values of the test solutions, and then calculate the contents of all to-be-determined elements in the samples after consulting relevant concentrations from the standard curves or regression equations. Blank experiments were conducted under the same analytical conditions, with blank interference deducted as required in the manual for instrument operation.

2.4. Selection of the Adsorption Model

Isothermal adsorption, isobaric adsorption and constant-volume adsorption are three types of adsorption, of which isothermal adsorption is the most widely used. When the temperature is constant, an isothermal adsorption balance will be obtained, at this moment, the curve of relationship between equilibrium adsorption quantity (q_e) and equilibrium concentration (Ce) is called the isothermal isotherm. The isothermal isotherms used in this paper are as below:

(a) Langmuir isothermal equation

Expressed linearly as:

$$\frac{C_e}{q_e}=\frac{C_e}{q_m}+\frac{1}{bq_m} \quad (1)$$

Expressed non-linearly as :

$$q_e=\frac{q_m bC_e}{1+bC_e} \quad (2)$$

Langmuir equation, as one of commonly used adsorption isotherm equations, was put forward by physicochemist Langmuir Itying in 1916 based on the molecular movement theory and some hypotheses, and now is widely applied to studies on adsorption, generally, to the single layer adsorption.

(b) Freundlich isothermal equation

Expressed linearly as:

$$ln\,q_e=\frac{1}{n}ln\,C_e+ln\,K \quad (3)$$

Expressed non-linearly as:

$$q_e=KC_e^{1/n} \quad (4)$$

The Freundlich equation, as one of commonly used adsorption isotherm equations, is suitable for multiple molecular layer adsorption. In this equation, the magnitude of n reflects how difficult the adsorption would be. Where q_e (mg/g) is the equilibrium adsorption quantity; *Ce* (mg/L) is the concentration of the solute while an adsorption balance is reached; q_m (mg/g) is the saturated adsorption quantity; *b* is Langmuir constant; K and l/n are empirical constants.

3. Results and Discussion

3.1. Optimization of the Removal Rate of Pb (II)

The effect of MAM on the rate of removal of Pb (II) by changing the initial conditions is as shown in Figs.1- 5.

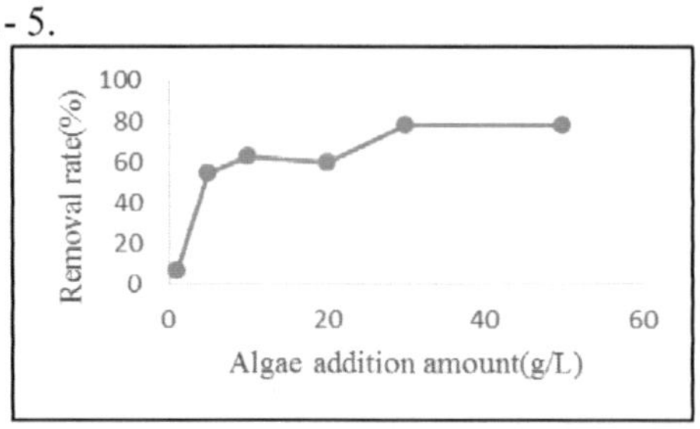

Figure1. Effect of the amount of MAM added

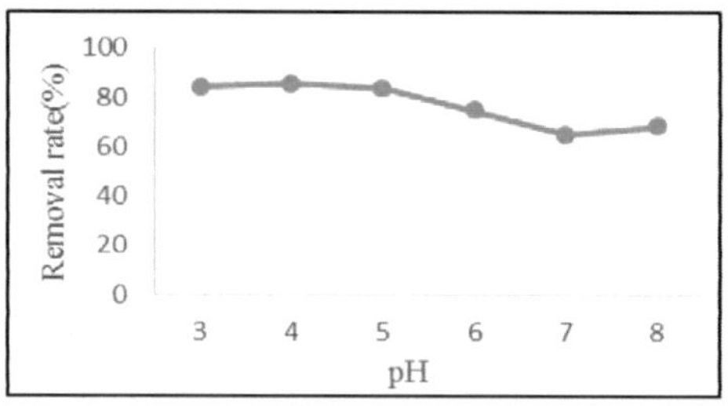

Figure 2. Effect of initial pH

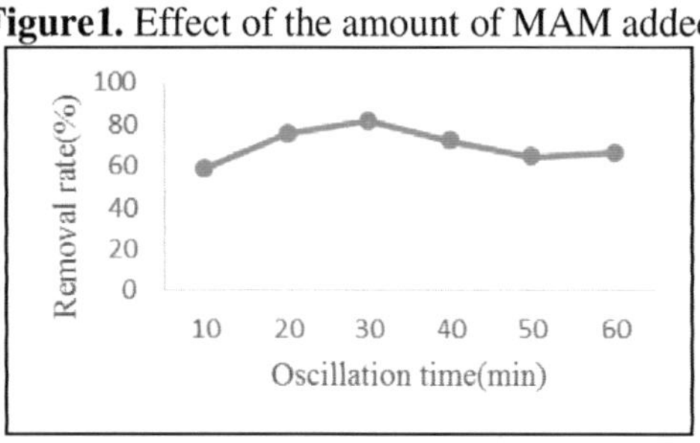

Figure3. Effect of shaking time

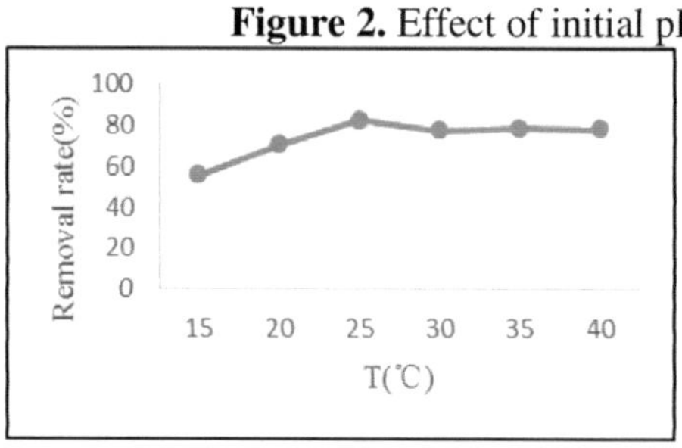

Figure 4. Effect of temperature

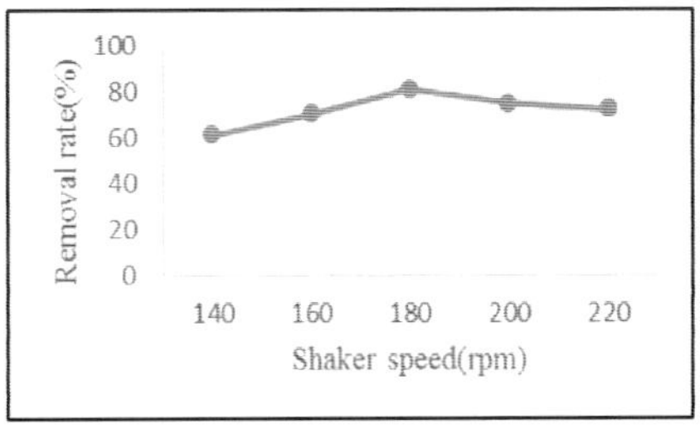

Figure 5. Effect of shaking speed on the removal rate of Pb (II)

The effects of the amount of MAM added, shaking time, pH, temperature, the concentration of Pb ions and the rotating speed of the shaking table on the performance of MAM to remove Pb(II) were determined by changing initial conditions, as shown in Figs. 1-5. As can be seen from Fig.1, with the increase in the amount of MAM added, the removal rate of Pb(II) rises gradually, till the amount of MAM added reaches 30g/L, at this moment, the removal rate reaches its maximum of 78%, after that, with the increase in the amount of MAM added, the removal rate no longer rises and tends to be steady. As can be seen from Fig.2, when raising the pH from 3 to 5, the removal rate of Pb(II) keeps at a high level, above 80%. As can be seen from Figs. 2-5, temperature, shaking time and rotating speed all have a certain effect on the removal rate of Pb(II), and with the increase in the temperature, shaking time and rotation speed, the removal rate presents an up-and-down trend. Before attaining the optimal conditions, the rise in temperature can improve the adsorbing rate, the increase in shaking time can make adsorption more even and complete, the increase in rotating speed can make sewage better contact with the porous structure of MAM and enlarge the surface area for adsorption, thus improve the removal rate. However, a drop in removal rate may appear when the temperature, shaking time and rotating speed attain the optical conditions, which may be the result of the desorption of Pb (II).

In summary, the maximum adsorption rate of Pb(II) can reach 85.51% at the optimal experimental conditions, that is, when the initial concentration of Pb(II) is 30mg/L, the amount of MAM used is 30g/L, the rotating speed of the shaking table is 180rpm, and the adsorption reaction proceeds in a thermo shaker at 25℃ for 30min.

3.2. Adsorption Isothermal Parameters of Pb (II)

Aqueous solutions were formulated with a Pb(II) concentration of 5, 10, 20, 30 and 50mg/L respectively, and adjusted with 0.1mol/L HCl and NaOH to keep their pH at 4. Add MAM to the solutions with an amount of 30g/L, sample after shaking the solutions at 25℃ for 30min under the condition that the rotating speed of the shaking table is kept at 108r/min, and then centrifuge the solutions for 10min under the condition that the rotating speed is kept at 180 r/min. Determine the concentrations of Pb(II) in the supernatants and calculate the removal rate (see Table 1). Experimental data was fitted using Langmuir and Freundlich isothermal equations[9] (see Table 2).

Table 1 Adsorption isothermal parameters of Pb (II) in MAM

Initial concentration(mg/L)	Equilibrium concentration(mg/L)	Removal rate (%)
4.29	1.36	68.28
9.38	2.21	76.41
18.46	4.58	75.17
26.90	4.77	82.26
45.20	9.79	78.35

Experimental data was fitted using Langmuir and Freundlich equations, and the results are as shown in Table 2. From the correlation efficient R^2 obtained by fitting the two model equations, it can be judged that the adsorption rate of Pb(II) has a quite high degree of fitting with both the Langmuir and Freundlich equations.

In general, the Langmuir model is intended to study single molecular layer adsorption, a type of physical adsorption; the Freundlich model is used to describe multiple molecular layer adsorption, but when n<1, this model describes a situation of single molecular layer physical adsorption, and adsorption is easy to happen when the value of n varies between 0.1 and 0.5 means, while n>2 means adsorption is difficult to happen. In this experiment, the value of n is 0.7970, which means adsorption is easy to happen, and the adsorption is of single molecular layer physical adsorption.

Table 2 Parameters of the Langmuir and Freundlich models

Model	Parameter	Correlation Coefficient	Fitting Equation
Langmuir	q=22.52mg/g, b=11.12/mg	R^2=0.9567	Y=0.4937x-0.0444
Freundlich	K=2.3287mg/g, n=0.7970	R^2=0.9578	Y=1.2547x+0.8453

4. Summary

Through single factor experiments, this paper concludes that MAM can remove the heavy metal Pb(II) out of water desirably, and the maximum adsorption rate of Pb(II) is 85.51% under the optical removal conditions, that is, when the amount of MAM is 30g/L, pH=4, the rotating speed of the shaking table is 180rpm, and the reaction time is 30min or above.

When MAM is used to adsorb Pb(II) in water, the maximum adsorption quantity in theory is 22.52mg/g, which means MAM is easy to adsorb Pb(II) in water, following a single molecular layer physical adsorption process. Therefore, it is advisable to use MAM to treat industrial waste water containing heavy metals.

5. References

[1] G.F.Shen.: Think about urban sewage treatment policy, China Environmental Protection Industry, (2004), No2, p. 16

[2] W.C.Gao:Optimization and Mechanism Study of Biosorption of Cu (II)Pb (II) from Aqueous Solutions by Microbial Flocculants of MBFGAI(MS, Hunan University, China 2011)

[3] W.F.Sun and D,Xiao: Current situation and management technology of heavy metal pollution in water body, Energy and Energy Conservation,(2012),No2,p.49

[4] X.Yue,K.Liu and X.L.Xia: Preventive Medicine Tribune,Vol.20(2014),No 3,p.209

[5] L.Li: Study on Heavy Metal Resistance and Biosoption of Lead (II) Ions from Aqueous Solution by WT6-5(MS, Northeastern University, China 2013)

[6] M.G.Geoffrey: Journal of Chemical Technology & Biotechnology,Vol.84(2009),No 1,p.13

[7] X.L.Hou,A.Q,Liu and L.P.Cai: Subtropical Agriculture Research,Vol.8(2009),No 2,p.106

[8] M.A.Ashraf,K.Mahmood and A.Wajid: Proceedings of International Conference on Food Engineering and Biotechnology(ICFEB 2011)[C](Bangkok, Thailand, May 7-9,2011),Vol,9,p.60

[9] I.Qureshi,S.Memon and M.Yilmaz: Comptes Rendus – Chimie,Vol.13(2010),No 11,p.1416

Acknowledgments

The present work is supported by funds from the National Natural Science Foundation of China (51278090) and the Major Science and Technology Program for Water Pollution Control and Treatment (2013ZX07202-010-05).

The New Developments Made in the Autotrophic and Heterotrophic Ammonia Oxidation

Mei Wang[1], Yurui Wu[2], Jiachao Zhu[1], Chenyi Wang[2], Yanbin Zhu[1], Qing Tian[1,*]

1 Department of Environmental Science and Engineering
DongHua University, 2999 Shanghai North people's Road,201620 P.R. China
2 Shanghai Caoyang No.2 High School, 160 Meichuan Road, Putuo District, Shanghai, 200333 P.R. China
E-mail address: tq2004@dhu.edu.cn

Abstract. Nitrification is defined as conversion of the ammonium(NH_4^+-N) to nitrate and it is generally divided into two categories: autotrophic nitrification and heterotrophic nitrification. With the metagenomes retrieved from the *Nitrospira* highly enriched biosamples, both the ammonium oxidation gens (full set of AMO and hydroxylamine dehydrogenase (HAO) genes) and nitrite oxidoreductase (NXR) coding genes necessary for nitrite oxidation were found in a completely nitrifying bacterium from the genus *Nitrospira*. Various species of *Nitrospira* (containing phylogenetically distinct ammonia monooxygenase and hydroxylamine dehydrogenase genes) had been found in many engineered systems and were proved to be comammox bacterium. Some anammox bacteria were found to use mono or dimethylamine and even methanol as electron donors, but the enzymes the employed in the special metabolism were not known. Many heterotrophic nitrifying bacteria have been isolated and characterized from soil, activated sludge, wastewater treatment system. A denitrification process was developed with poly(3-hydroxybutyrate) (PHB) granulesas source of carbon for denitrification. It is possible from metabolic foundations that the heterotrophic nitrifier can consume intracellular stored PHB as the alternative carbon source. This direction still waiting for further exploration.

1. Introduction

Nitrification, the oxidizing ammonia (NH_3) to nitrate, is a key process in nitrogen cycles in ecosystems. It is process that had been studied widely in natural and agricultural waters [1] (e.g. soil from farms or constructed wetlands) and been applied widely in the water and wastewater treatment processes [2]. In particular, the process of simultaneous nitrification-denitrification with complete or partial nitrification in order to achieve the most economic pattern for nitrogen removal [2]. E.g., the nitrification pathways in landfills leachate treatment process has aroused great attentions [3]. To date, studies on the artificial systems (such as aquaculture for the growing of fish and plants) in eliminating toxic ammonium and recycle nitrogen by plants via nitrate also made breakthrough [4]. The milestone breakthrough in recent research should attribute to complete oxidative ammonia into nitrate in one organism (name as complete ammoxidation or Comammox). A pure Comammox bacterium has been obtained from the genus *Nitrospira* [5].

In the conventional ammonia oxdication process, the biomass yield was low for the low energy yield during nitrification, and as autotrophic bacteria, nitrifiers only obtain carbon from autotrophic CO_2 fixation via the Calvin cycles. In this way, the autotrophic nitrifiers can only be predominant in the low organic carbon environment. However, recent researches uncovered that many autotrophic nitrifiers and mixotrophic bacteria can propagate in the high carbon source contained environment and involve actively in the process of nitrification and phosphorous recycle and accumulation [6]. It is possible these bacteria survive via the mixotrophs growth passages and using organic and inorganic compounds as the alternative carbon and energy sources. Among the organic carbon sources, polyhydroxyalkanoate (PHA), a biopolymer naturally synthesized by many bacteria and intracellularly accumulated as granules has aroused attention in the study of nitrification. These intracellular stored carbon source were found to be important in denitrification[7]. This paper is to summarize the progress of nitrifying microorganisms' researches in recent years. The culture conditions of synthesizing PHB by different microorganisms are introduced. Furthermore, the effect of varied factors on nitrifying microorganisms was also discussed and the outlook of future work was also proposed.

2. Autotrophic Nitrifying Bacteria

2.1. Complete Nitrification

Conventional nitrogen removal is a two-step process involving two groups of microbes: i) oxidize ammonia to nitrite by the ammonia oxidizing microorganisms (i.e., ammonia oxidizing archaea, AOA and ammonia oxidizing bacteria, AOB); ii), nitrite is oxidized to nitrate by the nitrite oxidizing bacteria (nitrite oxidizing bacteria, NOB). Two separate processes of nitrification lead to the coexistence of AOB with NOB. However, Costa et al [8] proposed that the ammonia is oxidized to nitrate by a microorganism completely is viable based on the kinetic theory of optimal metabolic processes. In 1999, Zheng Ping et al [9] proposed that the nitrification substrate (ammonia) is more conducive to supporting the growth of one kind of microorganism than two kinds of microorganisms (ammonia oxidizing microorganisms and nitrite oxidizing bacteria) theoretically because of lower energy. With the metagenomes retrieved from the *Nitrospira* highly enriched biosamples, both the ammonium oxidation gens (full set of AMO and hydroxylamine dehydrogenase (HAO) genes) and nitrite oxidoreductase (NXR) coding genes necessary for nitrite oxidation were found in a completely nitrifying bacterium from the genus *Nitrospira*. Daims H. et al. revealed that complete ammonia oxidation (Comammox) bacteria were successfully enriched in deep oil wells [5] and the pure culture were obtained in growing biofilms with oligotrophic conditions in an aquaculture system [10]. Meanwhile, Van Kessel, MAHJ[10] and Daims [5] reported three different cultivated bacteria and an uncultivated bacterium [11] that carry out complete oxidation of ammonia to nitrate. From then on, various species of *Nitrospira* (containing phylogenetically distinct ammonia monooxygenase and hydroxylamine dehydrogenase genes) had been found in many engineered systems including the drinking water systems [12] and were proved to be comammox bacterium. These findings fundamentally set up a new concept of nitrification. Up to date, the enormous potential of the completely nitrifying *Nitrospira* in the nature and engineered system still need to be deplored.

2.2. Anaerobic Ammonium Oxidation (ANAMMOX) Bacteria and Their Mixotrophic Growth

Most nitrifying bacteria need to survive low ammonium presence environments, some of the situation comes from the electron loss in the cyclic electron flow in the nitrification and simultaneous denitrification process. As a result, some nitrifiers can utilize the alternative external organics as the electron donors to replace ammonium. E.g. the new strain Pseudomonas sp. C27 that performs both autotrophic and heterotrophic denitrification in mixotrophic DSR medium was isolated by Chen Chuan[13]. Feray and Montuelle[14] revealed that several nitrifying strains were able to grow under mixotrophic conditions in response to environmental changes. These microorganisms were found to be prevalent in biofilms in wastewater treatment system. Another example is anaerobic ammonium oxidation (anammox) bacteria. Anammox was discovered in the early 1990s in a denitrifying fluidized bed reactor in the Netherlands[15]. Anaerobic ammonia-oxidizing bacteria (anammox)oxidize ammonium using nitrite as an electron acceptor and producing nitrogen under anoxic conditions and are considered to have tremendous potential in the treatment of rich-ammonia wastewater. These novel bacteria have a highly unusual physiology, in that they live by consuming ammonia in the absence of oxygen. Some anammox bacteria were found to use mono or dimethylamine and even methanol [16] as electron donors, but the enzymes the employed in the special metabolism were still waiting for uncover.

3. Heterotrophic Nitrifying Bacteria

Heterotrophic nitrification process implies heterotrophic microorganisms oxidize ammonia nitrogen to nitrite nitrogen and nitrate nitrogen under aerobic conditions. Most heterotrophic nitrifying bacteria have a phenomenal ability to reduce nitrate to N_2 or N_2O. In general, compared with autotrophic nitrifying microorganisms, heterotrophic nitrifying microorganisms gain several advantages. E.g., higher buffering capacity to organic loads, less oxygen demand and higher growth rates. In addition, these heterotrophic microorganisms presented higher growth rates.

3.1. Isolation of Heterotrophic Nitrifying Bacteria

In the last twenty years, many heterotrophic nitrifying bacteria have been isolated and characterized from soil, activated sludge, wastewater treatment system. It is shown that heterotrophic nitrifying bacteria wildly exist in all kinds of natural environments. More recent studies have highlighted the existence of heterotrophic nitrifiers such as *Pseudomonas*, *Acinetobacter*, *Bacillus* and *Alcaligenes* [17]. Since, the heterotrophic nitrification process implies heterotrophic microorganisms can oxidize ammonia nitrogen to nitrite nitrogen and nitrate nitrogen under aerobic conditions. It is possible, the nitrogen removal can occur under aerobic conditions with the process of simultaneous nitrification and denitrification. A denitrification process was developed with poly(3-hydroxybutyrate) (PHB) and copolymers with 3-hydroxyvalerate (PHBV) granules, as source of carbon for denitrification. Vock et al. [7] discovered that activated sludge could store a carbon source, by transforming organic carbon from wastewater to the form of PHA stored in bacteria cells. Wenjie Wan [18] reported that denitrifying phosphorus-accumulating bacterium(DPAB). Enterobacter cloacae HW-15 presented high heterotrophic nitrification–aerobic denitrification capacity. Barak and van Rijn [19] revealed that *P. denitrificans* was able to reduce nitrate With PHB and without an external carbon source. Since the autotrophic nitrifier (including the AOB and NOB and anammox bacteria) can use a broad range of inorganic and organic compounds as the energy and carbon sources. It is possible from metabolic fundations that the heterotrophic nitrifier can consume intracellular stored PHB as the alternative carbon source. Although, this concept is still hypothetical without experimental evidence. However, from the review of the recently published reference demonstrate this is fascinating direction that still waiting for further exploration.

3.2. Factors Affecting Heterotrophic Nitrification

The insufficient DO might be responsible for the low substrate removal, Huang [20] et al report that *Acinetobacter* sp. Y16 only showed an ammonium removal efficiency of 20.4% in static cultivation. However, an extremely high DO level may also inhibit the substrate removal, as described by Zhao [21], A. faecalis strain NR. performed well in removing ammonium and TOC with DO concentration of 1.89-7.71 mg/L.

As we know, the C/N ratio plays an important role in the Heterotrophic nitrification and aerobic denitrification process. C/N ratio is the main control factor to obtain efficient denitrification, the removal efficiency would be comparatively low when the carbon source was insufficient, or the C/N ratio was low. Heterotrophic microbes often require high concentrations of organic carbon to assimilate ammonium, but few have rapid ammonium removal rates within wide C/N ratios [22], [23]

Temperature and pH are all the important environmental factors as for microorganisms. Mike S.M. Jetten showed that anammox biomass was positively affected by the digester effuent. The pH (7.0-8.5) and temperature (30-37°C) is the best for the process

Many researchers screened heterotrophic nitrification-aerobic denitrifying bacteria, and in the exploration of the growth conditions in the optimal pH range. Duan et al. [24] have respectively reported two strains that have good nitrogen removal performance at alkaline conditions but low nitrogen removal rates at neutral conditions. *Aeromonas* sp. HN-02 grow at pH range of 6-9[25]. Chen [12] et.al found that optimal pH is 9.0 as for heterotrophic denitrifying strain *Pseudomonas* sp. C27. Previous studies revealed that Heterotrophic nitrifiers have a broader range of pH generally [26].

As described by Yan G [27], when the temperature elevated from 37°C to 50°C, the removal efficiency of nitrate decreased 53% for heterotrophic nitrification-aerobic denitrification strain *Halomonas Camposalis* ha3.

Carbon source is essential for bacterial growth, so it was considered to be an important factor influencing heterotrophic nitrification ability. Sodium succinate, citrate and acetate can be directly utilized by microorganisms because their molecular structures are low-molecular-weight. Therefore, microorganisms can quickly obtain energy by using those compounds and achieve higher efficiency of nitrogen removal. Zhang Shusong [28] isolated new strain of D. *polyhydroxybutyrativorans* SL-205 that can utilize nitrate and nitrite for denitrification and ammonium for nitrification using poly(3-hydroxybutyrate-co-3-hydroxyvalerate) (PHBV) as the sole carbon source under aerobic conditions from a denitrification reactor.

4. Conclusions

More and more studies reported that comammox are all autotrophic microorganisms that coexists with AOA and AOB in the soil, and utilize ammonia-nitrogen as the nitrogen source and carbon dioxide as the sole carbon Source. However, the relationship of heterotrophic nitrifying bacteria and comammox should be further investigated. Heterotrophic nitrifying bacteria can synthesize PHB, but its contribution to the biological nutrient removal (BNR) system is unknown. With the continuous popularization of new detection methods and the decrease of cost, it is easier to obtain new isolated species and their metabolic pathways that change their growth pattern in the nature and engineered environments. In this way, people have more opportunities to find out metabolic pathways of new species that change their lifestyle with the change of environment and being actively pursued for engineering application.

5. References.

[1] Grunert O, Reheul D, Van Labeke M-C, Perneel M, Hernandez-Sanabria E, Vlaeminck SE, et al. Growing media constituents determine the microbial nitrogen conversions in organic growing media for horticulture. *Microbial Biotechnology*. 2016;9(3):389-99.

[2] Ge S, Wang S, Yang X, Qiu S, Li B, Peng Y. Detection of nitrifiers and evaluation of partial nitrification for wastewater treatment: A review. *Chemosphere*. 2015;140:85-98.

[3] Sun X, Zhang H, Cheng Z, Wang S. Effect of low aeration rate on simultaneous nitrification and denitrification in an intermittent aeration aged refuse bioreactor treating leachate. *Waste Management*. 2017;63:410-6.

[4] Zou Y, Hu Z, Zhang J, Xie H, Guimbaud C, Fang Y. Effects of pH on nitrogen transformations in media-based aquaponics. *Bioresource Technology*. 2016;210:81-7.

[5] Daims H, Lebedeva EV, Pjevac P, Han P, Herbold C, Albertsen M, et al. Complete nitrification by Nitrospira bacteria. *Nature*. 2015;528(7583):504-+.

[6] Tian Q, Zhuang L, Ong SK, Wang Q, Wang K, Xie X, et al. Phosphorus (P) recovery coupled with increasing influent ammonium facilitated intracellular carbon source storage and simultaneous aerobic phosphorus & nitrogen removal. *Water Research*. 2017;119:267-75.

[7] Vocks M, Adam C, Lesjean B, Gnirss R, Kraume M. Enhanced post-denitrification without addition of an external carbon source in membrane bioreactors. *Water Research*. 2005;39(14):3360-8.

[8] Costa E, Perez J, Kreft JU. Why is metabolic labour divided in nitrification? *Trends in Microbiology*. 2006;14(5):213-9.

[9] Zheng P, Feng X S. Biochemical Principles of Nitrification. *Microbiology Bulletin*. 1999(03):215-7.

[10] van Kessel MAHJ, Speth DR, Albertsen M, Nielsen PH, Op den Camp HJM, Kartal B, et al. Complete nitrification by a single microorganism. *Nature*. 2015;528(7583):555-+.s

[11] Santoro AE. The do-it-all nitrifier. *Science*. 2016;351(6271):págs. 342-3.

[12] Wang Y, Ma L, Mao Y, Jiang X, Xia Y, Yu K, et al. Comammox in drinking water systems. *Water Research*. 2017;116:332-41.

[13] Chen C, Ho K-L, Liu F-C, Ho M, Wang A, Ren N, et al. Autotrophic and heterotrophic denitrification by a newly isolated strain Pseudomonas sp C27. *Bioresour Technol*. 2013;145:351-6.

[14] Feray C, Montuelle B. Chemical and microbial hypotheses explaining the effect of wastewater treatment plant discharges on the nitrifying communities in freshwater sediment. *Chemosphere*. 2003;50(7):919-28.

[15] Mulder A, Vandegraaf AA, Robertson LA, Kuenen JG. ANAEROBIC AMMONIUM OXIDATION DISCOVERED IN A DENITRIFYING FLUIDIZED-BED REACTOR. *Fems Microbiology Ecology*. 1995;16(3):177-83.

[16] Kartal B, van Niftrik L, Rattray J, de Vossenberg JLCMv, Schmid MC, Damste JSS, et al. Candidatus 'Brocadia fulgida': an autofluorescent anaerobic ammonium oxidizing bacterium. *Fems Microbiology Ecology*. 2008;63(1):46-55.

[17] Shoda M, Ishikawa Y. Heterotrophic nitrification and aerobic denitrification of high-strength ammonium in anaerobically digested sludge by Alcaligenes faecalis strain No. 4. *Journal of Bioscience and Bioengineering*. 2014;117(6):737-41.

[18] Wan W, He D, Xue Z. Removal of nitrogen and phosphorus by heterotrophic nitrification-aerobic denitrification of a denitrifying phosphorus-accumulating bacterium Enterobacter cloacae HW-15. *Ecological Engineering*. 2017;99:199-208.

[19] Barak Y, Rijn JV. Atypical Polyphosphate Accumulation by the Denitrifying Bacterium Paracoccus denitrificans. *Applied & Environmental Microbiology*. 2000;66(3):1209-12.

[20] Huang X, Li W, Zhang D, Qin W. Ammonium removal by a novel oligotrophic Acinetobacter sp Y16 capable of heterotrophic nitrification-aerobic denitrification at low temperature. *Bioresource Technology*. 2013;146:44-50.

[21] Zhao B, Tian M, An Q, Ye J, Guo JS. Characteristics of a heterotrophic nitrogen removal bacterium and its potential application on treatment of ammonium-rich wastewater. *Bioresour Technol*. 2017;226:46-54.

[22] Joo HS, Hirai M, Shoda M. Characteristics of ammonium removal by heterotrophic nitrification-aerobic denitrification by Alcaligenes faecalis no. 4. *Journal of Bioscience and Bioengineering*. 2005;100(2):184-91.

[23] Zhang Q-L, Liu Y, Ai G-M, Miao L-L, Zheng H-Y, Liu Z-P. The characteristics of a novel heterotrophic nitrification-aerobic denitrification bacterium, Bacillus methylotrophicus strain L7. *Bioresour Technol*. 2012;108:35-44.

[24] Duan J, Fang H, Su B, Chen J, Lin J. Characterization of a halophilic heterotrophic nitrification-aerobic denitrification bacterium and its application on treatment of saline wastewater. *Bioresour Technol*. 2015;179:421-8.

[25] Chen Q, Ni J. Ammonium removal by Agrobacterium sp LAD9 capable of heterotrophic nitrification-aerobic denitrification. *Journal of Bioscience and Bioengineering*. 2012;113(5):619-23.

[26] Yang L, Ren Y-X, Liang X, Zhao S-Q, Wanga J-p, Xia Z-H. Nitrogen removal characteristics of a heterotrophic nitrifier Acinetobacter junii YB and its potential application for the treatment of high-strength nitrogenous wastewater. *Bioresour Technol*. 2015;193:227-33.

[27] Guo Y, Zhou X, Li Y, Li K, Wang C, Liu J, et al. Heterotrophic nitrification and aerobic denitrification by a novel Halomonas campisalis. *Biotechnology Letters*. 2013;35(12):2045-9.

[28] Zhang S, Sun X, Fan Y, Qiu T, Gao M, Wang X. Heterotrophic nitrification and aerobic denitrification by Diaphorobacter polyhydroxybutyrativorans SL-205 using poly(3-hydroxybutyrate-co-3hydroxyvalerate) as the sole carbon source. *Bioresour Technol*. 2017;241:500-7.

Acknowledgments.
This study was financially supported by National Natural Science Foundation of China (21777024), National Natural Science Foundation of China (51478099) and Natural Science Foundation of Shanghai, China (16ZR1402000), and Shanghai Adolescence Science Innovation Practice Funds.

Experimental Study on the Integrated Performance of Dust Removal and Water Vapour Recovery in Catalytic Cracking Gas

LIU An-Lin, PENG Lv, ZHANG Yan-Hong, MA Liang* and HE Meng-Ya

State Key Laboratory of Chemical Engineering, East China University of Science and Technology, Shanghai 200237, China.
E-mail: maliang@ecust.edu.cn

Abstract. Industrial combustion flue gas contains fine dust, water vapor, in order to remove micro dust, and to recover water vapor at the same time. we have put forward a research on the technology of recovering water vapor and dust removal by using a cyclone separator for the first time. By studying the pressure drop, the separation efficiency and the efficiency of recovering condensed water of the flue gas, we can get the optimum technological parameters for the actual working conditions. The results show that the pressure drop increases with the increase of inlet velocity and temperature, and the rate of increase is increasing. When under the same cooling water flow rate, the dust separation efficiency increases first and then decrease with the inlet velocity increase. When the inlet velocity is near the 50 m^3/h, the separation efficiency reaches the maximum, 96%. The amount of vapor recovery increases with the increase of cooling medium flow rate in 15 minutes. When at the same cooling medium flow rate, the recovery increases first and then decrease with the inlet velocity increase, and reaches the maximum when the inlet velocity is 60-80 m^3/h.

1. Introduction

A large amount of dust and vapour water in the industry are directly discharged from the incineration flue gas, resulting in serious waste of resources and environmental pollution [1]. The recovery of water vapour in flue gas is not only beneficial to the recovery of latent heat of flue gas, but also can save industrial water and reduce production cost [2]. The removal of dust in the flue gas will also significantly alleviate the increasingly severe fog and haze [3]. Therefore, the research on the integration of vapour water recovery and dust removal is of great significance in exhaust gas.

In recent years, scholars have conducted a series of studies on the heat transfer and dust removal performance of cyclone separators by experiments and simulation [4] [5], Chang-Eon Lee found that condensate recovery is not only beneficial to the latent heat recovery of flue gas, but also can absorb acid gas, which had obvious significance of environmental protection [6] Jolius Gimbun et al. attempted to enhance rotating centrifugal separation by improving the separable factors of materials, for example, to reduce the viscosity of materials by adjusting the temperature [7]. Nwabunwanne Nwokolo used a cyclone to recovery the heat from the downdraft biomass gasifier [8]. R., Parke et al. had studied the separation performance of cyclone separator under the condition of high temperature and high pressure, which is of great importance to the application of cyclones in industry [9] [10].

The above research has laid a foundation for the application of cyclone separator, but there is no cyclone separator used as the tube path in the integration of flue gas condensed water recovery and dust removal. As a common separation equipment in the modern chemical industry, cyclone separator has the characteristics of simple structure, high separation efficiency and mature technology, and is especially suitable for use in harsh environments such as high temperature. In this paper, a self-designed cyclone separator is used as a tube to study the technology of dust separation and condensation water recovery in catalytic cracking incineration flue gas, the high temperature dust laden gas is separated in the cyclone separator, and the heat is transferred with the cooling medium in the casing wall, and the flue gas temperature is reduced, so that the condensed water is precipitated. The influence of the gas velocity, cooling medium flow rate and flue gas temperature on the pressure drop, separation efficiency and vapor water recovery was experimentally determined.

2. Experiment

2.1. *Design of cyclone*

As shown in Figure 1, The cyclone used in this experiment reduced the diameter of the cylindrical section to enhances the centrifugal effect. In order to make the airflow steady, a parallel deflector is added at the entrance, increase the length of the cone properly, and the size parameters of the cyclone are shown in table 1. In this experiment, the cyclone separator, the tank body and the pipe are all 304 stainless steel, all of which have good heat transfer performance and corrosion resistance.

Table 1. Dimensions(mm) of the cyclone

D	a	b	d_x	d_c	H_1	H_2	H_3	S	I_n
75	60	30	30	30	300	140	200	200	50

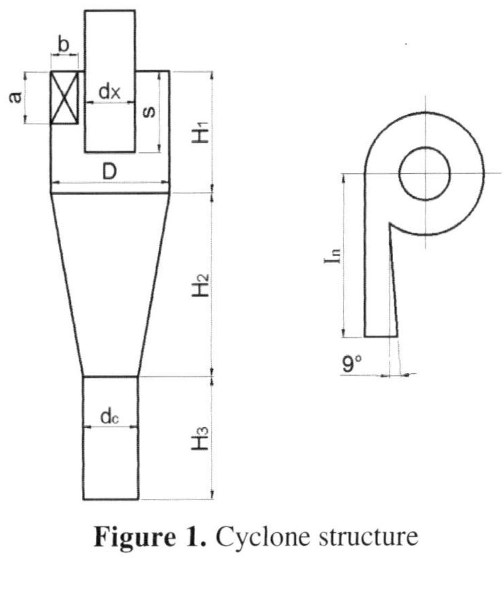

Figure 1. Cyclone structure

2.2. *Experimental process*

As shown in Figure 2, CO and air are burned in an incinerator (1), and the dust laden flue gas with high temperature is sucked into the cyclone (8) under the action of a fan (10). At the same time, under the action of the centrifugal pump (6), the liquid in the cooling medium buffer tank (2) enters the shell path (9) along the tangential direction at the lower part of the tank. The high temperature dust gas in the cyclone (8) separator for gas-solid separation, at the same time, the high temperature gas transfers the heat with the cooling medium, so that the temperature of the flue gas drops rapidly, then water vapor in the flue gas condensed rapidly into small droplets, the small droplets in the cyclone field capture dust, which further strengthen the separation efficiency. The rotating direction of the smoke in the cyclone (8) separator is opposite to the rotation direction of the cooling medium when entering the shell path (9), thereby improving the heat transfer efficiency. The separated dust condensed water flows out from the bottom of the tank body, and the flue gas after separating and cooling enters into the lower part of the desulfurization tower (11) under the action of the fan (10). The alkali liquor (12) is transported to the upper end of the desulfurization tower (11) under the action of an anticorrosive pump (13) and is sprayed downward, and finally the purified gas is discharged after the desulfurization. A thermometer (7) is arranged at the inlet and the outlet of the high temperature flue gas for measuring the temperature before and after the heat exchange of the flue gas, as well as the cooling medium. A flow meter (5-1, 5-2) is provided on the inlet pipe of the high temperature gas and the cooling medium to measure the flow velocity of the smoke and the cooling medium. The bypass valve (3-1, 3-2) is used to regulate the flow rate and concentration of the flue gas.

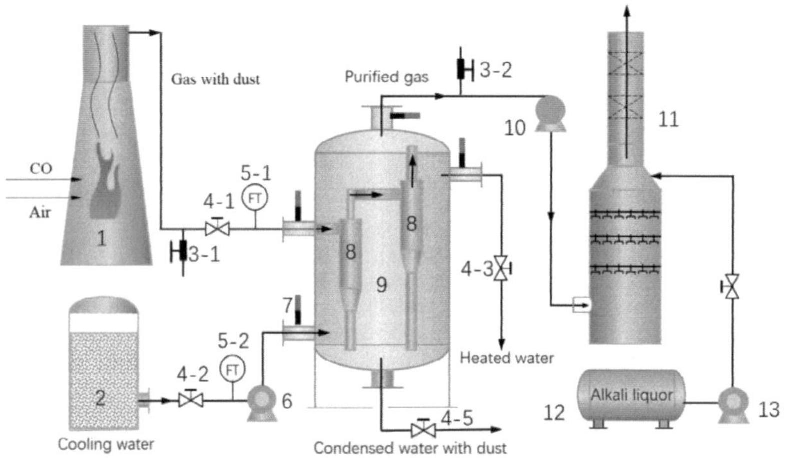

1. Incinerator 2.Cooling water 3. Bypass valve 4. Cut-off valve 5. Flowmeter 6. Centrifugal pump 7. Thermometer 8. Cyclone separator 9. Shell path 10. Fan 11. Desulfurization tower 12. Alkali liquor 13. Anticorrosive pump 14. U-tube manometer 14

Figure 2. Flow chart of experiment

3. Experimental results and discussions

3.1. Pressure drop

When the cooling medium velocity is 2.1 m^3/h and the flue gas temperature is 120, 100, 80 C°, respectively, the pressure drop between the inlet and outlet of the experimental equipment is measured. As shown in Figure 3, the pressure drop increases as the inlet velocity increases, and the rate of increase is accelerating. When at the same inlet velocity, the pressure drop will increase as the inlet temperature increases. Generally, the pressure loss of the cyclone mainly consists of the overflow pipe, the air inlet, and the loss caused by the vortex flow. But the higher the flue gas temperature is, the greater the viscosity of the dust in the flue gas is, the friction resistance between the flue gas and the side wall of cyclone separator increases, so that the pressure drop increases.

3.2. Separation efficiency

When the temperature of inlet gas is 120C°, we can see from Figure 4, In the same cooling medium flow rate, with the inlet velocity increased, the dust removal efficiency increased first and then decreased rapidly, and the air velocity in the vicinity of 50 m^3/h dust removal efficiency reached the maximum. This is because the proper increase in inlet velocity can increase the centrifugal force of the dust in the rotation, When the inlet velocity is too high, it is easy to cause a sharp increase in the resistance in the cyclone separator, and the dust rebound and the small dust droplets are broken down, so as to reduce the dust removal efficiency. When under the same air inlet velocity, the separation efficiency increases with the increase of cooling medium flow rate. This is mainly because the larger the cooling medium flow is, faster the high temperature smoke is cooled, small droplets are faster condensed, which strengthen the separation efficiency. When the flow rate of cooling medium is 2.1 m^3/h and the inlet gas velocity is 50 m^3/h, the removal efficiency reach the maximum is 96%. Under the different cooling medium flow rate, the maximum change rate of dust removal efficiency is more than 18% at the same gas inlet velocity.

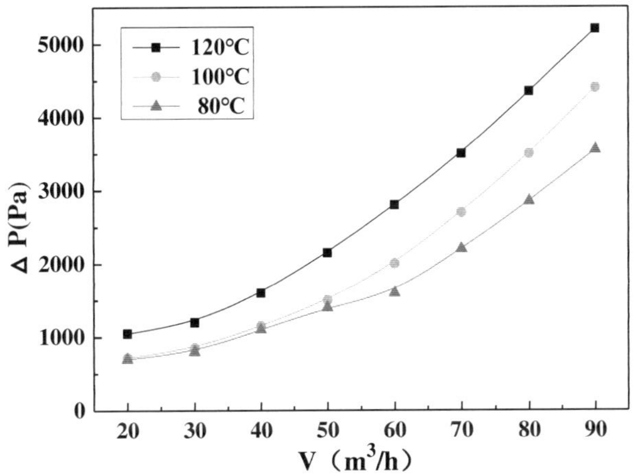

Figure 3. Relationship of V and ΔP

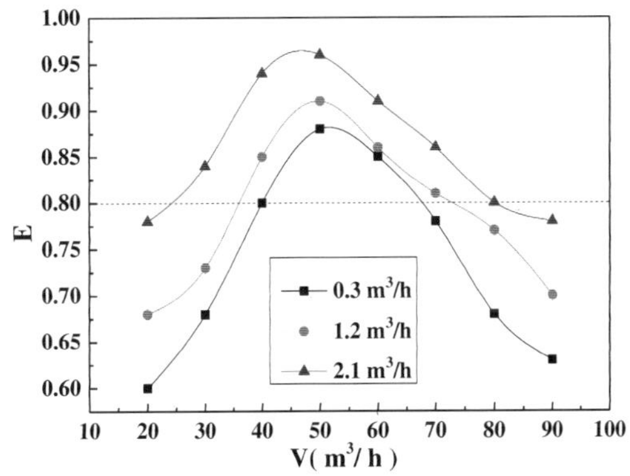

Figure 4. Relationship of V and E

3.3. Recovery efficiency of vapor water

As shown in Figure 5, the inlet temperature is 120C°, when the flow rate of the cooling medium is 0.3, 0.9, 1.5, 2.1 m^3/h and the inlet gas velocity is 20-100 m^3/h. The recovery of condensed water is studied with the change of inlet gas velocity and cooling medium flow rate in 15 minutes. The results show that under the same cooling medium flow rate, with the increase of inlet velocity, the recovery of condensed water increases first and then decreases, and greater the flow velocity of cooling medium is, the faster the increase rate of condensed water. This is because when the cooling water flow rate is lower, the cooling effect is poor, and the amount of condensation water is less. With the increase of cooling medium flow rate, the cooling effect is enhanced, and the water vapour in flue gas is condensed easily. When the flue gas velocity is less than 80 m^3/h, higher the gas velocity is, larger the volume of condensed water we can get in the unit time. When the flue gas velocity is larger than the

80 m³/h, gas residence time is too short, so the water vapor is not condensed in time, and is discharged before precipitation, which result in a sharp drop in the amount of condensed water recovery.

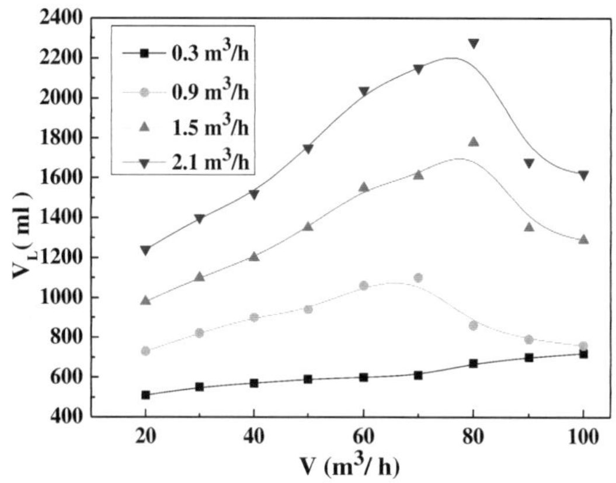

Figure 5. Relationship of V and V_L

4. Conclusion

Study on the integration of dust removal and condensed water recovery of catalytic cracking flue gas by cyclone separator. Through the study of pressure drop, dust removal efficiency and recovery efficiency of vapor water, the optimum parameters under industrial conditions are obtained.

(1) When under the same inlet speed, the pressure drop increases with the increase of inlet gas' temperature; when at the same inlet temperature, the pressure drop increases with the increase of the inlet velocity, and the rate of increase is accelerating.

(2) Under the same cooling medium flow rate, the separation efficiency increases first and then decrease with the increase of inlet velocity. When the inlet velocity is near the 50 m^3/h, the dust removal efficiency is the highest. Under the experimental conditions, when the flow rate of cooling medium is 2.1 m^3/h, the separation efficiency of flue gas is the highest. When the inlet velocity near the 50 m^3/h, the separation efficiency is highest of 96%.

(3) When under the same cooling medium flow rate, the recovery efficiency of condensed water first increases and then decreases with the increase of inlet velocity, and when the inlet velocity is 60-80 m^3/h; the recovery of condensed water is the maximum. When at the same inlet velocity, the recovery of condensed water increases with the increase of cooling medium flow rate.

5. Reference

[1] Chen M Q 2014 *Sci. Tech. Innova. Herald.* 32 12.
[2] P. Szulc. T, Tietze. K and Wójs 2015 *Arch. Civ. Mech. Eng.,* 15 1017.
[3] Fu P, Wang F, Yang XJ, Ma L, Cui X and Wang H 2017 *Environ. Sci. Technol.*, 15 1587.
[4] Nag P. K. and Gupta A. V. S. S. K. S. F. 1999 *Heat Transfer Eng*, 20 28.
[5] Park K, Hong C. H. Han J. W., Kim B.S., Park C. S. and Kwon O. K. 2012 *Int. J. Mech., Aerosp., Indus., Mech. and Manuf. Eng.*, 6 217.
[6] Leea C. E., Yua B. and Leeb S. 2015 *Energy*, 86 267.
[7] Jolius G., Chuah T. G., Fakhru R. A. and Choong T. S. Y. 2005. *Che. Eng. Process.*, 44 7.
[8] Nwokolo N., Mamphweli S. and Makaka. G. 2016 *Appl. Therm. Eng.*, 98 1158.
[9] Patterson P. A and Munz R. J. 1989 *The Canadian J. Che. Eng.*, 67 321.
[10] Parker R., Jain R., Calvert S., Drehmel D. and Abbott J. 1981 *Environ. Sci. Tech.*, 15 451.

IOP Conf. Series: Earth and Environmental Science **178** (2018) 012018 doi:10.1088/1755-1315/178/1/012018

Preparation and Performance Analysis of Lignocellulose Water Retention Agent and Combination with Coal Gangue Ceramsite

Xiaoyu Fang, Shengjie Fu and Yonggang Li[1]

1 College of Chemical Engineering, Qinghai University, Xining 810016,PR China
E-mail: li.yonggang@126.com

Abstract. Environmental friendly water retention agent has been widely used in agriculture, since its ability to absorb water, hold in water, resist salt. In this study, the water rentention agent was polymerized with using wood cellulose as raw material, acrylamide as monomer, N, N - methylene double acrylamide as crosslinking agent, ammonium persulfate as the initiator. The optimized technological conditions was obtained by the orthogonal experiments: the quality ratio of cellulose and acrylamide was in 1:6, ammonium persulfate was 1.5 wt%(of the total amount of cellulose and monomer), *N, N* - methylene double acrylamide 0.25 wt%(of the total amount of cellulose and monomer). The maximum water absorbency of water retention agent was 330.07g/g in distilled water and 117.48g/g in 0.9 wt% KCl solution. On the basis of this, using water retention agent as the coating for coal gangue ceramsite, and the maximum water absorbency of water retention agent was 332.15g/g in distilled water and 118.91g/g in 0.9wt% KCl solution. The results show that adopting above process not only can obtain good water retention agent, but also can enhance the water retention and salt resistance by using with ceramsite.

1. Introduction

Lignocellulose is plentiful in nature and is one of the most promising renewable resources[1]. The structure of lignocellulose is similar to super absorbent resin, both of them are natural reticular formation containing a large number of hydrophilic group[2]. Using lignocellulose as raw material to carry on the graft copolymerization reaction to prepara super absorbent resin, will greatly increase the additional value. It's one of the important way to use lignocellulose. At the same time, comparing to other raw materials of water retention agent ,such as starch, protein, etc., lignocellulose can widely replace chemical raw materials and will not affect the human food supply[3]. At present, the research on the super absorbent resin of cellulose has been paid attention to, but the research on the salt resistance is not sufficient. In the western regions of China, such as Qinghai's ecological environment is fragile because the salination in some areas is serious, and enhancing the salt resistance of water retention agent has profound guiding significance for the improvement of saline and alkaline land[4]. This study mainly considered the following factors:the quality ratio of cellulose and acrylamide, the dosage of initiator, the dosage of crosslinking agent[5]. Experimenting with each influence gradually and on that basis designing orthogonal experiment to obtain the optimal technological condition. Coal gangue ceramsite is beneficial to loose soil and has absorbability[6]. Using it with water retention agent in agriculture soil, not only can improve soil porosity and permeability but also can take full of advantage of water retention agent absorbent and salt resistance. On the basis of preparing water retention agent, this paper used the water retention agent as the coating for coal gangue ceramsite and research performance of them closely combined.

2. Experiment

2.1. Experimental Material

lignocellulose SD - 300; ammonium persulfate, AR; acrylamide, AR; *N, N* - methylene double acrylamide, AR; sodium hydroxide, AR ; absolute ethyl alcohol, AR; coal gangue ceramsite(made in our laboratory).

2.2. Preparation of Water Retention Agent

In a flask, dissolve a certain amount of cellulose in distilled water, N_2 atmosphere. Cool to 20°C, add the initiator ammonium persulphate and stir to dissolve about 5mins. Then add acrylamide and *N, N* - methylene double acrylamide at 70°C, N_2 atmosphere. Stir the mixture until it becomes uniform slurry. Add sodium hydroxide to hydrolyze. After the gel is formed, remove it from the reaction vessel. Then, wash it with distilled water and absolute ethyl alcohol. Dry it in a constant temperature oven. After a while, take out the product and snip into small chunks.

2.3. Combination with Coal Gangue Ceramsite

Combine water retention agent made by experiment with coal gangue ceramsites, which are about 5-7 mm in diameter and 423 kg/m³ in bulk density. Research the performance of the product after the combination.

Take a group of ceramsite, coat them with water retention agent. Ceramsites should be wrapped in resin totally and the thickness of the coating layer is about 3 mm.

2.4. Performance of Product

2.4.1. Water Absorption

Water absorption (ml/g or g/g) can be defined as the volume or mass of the liquid absorbed by the unit mass water-retaining agent in unit time.

First, cut a dry sample and weigh accurately with the analytical balance, and the mass of the dry sample was m_1(g). Then, add the sample and 300ml distilled water into 500ml beaker, mix well. Static for 24h, the water retention agent becomes hydrogel after absorbing water. At this point, use absorbent paper absorb the surface water. Then, weigh the hydrogel accurately with the analytical balance, the mass is m_2 (g). Repeat three times to take an average. Thus, the water absorption rate of the sample is Q_1 (g/g). The formula of water absorption ration is following:

$$Q_1 = (m_2-m_1)/m_1 \quad (1)$$

2.4.2. Salt Resistance

The salt resistance of water retention agent can be defined as the ability to absorb water in saline solution. Experiment with nonelectrolyte urea and electrolyte potassium chloride to research the absorption rate respectively, experimental method is similar with the water absorption rate.

Take a polymer and weigh it accurately with the analytical balance, the mass is m_3 (g). Then place the sample in 500ml beaker and add in 300ml of 0.9%wt salt solution and mix well. Static 24h, the surface water was absorbed by absorption paper, accurately weigh the hydrogel. The mass is m_2 (g). Next, take the average of three repeated measurements. The salt absorption rate of the sample is Q2(g/g). The calculation formula of salt resistance is:

$$Q_2 = (m_2 - m_1)/m_1 \quad (2)$$

The salt resistance of water retention agent is usually researched by using salt solution which can ionize a valence ion[7], such as KCl, which is also a kind of common potassium fertilizer. Therefore, the salt absorption rate in the results discussion and analysis was measured in 0.9wt% KCl solution. The salt absorption rate in urea solution was studied in the combination of water retention agent and ceramsite.

3. Results Analysis and Discussion

3.1. Effect of the Mass Ratio of Cellulose and Monomer

The biggest factor is the mass ratio of cellulose and acrylamide[8]. We chose the mass ratio : 1： 3、1： 4.5、1： 6、1： 8、1： 10、1： 12. Experimental conditions: ammonium persulfate 2.5 wt%(of the total amount of cellulose and monomer), *N, N* - methylene double acrylamide 0.375 wt% (of the total amount of cellulose and monomer).

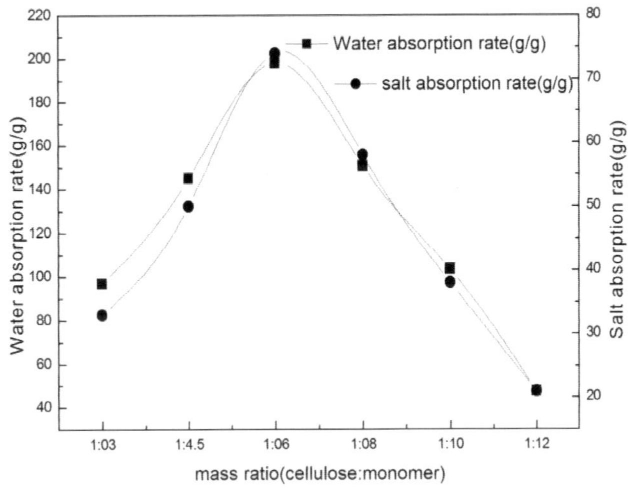

Figure 1. The effect of the mass ratio of cellulose and monomer on water absorption and salt resistance.

Figure 1 shows that when the mass ratio of cellulose and monomer is in the range of 1:3 to 1:6, the absorption rate of products increases with the increase of mass ratio. Once ratio exceed 1:6, the absorption rate of the product decreases with the increase of mass ratio. Since acrylamide is nonelectrolyte, it hardly disassociates into ions in water[9]. With the increase of mass ratio, the number of grafting point increases, the absorbent and salt resistance also incre. But when ratio exceed 1:6, the resin is self-linking highly, and the grafting rate decreases, so the absorption rate decreases[10]. Thus, when the mass ratio of cellulose and monomer is 1:6, water retention agent has the maximum absorption rate.

3.2. Effect of Ammonium Persulphate

The mass ratio of cellulose and monomer is 1:6, *N, N* - methylene double acrylamide 0.375% (of the total amount of cellulose and monomer). Experimental conditions keep unchanged, change ammonium persulfate mass, the results are shown in Figure 2.

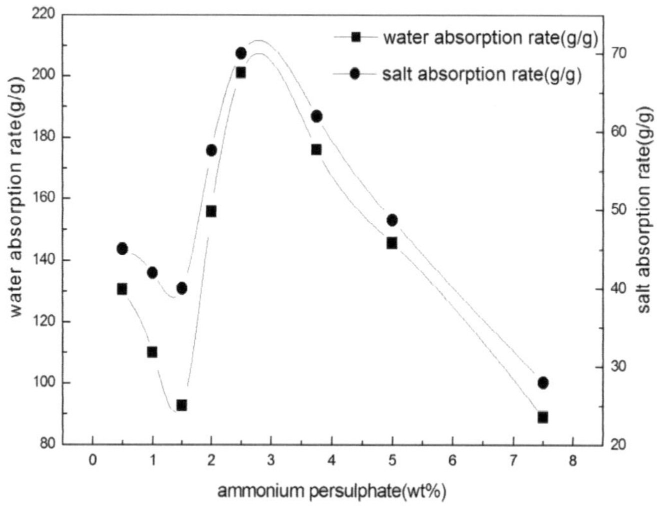

Figure 2. The effect of ammonium persulphate on water absorption and salt resistance.

The reason why initiator can accelerate the reaction rate is producing new free radical[11]. From Figure 2: the evolution of the effect can be broadly divided into three phases.

When ammonium persulfate dosage below 1.5 wt% of the total amount of cellulose and monomer, product absorbability decreases with the increase of ammonium persulfate dosage.

When ammonium persulfate dosage in the range of 1.5 wt% to 2.5wt%, product absorbability increases with the increase of ammonium persulfate dosage.

When ammonium persulfate dosage exceed 2.5 wt%, product absorbability decreases with the increase of ammonium persulfate dosage.

Figure.2 shows that when ammonium persulfate dosage is 2.5 wt%, water retention agent has maximum absorption rate. Under low dosage, initiator produces new free radical at a slower rate.So water retaining agent has slower aggregation rate, small crosslinking degree, poor absorbability[12]. As ammonium persulfate dosage increases, formation rate of free radical speeds up. It will accelarate aggregation rate, increase crosslinking degree, also increase product absorbability. However, when dosage exceeds the limits, although add moreinitiator will accelarate formation rate of free radical and aggregation rate, chain termination reaction rate speeds up , add percent of oligome, decrease product absorbability totally[13].

3.3. Effect of N, N - Methylene Double Acrylamide

Experimental conditions: the mass ratio of cellulose and monomer is 1:6, ammonium persulfate 2.5%(of the total amount of cellulose and monomer). Experimental conditions keep unchanged, change cross linking mass, the results are shown in Figure 3.

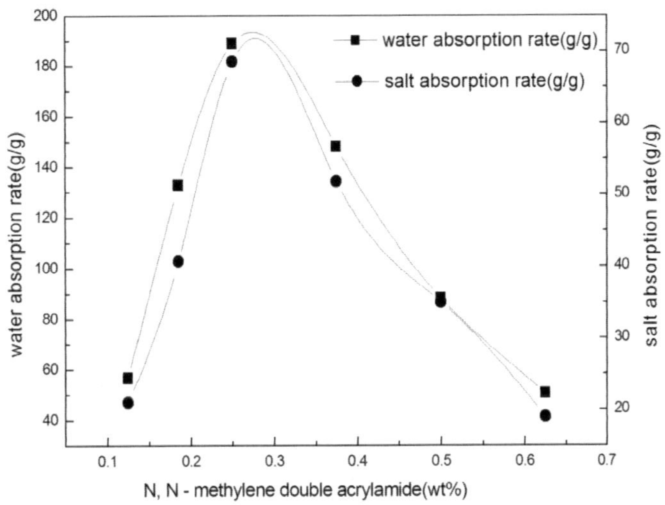

Figure 3. The effect of *N, N* - methylene double acrylamide on water absorption and salt resistance.

With increase of crosslinking agent dosage, product water absorption and salt absorption increase first and then decrease. Because in grafting reaction, the function of crosslinking agent are forming water uptake networks by cross-linking absorbent groups and make lignocellulose framework stronger to improve product absorbability[14]. When crosslinking agent dosage is too little, resin density lower, the three-dimensional net structure of polymer bigger, more dissolvable parts, higher water-solubility, so product absorbability is poor[15]. On the contrary, when crosslinking agent dosage is too much, the net-structure of polymer has more cross-links. However, crosslinking points associating with one another strongly will limit the space, and it would be bad for water retention agent swelling in water, so product absorbability is poor as well[16]. In conclusion, when *N, N* - methylene double acrylamide dosage is 0.25 wt%, water retention agent has maximum absorption rate.

3.4. The Optimized Technological Conditions

By analyzing the mass ratio of the monomer, the dosage of initiator and crosslinking agent, we get that when the mass ratio of cellulose and monomer is 1:6, ammonium persulfate 2.5 wt%, *N, N* - methylene double acrylamide 0.375 wt%, water retention agent has slightly higher absorption rate. On this basis, run the optimum conditions to obtain the optimized technological conditions. Factor levels are shown in Table 1.

Table 1. Factor levels.

Factor level	the mass ratio of the monomer	Ammonium persulfate dosage %	N,N-methylene double acrylamide dosage%
Level 1	1：5.5	2.25	0.25
Level 2	1:6	2.5	0.3125
Level 3	1:6.5	2.75	0.375

Through orthogonal experiment, we obtained the process parameters and the optimal formula. Experimental conditions: the mass ratio of cellulose and monomer is 1:6, ammonium persulfate 2.25%, *N, N* - methylene double acrylamide 0.375%. In this case, water retention agent has maximum absorption rate and the maximum water absorbency of water retention agent was 332.15g/g in distilled water and 118.91g/g in 0.9wt% KCl solution. The performance of water retention agent made in experiment is better than other water retention agent with cellulose reported in literatures[17].

3.5. The Absorbability after Water Retention Agent and Ceramsite Combining

Take water retention agent, water retention agent used with ceramsite, measure water absorption rate and salt absorption rate(in 0.9 wt% urea solution and 0.9 wt% KCl solution). Results are shown in Table 2. Analyzing Table 2, water retention agent used with ceramsite has no influence on water absorption rate and salt absorption rate, nonelectrolyte urea has little affect on absorbability.

Table 2. Comparison of absorbability

Absorb media（0.9%wt salt solution）	distilled water	urea	KCl
absorbability（g/g）	330.07	310.16	117.48
absorbability combined with ceramsite （g/g）	332.15	458.54	118.91

Based on the theory of Flory-Huggins water uptake model[18], water retention agent absorbability relates to ion concentration of electrolyte solution. Urea is nonelectrolyte, exists as molecules in water, can not dissociate into ions. So it has little affect on absorbability. On the contrary, potassium chloride dissociates in water, absorbability declines precipitously. But in this experiment，water retention agent is good in saltresistance，leading to its strong applied value in saline soils. The existing literatures have not reported cellulose water retention agent absorbability in 0.9 wt% KCl solution.

4. Conclusion

This paper synthesized a super absorbent resin by using a lignocellulose SD-300, and measured its water absorption ratio and the salinity resistance. The optimized technological conditions are obtained by analyzing the experimental results: When the mass ratio of lignocellulose and monomer is 1:6, the dosage of initiator is 2.25%, the dosage of crosslinking agent is 0.375%, water retention agent can reach 330.07g/g in distilled water and 117.48g/g in 0.9 wt% KCl solution. After combining with coal ceramsite, the water absorption rate is 332.15 g/g, and the salt absorption rate is 118.91g/g. It can be seen that the water retention agent absorbent capacity and salt tolerance are not affected after combining，using them together is promising in soil improvement.

5. References

[1] Santos, Ricardo B, Treasure, Trevor, Min, Douyong, Jameel, Hasan, Phillips, Richard, Chang and Hou-Min 2013 Impact of hardwood lignin characteristics on biorefinery International *Bioenergy and Bioproducts Conference.* V 1, pp 681-725

[2] Edward M and Rubin 2008 Genomics of cellulosic biofuels. *Nature* 454, pp 841-845

[3] Reddy N and Yang Y 2005 Biofibers from agricultural byproducts for industrial applications *Trends Biotechnol.* 23, pp 22–27

[4] Li Bin, Wang Zhichun, Sun Zhigao, Chen Yuan and Yang Fu 2005 Resources and sustainable resource exploitation of salinized land in China *Agricultural Reaserch In The Arid Areas.* 23（2）, pp 154-158

[5] GabrieliiaI, GatenholmP and Glasser WG 2000 Separation, characterization and hydrogel—for mation of hemicellulose from aspenwood *Carbohydrate Polymers.* 43, pp 367—374

[6] Omidian H and Zohuriaan-Mehr M J 2002 DSC studies on synthesis of superabsorbent hydrogels *Polymer.* 43, pp 269-277

[7] Kreps and J. A 2002 Transcriptome changes for Arabidopsis in response to salt, osmotic, and cold stress *Plant Physiol.* 130, pp 2129–2141

[8] Davis, SC, Hellens, WV and Zahalka 1996 *HA Polymeric Materials Encyclopedia.* Vol 4，2264

[9] Bakass M, Mokhlisse A and Lallemant M 2001 Absorption and desorption of liquid water by a superabsorbent polyelectrolyte: role of polymer on the capacity for absorption of a ground *Poly Sci.* 82(6), pp 1541-1548

[10] Tiwart A and Singh V 2008 Microwave induced synthesis of elect rical conducting gum acacia graft polyaniline *Carbohydrate Polymers.* 74(3), 4272434

[11] Singh V and Tiwart A 2008 Microwave promoted hydrolysis of plant seed gumson alumina support *Carbohydrate Research*. 343, 1512154.
[12] A.Pourjavadi, H.Salimi and M.S.Amini-Fazl 2006 Optimization of synthestic conditions of a novel collagen-based superabsorbent hydrogen by taguchi method and investigation of its metal ions absorption *Journal of Applied Polymer Science*. 102, pp 4878-4885
[13] Thimma, Reddy T, Reddy, N Subbarami, Tammishetti and Shekharam 2003 Synthesis and characterization of guar gum-graft-polyacrylonitrile *Polymers for Advanced Teehnologies*. 14(10), pp 663-668
[14] A. Sannino, S Pappada and M. Madaghiele 2005 Crosslinking of derivatives and hyaluronic acid with water-soluble carbodiimide *Polymer*. 46, pp 11206-11212
[15] H.Andry and T.Yamamoto 2009 Water retention, hydraulic conductivity of hydrophilic polymers in sandy soil as affected by temperature and water quality *Journal of Hydrology*. 273, pp 177-183
[16] Jianbin Zhao and Shibing Wang 2015 Study of synthesis and performance of lignin-based acrylic forestry aquasorb *Applied Chemical Industry*. 44(11), pp 2051-2054
[17] Sufei Yu 2013 Preperation and performance of a soil aquasorb modified by cellulose Shanghai. Donghua University. D, pp1-59
[18] Jens Voepe, UlricaEdlund and Ann-ChristineAl bertsson 2009 Alkenyl-functionalized precursors for renew able hydrogels design *Polymer Chemistry*. 47, pp 355-3606

Acknowledgments
This work was supported by the "Chunhui Plan" of Ministry of Education PRC (z2016096) and the "Principal Cup" innovation entrepreneurship incubation project of Qinghai University (2017-xzb-05k)

Study on the Preparation of Corn Starch Salt-resistant Water Retention Agent and Properties of Retention Agent Combined with Coal Gangue Pottery

Shengjie Fu, Xiaoyu Fang and Yonggang Li[1]

1 College of Chemical Engineering, Qinghai University, Xining 810016, PR China
E-mail: li.yonggang@126.com

Abstract. Environmental protection agent is widely used in agriculture because of its ability to absorb water, water retention and salt resistance. In this study, Qinghai corn starch was used as raw material. A high absorbent resin was prepared using acrylic acid and acrylamide as monomer, N, N-methylene diacrylamide as cross-linking agent and ammonium persulfate as initiator for graft copolymerization. The Optimized experimental conditions as follows: monomer (acrylamide + acrylic) : Qinghai corn starch = 12 (mass ratio), acrylamide content was 12% (wt %).The concentration of the initiator was *a*/2. The absorption rate of the product reached 1548g/g, and the salt absorption rate was 96g/g (0.9 wt% KCl solution). Then on this basis, the water absorption rate of retention agent combined with coal gangue pottery is 1549g/g, and the salt absorption rate is 97g/g. The results showed that the water retention agent not only has good water retention and salt resistance, but also its water retention and salt resistance are not affected by the combination with Coal gangue pottery.

1. Introduction

Corn starch (CS) is considered to be a kind of environmental friendly material because of its low cost, safety and health and good biodegradation performance, often as a graft polymer material to research and product super absorbent polymers [1]. In spite of retention agent of CS showed good water imbibition and begin to be reserached and used widely, its salt resistance didn't get enough attention[2]. The retention agent would be further limited to the use of salinity soil remediation, such as saline-alkali land in Qinghai salt lake area[3]. At the same time, the governance of desertification needs that water retaining agent has a good performance of water keeping and salt resistance, in order to achieve the effect of water imbibition and fertilizer imbibition[4].

Ceramsite has larger specific surface area, relatively stable chemical properties and good adsorption, which ensure the soil is loose and breathable[5]. So ceramsite is widely used in soil improvement [6].Our lab had done a lot of research on ceramsite preparation of coal gangue of Qinghai Reshui coal industrial park. The coal gangue ceramsite can be used in soil repairing and water retaining and fertilizer keeping[7].

Considering the characteristics of water retention agent and adsorbent pottery. This Study used CS as main material, acrylic acid (AA) and acrylamide (AM) as graft monomer, ammonium persulfate as initiator on graft copolymerization, prepared injectivity resin and test its water retention and salt resistance. In the end, combine the water protection agent with the coal gangue ceramic granule and test combination's water retention and salt resistance.

2. Experimental Part

Please follow these instructions as carefully as possible so all articles within a conference have the same style to the title page. This paragraph follows a section title so it should not be indented.

2.1. Experimental Materials
Qinghai CS(industrial grade); acrylic acid(AR); acrylamide (AR), ammonium persulphate(AR);*N,N*-methylene diacrylamide(AR); NaOH(AR).

2.2. Preparation of Retention Agent

2.2.1. Starch gelatinization. Five grams of CS was dissolved in 50 mL distilled water in a 100mL beaker with magnetic stirring at 65°C. Cooled to 20°C after 2 h.

2.2.2. Configuration of monomer solution. Added 43mL of acrylic acid into the 4 mouth flask, 1mol/L NaOH solution was added for partial neutralization. Then, ammonium persulfate was added as initiator. The mixture was stirred for 5mins under nitrogen and then added acrylamide.

2.2.3. Graft copolymerization. Added 0.01 g N,N'-methlenebisacrylamide in the 4 mouth flask under nitrogen. Then, gelatinized starch was added after 5 mins. Until it became a paste of polymerization at 70°C. After standing for 2h, the poduct was washed with ethanol and deionized water.

2.3. Retention Agent Combined with Coal Gangue Pottery
The coal gangue is prepared in the laboratory (coal gangue is from Qinghai hot water coal industry park). The diameter of the coal gangue is 5-7mm, the accumulation density is 423 kg/m³, and the volume density is 889 kg/m³. Coat the coal gangue ceramic granule with the water protection agent. After drying, the thickness is 3mm, and the ceramic particles are not exposed directly.

2.4. Performance Test of Water Protection Agent

2.4.1. Determination of Water Absorption Rate. Weigh 0.3g water retention agent accurately. Add it to 1kg deionized water in the mixing conditions. Stand for three hours and take out the product after absorbing water. Drain its surface with water absorbent paper and its mass is m_1. Thus, the water absorption rate of the sample is S_1(g/g). The calculation formula of water absorption rate [8] is following:

$$S_1=(m_1 -0.3)/0.3 \quad (1)$$

2.4.2. Determination of Salinity Resistance. The salt resistance of water retention agent is the water absorbent capacity in a certain concentration of salt solution. The determination of the salinity is usually using 0.9%(wt) NaCl solution [9]. Considering that the fertilizers used in agriculture mainly include nitrogen fertilizer and potash fertilizer. In this paper, the determination of salt resistance mainly used 0.9% urea solution(Nitrogen-containing non-electrolytes) and 0.9% KCl solution(contains potassium and a valence ion can be electrolyzed).
When measuring the salt resistance, weigh 0.5g water retention agent accurately at first. Then add it to the 500g target concentration of salt solution with mixing. Static 3h to absorb water, then remove the product after expanding. At last, absorb water from its surface with water absorbent paper and its mass is m_2. The salt absorption rate of the sample is S_2(g/g). The formula for calculating the salinity resistance of water retention agent [10] is following:

$$S_2 = (m_2-0.5) /0.5 \quad (2)$$

3. Results Discussion
The experimental conditions we chose: The content of AM is 10 wt% of the total (wt), monomer content (AM + AA) :CS= 10, acrylic neutralization is 70 wt%, and initiator concentration is 0.1 wt%(relative to monomer). The single factor variables are discussed .

IOP Conf. Series: Earth and Environmental Science **178** (2018) 012019 doi:10.1088/1755-1315/178/1/012019

3.1. Influence of the Content of AM on the Product's Absorbability

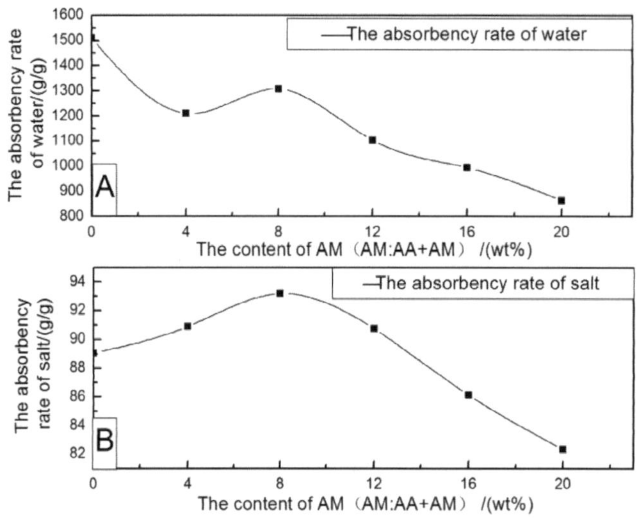

Figure 1. Influence of acrylamide content on the amount of liquid absorbed.

According to Figure 1, when AM accounted for 0% to 8%, product's salt absorption rate increases with the increase of the mass ratio. Bibulous rate drops in AM content from 0% to 4%, increases at 4 to 8% range. When the content of AM exceeds 8%, the absorption rate of the product decreases with the increase of mass ratio.

When AM is 4%, the water absorption rate decreases. Because the addition of AM reduces the hydrogen bonds formed with water and carboxyl. When the content of AM increases to 8%, the water absorption rate and salt absorption rate become higher. It is due to the synergistic effect of a certain number of different hydrophilic groups on the large molecular chain. Since -CONH2 is a non-ionic group, the degree of dissociation in water is small and the influence of ions is small. Therefore, the introduction of nonionic monomers enhances resin's anti-electrolyte ability and the salt absorption rate of the product. But there is a moderate amount of it[11]. When dosage more than 8%, due to the hydrophilicity of -CONH2 is not as good as - COOH, excessive dosage leads to water absorbent and electrolyte resistance decreases. Thus it can be seen that a mixture of different hydrophilic groups in the appropriate copolymerization is one of the effective ways to improve the ability of water-absorbing resin to resist salt. [12]

3.2. Influence of the Initiator Concentration on the Absorption Rate of the Product

As shown in Figure 2, with the increase of the trigger dose. The product absorption rate rises at first and there is the peak around a(a=1/1200). Then the absorbent rate shows a decreasing trend when the initiator concentration was greater than a.

When the dosage of initiator is low, the number of grafted active centers generated on the starch skeleton is correspondingly less, the polymerization rate is slow and the crosslinking degree is not enough, the absorbability performance is not good. However, when the dosage is large enough, it can improve the polymerization speed. But at the same time, it also increases the chance of the chain termination, and the monomer has more chance to increase which means the absorption performance is reduced. The test results show that the absorbent rate reaches its peak when the initiator concentration is a. [13]

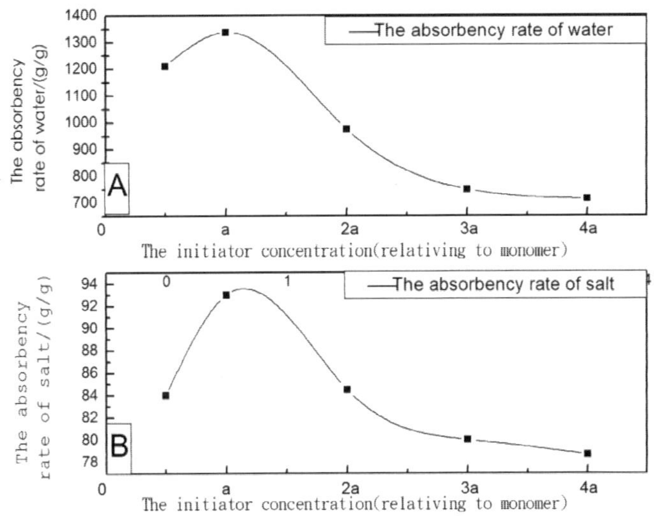

Figure 2. Relationship between the concentration of initiator and the amount of product absorbent (a = 1/1200).

3.3. Influence of the Neutralization of Acrylic Acid on Product Performance

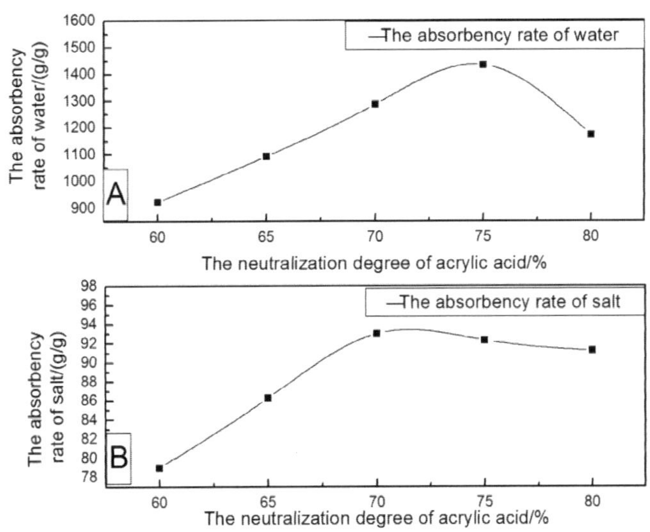

Figure 3. The effect of acrylic neutralization on product performance.

As shown in Figure 3, when the AA neutralization (Dn) is 75%, the absorption rate shows the upward trend. And at Dn=75%, the absorption rate reaches its peak. When the Dn is low, the -COO content is low, but the -COOH content is high. At the same time, the reaction is intense, and it is easy to form the highly crosslinked polymers. The concentration of ions in the structure of the polymer network is small, and the permeability of the polymer network is low, which leads to the low absorption rate. When the Dn is larger than 75%, the degree of cross-linking is decreased on the one hand and the strong hydrophilic -COO is increased on the other hand with the increase of Dn. It makes the increase of the inner osmosis pressure in the crosslinking network, which leads to the increase of the absorption rate. However, when the Dn is too high, the reactants are less interrelated and the product crosslinking is weaker, high content of the coo-content can also increase the waterborne of the resins. So the absorption rate shows a downward trend. [14]

3.4. Influence of Monomer and Qinghai Corn Starch Ratio on Product Performance

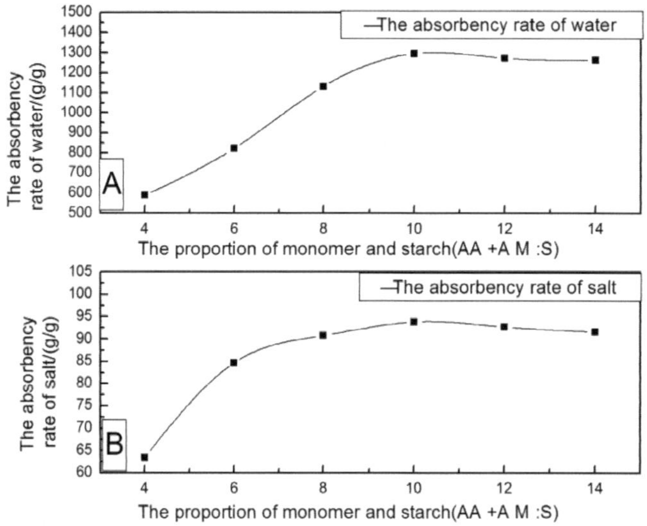

Figure 4. The effects of the monomer and starch ratio (AA +A M: S) on product performance.

Figure 4 is the relationship between AA + AM: S(w t) and product absorption rate. When AA + AM: S< 10, the absorption rate increases with AA + AM: S. When the mass ratio is 10, the absorbant rate reaches its peak. Starch is used as grafting skeleton in polymer, but it does not absorb water better. The starch content is too high which makes the grafting side chain is short, the network structure is small and the absorption rate is low. But as AA + AM: S ratio increases, starch content is not enough. When AA + AM: S> 10, the absorption rate of the product has decreased due to the increase probability of the monomer and the increase of the crosslinking degree of the graft side. [15]

3.5. Process Condition Optimization

In order to get the water retention agent with better salt resistance, we analyzed the single factor experimental results. Results show that when AM content at about 8%, initiator concentration between $a/2$ and $2a$, acrylic acid neutralization degree at around 75% and AA + AM: CS at around 10, the salt absorption rate water retention agent is highest. According to the above, the factors of orthogonal test are determined in Table 1.

Table 1. Factor levels

Factor	Content of AM（wt%）	Concentration of initiator (relative to monomer)	Neutralization of acrylic (%)	AA+AM :CS
Level 1	7	*a/2*	64	9
Level 2	9	*a*	70	11
Level 3	11	*2a/3*	76	13
Level 4	13	*2a*	82	15

IOP Conf. Series: Earth and Environmental Science **178** (2018) 012019 doi:10.1088/1755-1315/178/1/012019

The results of the orthogonal test are: when AM content is 12 %, initiator concentration is *a*/ 2, AA neutralization degree is 80 %, AA + AM: CS= 12, the product has better anti-salt performance. The absorption rate is 1548g/g, and the salt absorption rate is 96g/g. Dang Jing and her team[16] used corn starch as material, acrylic acid as monomer for grafting copolymerization to the water absorbent resin preparation. The water absorption rate is 730g/g, and they have no salt-resistant analysis. Comparing our preparation with theirs, the water absorption rate of the product was increased by 818g/g, and the salt resistance reached 96g/g.

3.6. Properties of Water Retention Agent and Its Using with Coal Gangue Pottery

By orthogonal experiment to obtain better formula to test the resistance to salt water retention agent. As shown in Table 2, the non-electrolyte urea has no influence on the salt resistance. But the electrolyte solution leads to a decrease in water absorbent capacity.

The data in the Table 1 prove that the water retention agent and the salinity resistance of the pellet are unchanged. Because of the simple adsorption of the clay, the amount of adsorption is very small and has no direct effect on the water retention agent. What's more, because the clay is porous inside, there is a certain adsorption to the plant's desired nutrients such as nitrogen and potassium fertilizer. Thus the water retention agent combined with coal gangue are in line with the agriculture's and forestry's actual needs.

Table 2. Water absorption rate and salt rate measurement data

Absorption medium	DI	Urea	KCl solution
Absorbent rate of water retention agent/(g/g)	1548	1481	96
Absorbent rate of water retention agent combined with Coal gangue pottery/(g/g)	1549	1484	97

4. Conclusion

This work used Qinghai corn starch as the raw material, grafted acrylic acid and acrylamide for the preparation of water retention agent with good salt resistance. The optimized process is: AM content is 12 %, initiator concentration is *a*/ 2, AA neutralization degree is 80 %, AA + AM: CS= 12, the product has better anti-salt performance. The absorption rate is 1548g/g, and the salt absorption rate is 96g/g. The water absorbent rate reaches 1549g/g and the salt absorption rate is 97/g after the water protection agent and the coal gangue are used together. At this time, the water retention ability and the ability to resist salt are not affected. In summary, it is of practical value that the water retaining agent uses together with the ceramic grain.

5. Reference

[1] Ying Ma, Jinlong Zhang. Current situation and development goals of corn processing industry [J]. *Heilongjiang grain and oil technology,* 2001,(02):10-13.

[2] An overview of the prevention and improvement of saline alkali land in the northwest region [J]. *Xinjiang agricultural science*, 1962,(08):295-296.

[3] Mingli Li, Yun Tong, Fang Li, Shunrong Yang. Development status and research progress of ceramic-clay concrete [J]. *Concrete world*, 2010,(09):16-18.

[4] Ming Zhou, Wanfen Pu, Yan yang . Optimization of the synthesis conditions of high-absorbent resin for anti-salt starch grafts [J]. *Petrochemical,* 2003,(04):314-316.

[5] Wang rui, Li qian, Qinyan Yue, Nana Peng, Baoyu Gao, Wang yan. Preparation of wheat straw/modified bentonite based high absorbent resin [J]. *Journal of shandong university*, 2014, 42(06):115-120.

[6] party jing, wang rumin, wang xiaojian, cheng lei. Preparation and performance of corn starch grafting acrylic acid high absorbent resin [J]. *China adhesive,* 2009,18(04):45-48.

[7] guo self-hua. Research on high absorbent resin [J]. *Science and technology information*, 2009,(01):2.

[8] LI Zongjin, ZHU Ding. Property Improvement of Portland Cement by Incorporating with Metakaolin and Slag[J]. *Cement and Concrete Research,* 2003, 33:579-584

[9] Z.Zhang, P.Chen, X.F. Du. et al. Effects of amylose content on prop-erty and microstructure of starch- graft- sodium acrylate copolymers[J].*Carbohydrate Polymers*,2014,102,453- 459.

[10] L.Huang, C.M. Xiao, B.X. Chen. A novel starch- based adsorbent for removing toxic Hg（II）and Pb（II）ions from aqueous solution[J]. *Journal of Hazardous Materials*,2011,192,823- 826.

[11] Asaka H,Yada A,Saeki S． Absorbents for salt — containing water: JP ,8016 016 ［P］． 1980 — 02 — 04.

[12] D D Vu, P Stroeven, V B Bui. Strength and Durability Aspects of Calcined Kaolin-blended Portland Cement Mortar and Concrete[J]. *Cement and Concrete Composites*, 2001,23:471-478

[13] Prinya Chindaprasirt, Chai Jaturapitakkul, Theerawat Sinsiri. Effect of Fly Ash Fineness on Compressive Strength and Pore Size of Blended Cement Paste[J]. *Cement and Concrete Composites*, 2005,27: 425-428

[14] KIATKAMJORNWONG S,CHOMSAKSAKUL W,SONSUK M. Radiation modification of water absorption of cassava starch by acrylic acid/acrylamide[J]. *Radiation Physics and Chemistry,*2000,59（4）:413-427.

[15] BHUNIYA S P,RAHMAN M S,SATYANAND A J,et al. Novel route to synthesis of allyl starch and biodegradable hydrogel by copolymerizing allyl -modified starch with methacrylic acid and acrylamide[J]. *Journal of Polymer Science:Part A: Polymer Chemistry,*2003,41（11）:1 650-1 658.

[16] J.R. Witono, I.W. Noordergraaf, H.J. Heeres, et al. Water absorp-tion, retention and the swelling characteristics of cassava starch grafted with polyacrylic acid ［J］. *Carbohydrate Polymers,*2014,102, 453- 459.

Acknowledgments

This work was supported by the “Chunhui Plan” of Ministry of Education PRC (z2016096) and the “Principal Cup” innovation entrepreneurship incubation project of Qinghai University (2017-xzb-05k)

Theoretical Structural Stability of Small Molecules Water

Pei Jiang [1], Qingshan Li [1、2], Jiajie Lu [2]

1 Key Laboratory of Metastable Materials Science and Technology, (Yanshan University) , Qinhuangdao, Hebei
2 Health and longevity Institute of traditional Chinese Medicine Research Institute ,Beijing, China

Abstract. It has been found that small molecules of water are fully consistent with our requirements for healthy drinking through the study of various types of water microstructure. And we also found that long-term consumption of small molecules of water can improve some lifestyle-related diseases. The current structure of small molecules of water drawing people most acceptable, The most stable structure of a water mass having n=2 to 5, which is more than two water molecules linked by a hydrogen bond, is ring-shaped and its basic building block is a water molecule containing a proton donor and a proton acceptor. It has been predicted almost like there are several low energy continuum energy of two or three dimensional when n=6, This is a cage-type three-dimensional structure of small molecules between groups and macromolecules turning point. Based on the principle of primary principle, this paper analyzes the structural stability of small molecular group water from the electronic level, and gives the optimal small molecular structure model.

1. Introduction

Water is the source of life. Since the day of existence, water is inextricably linked with life. Human beings can survive and develop with water. Water participates in the digestion and absorption of food and transports nutrition; participates in excretion of body metabolism and metabolites; participates in the regulation of body temperature; and keeps the lubrication of organs, joints and ligaments, which are necessary for maintaining the normal physiological functions of human body. Liquid water in nature does not exist as a single molecule (H_2O), but exists in the form of molecular clusters, so the size of the water molecules is divided [1]. Academic size of water molecules typically using NMR spectral line width to reflect the size of the water cluster structure, the wider the lines, the larger clusters; more narrow spectral lines, the smaller clusters. Specifically, different water samples are used to represent the half-width of ^{17}O-NMR, and the water below 90 Hz (quite less than 10 water molecules) is called small-molecule hydration, whereas the macromolecule hydration is the other. The smaller the water cluster, the water penetration, solvency, diffusion force, emulsifying power, metabolic force and detergency will increase, water activation is stronger and the oxygen content is higher [2].

In the process of free access to human cell membrane, the small molecule group can penetrate the nucleus and DNA quickly, improve the function of cell metabolism and participate in the activity of life. After entering the human body, these active water constantly activates human cells and can carry more beneficial nutrients, minerals and oxygen to the human body and enter every corner of the human body and cells to make human cells more energetic and energetic. This structure of water molecules is a dynamic combination of its stable existence of time is only about 10~12 s, both the water molecules have been added to some water molecules, and some water molecules leave the water molecules .

And the molecular mass is just an average. At room temperature, water molecules in general water are about 30 to 40 water molecules. The mystery of water lies in the water molecule. The molecular size of water is related to the temperature of water, the ion concentration and the changing experience. Electric field, magnetic field, sonic wave and infra-red ray can all influence the structural change of molecular groups [3]. The study found that the smaller the water molecules, the greater the activity, the better water tastes.

The molecular structure of water molecules is a dynamic combination, like a busy supermarket, coming and going, in and out. This means that the water molecules are constantly changing, and there are ways to change the amount of water molecules, such as magnetic fields, sound waves, rays, infrared, pressure, temperature, etc. One of the most convenient way is to heat, the faster the molecular motion, the more unfavorable to the formation of hydrogen bonds, heated to 100° C, resulting in a gaseous water, the water molecules close to single-molecule or small molecule structure, such as tap water after boiling The distilled water belongs to the small molecule group. Therefore, access to small molecules of water is not mysterious. However, after research, at present, the fowllowing is the most accept structure of small molecule water. The most stable structure of water molecules beyond n=2-5, which is beyond the two water molecules linked by a hydrogen bond, is ring-shaped. The basic structural unit is a water molecule containing a proton donor and a proton acceptor. By theoretical It is very important to calculate the structural stability of small molecule water.

Small molecular group water has an important role in human health [4] [5] .

Activate cells - enhance immunity. The basis of cell activation is due to the fact that small groups of weakly basic water molecules are very small, can pass through the water channel, into the cell, but also due to water surface tension increases tightly attached to the cell membrane, and will not produce positive and negative charge attracting effect, the water molecule cohesion of high energy charge, increase the cell membrane potential, the cell is full of vitality, promote cell metabolism and self-healing ability, thereby enhancing the body immunity.

Strong penetration - improve metabolic function. Small molecules weak alkaline water, molecular groups neatly arranged, high density, strong penetration, can rapidly penetrate human tissues and cells, to enhance nutrient replenishment and waste discharge, promote metabolism, thereby enhancing metabolic function.

Solubility - Enhanced electrolyte ionization. Because of its small molecular mass of small molecules, small group, dissolved greatly enhanced, easily absorbed by the body; can dissolve the vessel wall impurities, restore blood vessel elasticity, prevention of hypertension, arteriosclerosis and other diseases. And because of the strong solvency, so that minerals and trace elements easily be electrolyzed as ions to improve their absorption rate. Such as calcium ions can be well absorbed by the body, add calcium, increase alkaline substances, bone marrow supplementation and enhance the body's immune function.

High dissolved oxygen - the ability to increase the vitality and metabolism of cells. Small alkalescence Weak alkaline water, due to its extremely small molecular groups, thus increasing dissolved oxygen reached 9.6mg/L, which can greatly increase cell viability and metabolic capacity, play a role in longevity.

Alkaline - to regulate the body's acid-base balance, enhance the elimination of acidosis function. "Acidic constitution is the source of all diseases." Shinohara Shinohara, a world-renowned medical doctor, points out that when acidic toxins accumulate in the body, the disease will occur. Small groups of weak alkaline water PH value of stability between 7.20 ~ 7.80, can neutralize acid toxins in vivo and promote detoxification in the body, and keep the body fluid was alkaline, enhance digestion and nutrient absorption, increase appetite.

Replenish minerals needed by the body - balances human nutrients. Yu Ruochu, a famous Chinese nutritionist and president of China Trace Element Science Society, pointed out: The imbalance of trace elements is the root cause of chronic diseases. Professor Jin Riguang, a Chinese quantum chemist and water nutrition expert, pointed out: Water without the motive force of life is the invisible killer of people's health. Professor Li Fuxing, an expert on water nutrition in China and a director of the

Professional Committee of China Medical Association for Healthy Drinking Water, pointed out: Returning to natural water is healthy water. Professor Zhong Bing of Guangdong Province Health Food Association also believes that many of the chronic diseases that endanger human health are related to the imbalance of trace elements in the body.

Close to the human cell concentration - has the function of promoting human health and improving the quality of life, the small molecules of weak alkaline water called the life water, and the frequency of cell fluctuations similar to contact with the cells will have resonance, resonance effect, and to take the cells in the waste and pathogen dilution, discharge, and promote cell metabolism.

Elimination of free radicals - to reduce the damaging effects of free radicals and enhance immune function. Free radicals are the main factors of human aging. Due to the small alkalescent weak alkaline water, which has a very low oxidation-reduction potential and is rich in hydroxyl anion, it can quickly eliminate excess free radicals in the body and enhance the body's immune function. A good medicine for aging, long drink will lay a solid foundation for health and longevity [6].

Calculation method: In this paper, the CASTEP [7] software module based on density functional theory was used to calculate the structural stability of small molecular clusters of water. Vanderbilt [8] ultra-soft pseudopotentials are used and the cut-off can be set to 500 eV to handle the ion-electron interactions. For the description of the exchange-related interactions between electrons, we use the PBE [9] function in the generalized gradient approximation. The Brillouin zone was sampled using the Monkhorst-Pack [10] meshing method and the k-point spacing was set at 0.01/Å. In the process of geometric optimization, the relaxation of crystal structure adopts the energy minimization method of Broyden-Fletcher-Goldfarb-Shanno [11]. The convergence criterion of atomic structure relaxation is that the force acting on each atom is less than 0.01 eV/Å , The energy change is less than 5.0×10-6 eV/atom, the stress is less than 0.02 Gpa. The basic model was constructed by setting up a square grid of 10× 10×10Å, placing the small molecular water structure in the center of the system, respectively, fixing the unit cell volume, optimizing the structure, optimizing the structure parameters and forming enthalpies listed in Table 1.

Results Discussion: According to Table 1, we can see that the H-O bond length and the H-O-H bond angle keep the original water molecule structure unchanged for different groups of small molecules. The size of hydrogen bonds decreases with the increase of the system. From the perspective of enthalpy formation, the formation enthalpy refers to the energy difference between the product and the reactants in the chemical reaction. If the energy of the product is lower than that of the reactants Energy, then the product thermodynamically stable, otherwise unstable. Therefore, the enthalpies of formation of these four kinds of small molecules of water are negative and the compounds satisfy the thermodynamic stability, and the four systems can be stable from the theoretical point of view. In order to further analyze the structural stability of small molecule water, we calculated the total state density and the density curve of the four systems, as shown in Figure 1. In the figure, the blue curve is the total state density, and the black and red curves represent the values of s and p orbitals. The green dot line represents the Fermi energy level of 0. The Fermi level is the valence band, the Fermi energy level is the guide band. The higher the value of the Fermi energy level indicates that the system is more unstable. The corresponding state density values of these four small molecules in the Fermi energy level are: 7.7,10.5,12.2 and 17.8 respectively. As can be seen from Figure 1, 3 ring water and 5 ring water have relatively high stability and are more stable in nature or experimental study.

Table 1. Small molecular cluster water structure parameters and the formation of enthalpy.

structure	composition	H-O bond length	H-O-H bond angle	Hydrogen bond	Enthalpy
Tricyclic	H6O3	0.976Å	105.92°	1.83Å	-1.41eV
Tetracyclic	H8O4	0.976Å	106.43°	1.69Å	-1.87eV
Pentacyclic	H10O5	0.975Å	106.29°	1.65Å	-2.34eV
Hexyclic	H12O 6	0.974Å	106.83°	1.67Å	-2.56eV

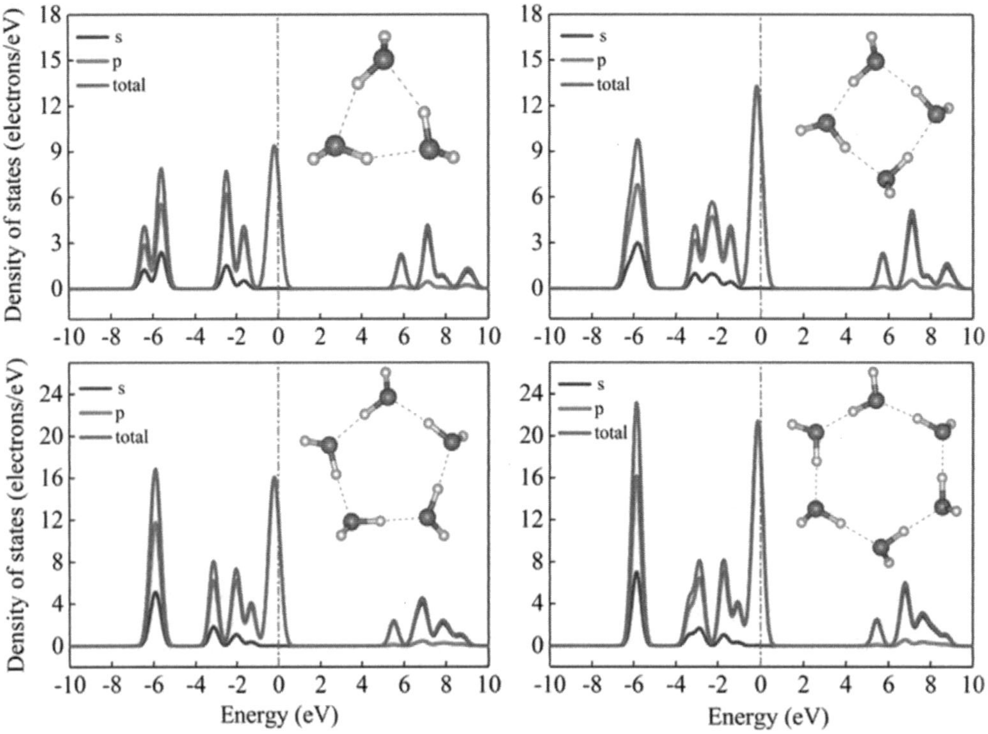

Figure 1. The total and partial wave density of small molecule water

Summary Outlook: It was found that the smaller the water molecules, the greater the activity, the better to drink; and the larger the water molecules, the less active, nor good for drink. Many experiments show that changing the structure of water has a great influence on the biological and physiological responses. Many studies also show that the fermentation of food, ripe, flavor and water structure is also closely related. 1. Small molecular group water does exist in nature. But it can be cognitively and discriminating. It is generally used to detect the safety barrier index, mineral content and proportional index. Then there are the "water molecules" indicators. 2. Clean and safe natural "small molecular mass water" is nature's "longevity fountain". People in one of the world's few longevity villages have also put up their staff to "live a long and healthy life".3. Natural "glacial mineral water" is a rare and expensive to ordinary people, but the minerals they contain are in line with the needs of the body. Therefore, it can be used as a healthy water to drink regularly, so as to supplement the nutrition that cannot be absorbed from food.

In order to solve the problem of drinking water, we can choose to install a water treatment device that solves the problem of water quality, hygiene and safety, and restores the water quality to "small molecular clusters." Water gave birth to life, the mystery of water is endless, but also for us to explore more step by step!

2. References

[1] Linshuang Wang, Rongshu Wang, Research Progress of Water Clusters. Chemical progress, 2001. 13(2): p. 81-86.

[2] Min Zhang, etal. Preparation and Evaluation of a High Quality Small Group of Drinking Water. National Infrared Heating and Infrared Medical Development Symposium Proceedings. 2011.

[3] Mingyang Lian, etal. Far infrared and small molecules of water.National Infrared Heating and Infrared Medical Development Symposium Proceedings 2015.

[4] Zhaoqing Wang, etal. Water function and application. World Alliance Medication Diet Research Committee International Medicinal Diet Therapy Symposium. 2014.

[5] Jiang Lu, Small molecules of water treatment of hypertension, coronary heart disease and diabetes significantly. Wish you good health, 2014(5): p. 56-56.

[6] Zhiwen Li, etal.A water treatment method and system for preparing stable small molecule water. 2014.
[7] Segall, M.D., etal., First-principles simulation: ideas, illustrations and the CASTEPcode. 2002. 14(11): p. 2717.
[8] Vanderbilt, D, Soft self-consistent pseudopotentials in a generalized eigenvalue formalism. Physical Review B, 1990. 41(11): p. 7892-7895.
[9] Perdew, J.P, K. Burke, and M. Ernzerhof, Generalized Gradient Approximation Made Simple. Physical Review Letters, 1996. 77(18): p. 3865-3868.
[10] Monkhorst, H.J. and J.D. Pack, Special points for Brillouin-zone integrations. Physical Review B, 1976. 13(12): p. 5188-5192.
[11] Pfrommer, B.G., etal., Relaxation of Crystals with the Quasi-Newton Method. Journal of Computational Physics, 1997. 131(1): p. 233-240.

IOP Conf. Series: Earth and Environmental Science **178** (2018) 012021 doi:10.1088/1755-1315/178/1/012021

Factors for Promoting Polyhydroxyalkanoate (PHA) Synthesis in Bio-Nutrient-Removal and Recovery System

Jiachao Zhu[1], Qing Tian[1], Yanbin Zhu[1], Jianpeng Yang[1], Mei Wang[1]

1 Department of Environmental Science and Engineering, Dong Hua University, 2999 Shanghai North people's Road, 201620 P.R. China
E-mail address: tq2004@dhu.edu.cn

Abstract. In the near future, the bio-nutrient-removal (BNR) system is no longer confined to meet the nitrogen and phosphorus removal requirement, but operated for sequestering and recovering carbon, nitrogen and phosphorus from wastewaters and generate more high-value-added by-products. Polyhydroxyalkanoate (PHA) is a metabolite of microorganisms under unfavourable growth conditions for their own energy storage and utilization. More and more researches have gradually uncovered the impacting factors on the microbial synthesis of PHA in the BNR and none-BNR system. This mini-review summarizes the impacting factors of carbon source types and dosage, nitrogen concentration and phosphorus recovery and temperature variation on the enrichment of PHA by microbial populations. Furthermore, the potential genomic reasons for promoting PHA synthesis are introduced. It provides a new perspective to stimulate the PHA synthesis by the functional microorganisms and its potential benefit for bio-nutrients removal and recovery in BNR system.

1. Introduction
In general sewage treatment plants, the lack of influent readily biodegradable chemical oxygen demand (rbCOD) will limit nitrogen and phosphorus removal efficiencies in the BNR systems. For example, the rbCOD / TN and rbCOD / TP ratios of urban sewages are confined within 4 and 16, respectively. Therefore, to achieve high efficient N and P removal and to meet the increasing stringent TN and TP discharging standard, external carbon sources are needed to compensate for the rbCOD deficiency. In the case of excessive carbon (C) and nutrient limitation, the bacteria tend to turn their metabolism into producing PHA, which is mainly consist of poly-3-hydroxybutyrate (PHB), poly-3-hydroxyvalerate (PHV) and poly-3-hydroxy-2-methylvalerate (PH2MV). As early as 1974, it has been reported that the accumulation of PHA was found in some specific microorganisms (e.g., polyphosphate accumulating organisms (PAOs) and glycogen accumulating organisms (GAOs) in the BNR system. Some species of PAOs can transform external carbon source into PHA for simultaneous denitrification and P accumulation. Some GAOs are also capable of absorbing VFAs and converting them into PHA and perform denitrification with the stored PHAs. PHAs are also an important carbon source for N removal (intracellular carbon source based denitrification) and P removal and recovery. The effect of carbon source type as well as dosage on the microbial synthesis of PHA is important fields to be explored. Temperature is another factor also worth considering in practical applications.

2. Carbon Source Type and Dosage
The researches on carbon source types have been extensively explored. Most of the studies applied acetate or propionate as external carbon source. Ethanol can also be used as external carbon source

in the EBPR system during long-term operation and achieve considerable phosphorus removal [1]. Many studies suggest that acetic acid is beneficial to the growth of PAO, yet some studies have shown that when acetic acid is used as a carbon source, the overproduction of polysaccharides would result in unstable phosphorus removal [2]. In the treatment of landfill leachate via endogenous denitritation (ED), using acetate as the sole carbon source is more likely to stimulate denitrifiers to store PHA than using propionate. In this way, higher nitrogen removal efficiency was achieved. As the comparison, the lowest PHA amount was observed when using glucose as the carbon source [3]. The study also investigated the composite carbon sources (acetate + propionate, acetate + propionate + glucose) and achieved similar denitrification efficiencies. The type of carbon source also affected the composition of PHA. It was reported that propionate mainly induced the synthesis of PHV, whereas the use of acetate induced the synthesis of PHB, the ratio of PHB / PHV synthesized under the mixture of acetate and propionate was directly dependent on the microbial uptake rate of acetate and propionate [4]. In the moving bed biofilm reactor (MBBR) system, the feasibility of using two common carbon sources, methanol and ethanol, was evaluated on the removal of micro-pollutants from biofilm. The methanol-fed MBBR contained higher microbial abundance than the ethanol-fed MBBR, suggesting that improved biotransformation of the target compound may be associated with higher microbial abundance [5]. In terms of the production of PHA using a mixed microbial culture, the effect of substrate composition on the PHA accumulation capacity was investigated. Six organic acids, malate, citrate, lactate, acetate, propionate and butyrate were tested. Only three volatile fatty acids (VFAs) were found to cause PHA production, with the feeding of acetate and butyrate resulting in PHB production, propionate promoting the formation of PHB / PHV copolymer (51% PHV) [6]. In addition to carbohydrates and VFAs, other feeding organics, such as sugars, agricultural wastes and whey, should mainly promote the synthesis of short-chain-length PHA (scl-PHA), while the synthesis of medium-chain-length PHA (mcl-PHA) should use long carbon chain VFAs or oily waste [7]. This provides a new idea for the future carbon source selection in the BNR system.

Besides these external carbon sources, the sludge generated within the system can also be recycled and reused as a carbon source. For example, primary sludge in the system after sonication was reused as an alternative carbon source and brought out high PHV containing biomass as well as ideal nitrogen and phosphorus removal [8]. Undoubtedly, such experiment would inspire an innovation of source recovery in the BNR system.

The dosing mode of carbon source will also affect the PHA synthesis. In Tian et al.'s study, an intermittent dosing of carbon source was employed and it was found that phosphorus uptake was significantly increased while increasing PHA content in biomass. In the process of phosphorus recovery, excess PHA stored in the biomass facilitated subsequent simultaneous nitrification / denitrification and phosphorus removal. And this operating strategy helps to choose some special functional microbial communities [9]. Some studies also use protein as a carbon source alternative. The protein itself contains C, N, P, K, Mg and other elements to meet the needs of microbial growth, and combined the release of P with magnesium ammonium phosphate for phosphorus recovery [10]. However, protein caused the introduction of a certain amount of ammonia nitrogen, this problem has not yet been solved. The above researches provide ideas and directions of how to improve the content of PHA and how to enrich functional population in the BNR system.

3. N and P Concentration

In addition to the carbon source type, concentrations of N and P will directly affect the amount and efficiency of PHA synthesis. Numerous studies have demonstrated that higher PHA levels are available at limited nitrogen or phosphorus concentrations [11]. In anaerobic-anoxic/nitrification sequencing batch reactor (A_2N-SBR) with COD / P and COD / N of 20.6 and 6.9, respectively, the high abundance of DPAO was confirmed. With limited N or P, there is more PHA synthesis and an increased proportion of PHV. When the P load is maintained at a constant level, the biodiversity of the population decreases as the carbon load increases, and returns to its original level after P load declined [12]. PHA based denitrification (PBD) treating municipal wastewater in a biofilm reactor was reported to be capable of removing high nitrate concentrations (39-52 mg N / L) with lower COD requirements (COD / N = 4.6-5.3). Particulate organisms intercepted in the denitrification stage were found to have the denitrification capacity solely based on PHA [13]. At a very high C / P ratio, P is depleted before C so that a microbial community with low PHA storage capacity but capable of rapidly accumulating P is selected. The rate of biomass uptake for a specific substrate is a linear function of the cell's P content [14].

4. Temperature Variation

Temperature ranges (35, 42 and 55 °C) were investigated using sewage sludge to produce PHA and reached the highest amount up to 0.39 gPHA / gVSS at 42 °C [15]. Short-term temperature changes (15-30°C) affect only the PHB synthesis rate but do not affect the maximum PHB storage capacity of the mixed cultures. Changes in PHB synthesis rate over long-term temperature changes are also affected by the adaption ability of the microbial culture. The temperature coefficient associated with PHA metabolism is higher than the short-term temperature change. On the other hand, the maximum growth rate of acetate-based mixed bacteria decreased with increasing temperature. At higher steady-state temperatures (30 °C), the mixed population still had a very high PHB storage (84 wt%). The rate decreased at lower temperatures (15 °C) and therefore required longer growth times [16].

5. Horizontal Gene Transfer

In recent years, apart from the researches on the production of PHA using a single strain, more researches have concentrated on the production of PHA using mixed microbial culture and the effects of related factors on PHA synthesis. In the BNR system, the ability to produce and store PHA leads to changes in the microbial community and determines the selection of dominant functional species. It is very important to explore the synthesis and storage of PHA in functional microbial communities under the changes of carbon, nitrogen and phosphorus ratio, temperature and the types and dosage of carbon source. In addition, with the development of multi-omics studies, the contribution of microbial community growth patterns and the specific horizontal translocation of PHA-encoding genes to the mutual aid microbial community would be discovered.

The ability of PHA synthesis can be obtained or enhanced through horizontal gene transfer between different species [17]. The genome analyais of *Pseudomonas extremaustralis* revealing the existence of another PHB-related gene cluster, *phbFPX*, that has a high degree of similarity to the gene belonging to *Burkholderiales* and a gene cluster *phaC1ZC2D* that encodes mcl-PHA. All mcl-PHA genes showed high similarity to genes from *Pseudomonas*, and interestingly, this gene cluster also showed native insertion of seven ORFs unrelated to mcl-PHA metabolism [18]. *Halomonas sp. TD01*, a commercially viable halophilic strain of scl-PHA production, was found to employ PHA synthase, depolymerase, and regulators / repressor proteins all participating in the PHA metabolic pathway, They found different levels of horizontal gene transfer among the genomes of different strains [19]. The results of the research results could enlighten the improvement of the ability of PHA in the mixed microbial culture of BNR system from genomics level.

6. Conclusion and Outlook

The above three factors are the most common factors in the current studies. In addition to carbon source types and dosage, N and P content, temperature, pH, sludge retention time (SRT) and mean cell residence time (MCRT) and many other factors will also affect the ability to synthesize PHA of the microbial functional communities. The change of quantity, composition and rate of PHA synthesis under the influence of many factors should be taken into account, especially the harsh environment such as low-temperature environment which often encounter in practical applications. Furthermore, selecting the economical carbon source for application accompanied with the P accumulating and recovery is one of the future research directions. More deeply, the availability of genomic information of the microbial community in the BNR system not only provides a theoretical basis for genetic modification of microorganisms, but also provides an in-depth approach on selection and enrichment of the functional population in the BNR system, on improving the nitrogen and phosphorus removal efficiency of the reactor removal efficiency and on recovering N, P, carbon sources and other high-valued by-products simultaneously.

7. References

[1] Puig S, Coma M, Monclus H, van Loosdrecht MCM, Colprim J, Balaguer MD. Selection between alcohols and volatile fatty acids as external carbon sources for EBPR. *Water Res.* 2008;42(3):557-66.

[2] Wong MT, Liu WT. Microbial succession of glycogen accumulating organisms in an anaerobic-aerobic membrane bioreactor with no phosphorus removal. *Water Sci. Technol.* 2006;54(1):29-37.

[3] Miao L, Wang S, Li B, Cao T, Zhang F, Wang Z, et al. Effect of carbon source type on intracellular stored polymers during endogenous denitritation (ED) treating landfill leachate. *Water Res.* 2016;100:405-12.

[4] Jiang Y, Hebly M, Kleerebezem R, Muyzer G, van Loosdrecht MCM. Metabolic modeling of mixed substrate uptake for polyhydroxyalkanoate (PHA) production. *Water Res.* 2011;45(3):1309-21.

[5] Torresi E, Casas ME, Polesel F, Plosz BG, Christensson M, Bester K. Impact of external carbon dose on the removal of micropollutants using methanol and ethanol in post-denitrifying Moving Bed Biofilm Reactors. *Water Res.* 2017;108:95-105.

[6] Fradinho JC, Oehmen A, Reis MAM. Photosynthetic mixed culture polyhydroxyalkanoate (PHA) production from individual and mixed volatile fatty acids (VFAs): Substrate preferences and co-substrate uptake. *J. Biotechnol.* 2014;185:19-27.

[7] Singh M, Kumar P, Ray S, Kalia VC. Challenges and Opportunities for Customizing Polyhydroxyalkanoates. *Indian J. Microbiol.* 2015;55(3):235-49.

[8] Tian Q, Ong SK, Xie XH, Li F, Zhu YB, Wang FR, et al. Enhanced phosphorus recovery and biofilm microbial community changes in an alternating anaerobic/aerobic biofilter. *Chemosphere.* 2016;144:1797-806.

[9] Tian Q, Zhuang LJ, Ong SK, Wang Q, Wang KW, Xie XH, et al. Phosphorus (P) recovery coupled with increasing influent ammonium facilitated intracellular carbon source storage and simultaneous aerobic phosphorus & nitrogen removal. *Water Res.* 2017;119:267-75.

[10] Puyol D, Batstone D, Hulsen T, Astals S, Peces M, Kromer JO. Resource Recovery from Wastewater by Biological Technologies: Opportunities, Challenges, and Prospects. *Front. Microbiol.* 2017;7.

[11] Montiel-Jarillo G, Carrera J, Suarez-Ojeda ME. Enrichment of a mixed microbial culture for polyhydroxyalkanoates production: Effect of pH and N and P concentrations. *Sci. Total Environ.* 2017;583:300-7.

[12] Wang Y, Geng J, Ren Z, Guo G, Wang C, Wang H. Effect of COD/N and COD/P ratios on the PHA transformation and dynamics of microbial community structure in a denitrifying phosphorus removal process. *J. Chem. Technol. Biotechnol.* 2013;88(7):1228-36.

[13] Krasnits E, Beliavsky M, Tarre S, Green M. PHA based denitrification: Municipal wastewater vs. acetate. *Bioresour. Technol.* 2013;132:28-37.

[14] Korkakaki E, van Loosdrecht MCM, Kleerebezem R. Impact of phosphate limitation on PHA production in a feast-famine process. *Water Res.* 2017;126:472-80.

[15] Morgan-Sagastume F, Hjort M, Cirne D, Gerardin F, Lacroix S, Gaval G, et al. Integrated production of polyhydroxyalkanoates (PHAs) with municipal wastewater and sludge treatment at pilot scale. *Bioresour. Technol.* 2015;181:78-89.

[16] Johnson K, van Geest J, Kleerebezem R, van Loosdrecht MCM. Short- and long-term temperature effects on aerobic polyhydroxybutyrate producing mixed cultures. *Water Res.* 2010;44(6):1689-700.

[17] Ciesielski S, Cydzik-Kwiatkowska A, Pokoj T, Klimiuk E. Molecular detection and diversity of medium-chain-length polyhydroxyalkanoates-producing bacteria enriched from activated sludge. *J. Appl. Microbiol.* 2006;101(1):190-9.

[18] Catone MV, Ruiz JA, Castellanos M, Segura D, Espin G, Lopez NI. High Polyhydroxybutyrate Production in Pseudomonas extremaustralis Is Associated with Differential Expression of Horizontally Acquired and Core Genome Polyhydroxyalkanoate Synthase Genes. *Plos One.* 2014;9(6).

[19] Cai L, Tan D, Aibaidula G, Dong X-R, Chen J-C, Tian W-D, et al. Comparative genomics study of polyhydroxyalkanoates (PHA) and ectoine relevant genes from Halomonas sp TD01 revealed extensive horizontal gene transfer events and co-evolutionary relationships. *Microb. Cell Fact.* 2011;10.

IOP Conf. Series: Earth and Environmental Science **178** (2018) 012022 doi:10.1088/1755-1315/178/1/012022

Treatment of Chemical Wastewater by O_3-BAC Method

Le Luo

Chongqing Vocational Institute of Engineering Chongqing, China

Abstract. Aiming at the problem of lower biodegradability and higher chromaticity of two stage biochemical treatment of petrochemical industries wasterwater treatment plant, O_3-BAC is used to treat the water outlet of sewage treatment plant and the $COD_{Cr,}$ and Ammonia-nitrogen and sewage chroma are analyzed as well. The results show that the treatment efficiency could be achieved through O_3-BAC method, that is, the COD_{Cr} is 26.7mg/L, the Ammonia-nitrogen is 0.18mg/L and the sewage chroma is 5 times.

1. Introduction

O_3-BAC (O_3-Biological Activated Carbon) is a wastewater treatment based on organic matters in ozonation. This process integrates O_3 pre-oxidation, activated carbon adsorption, biological oxidation and activated carbon regeneration. The O_3 pre-oxidation means that the 0.002~0.08μm particles are oxidized into particles less than 0.002μm so that both of adsorption efficiency of activated carbon and degradation efficiency of microbe are improved. The absorption of activated carbon could easily lead to a enrichment of higher concentration pollutants in local area, which is beneficial to the diffusion of pollutants into cell membrane and improves the treatment efficiency of low concentration wastewater[1]. Activated carbon has developed internal pores which is beneficial to microbe reproductions [2]. In addition, the unit area of activated carbon is too large to accumulate microbes, which is beneficial to improve the removal effieciency of organic matters [3]. At last, the microbial degradation of activated carbon could decompose organic pollutants into CO_2 and H_2O. When completing the degradation, the activated carbon could recover its adsorbability[4]. The former treatment of coagulation air is to remove pollutants larger than 0.08μm so that the oxidation efficiency of O_3 could be improved. The O_3-BAC method solve the problem of high cost of O_3 oxidation and the difficulty of activated carbon adsorption.

2. The Development of O3-BAC

After 1960s, with the development of industry, the source water was generally polluted by organic matter, such as pesticides, insecticides, herbicides, various additives, endocrine interferons, etc.. In 1970s, chlorine disinfection was found to react with organic compounds to produce disinfection by-products, which is harmful to human health. At the same time, the conventional treatment process had no removal ability for these organic compounds, so that the impact of drinking water on human health has been a great concern, forcing people to seek new drinking water purification technology to make up for the shortcomings of conventional treatment processes. In this context, O_3-BAC has been developed and applied.

The first use of O_3-BAC was a water plant in Dusseldorf, Germany, in 1961 [5]. On the basis of the original ozone+filtration, the activated carbon adsorption was added, by which the effluent water quality was obviously improved. Its success has attracted the attention of the German and Western European water treatment engineering circles. In 1967, Parkhurst et al [6] affirmed the beneficial effect of microorganisms on activated carbon, and finally established the O_3-BAC Technology. Since the beginning of 1970s, O_3-BAC technology has been extensively studied and applied [7].

IOP Conf. Series: Earth and Environmental Science **178** (2018) 012022 doi:10.1088/1755-1315/178/1/012022

3. Case Study

3.1. Case description

The design scale of sewage treatment plant in Petrochemical Industry is $5.5\times10^4 m^3/d$, and the operating scale is 2.40×10^4~$3.36\times10^4 m^3/d$. The process of O_3-BAC includes oil separation, average adjustment, hydrolyzation, A/O and the treatment of aeration tank and clarifier. The sewage comes from refinery, addictive plants, auxiliary plants and fine chemical plants. The composition of sewage could be divided into naphthalene, quinoline, pyridine, polyvinyl alchohol, styrene, acrylonitrile, nitrobenzene and two vinylichloride, which cause the two stage biochemical treatment effluent has lower biodegradability and higher chromaticity, and the sewage quality cannot meet the requirement of wastewater reuse pretreatment after the removal of COD_{Cr}. Therefore, the O_3-BAC method could treat the two stage biochemical effluent of sewage treatment plant properly.

3.2. The processing installation of O3-BAC method

The processing installation of O_3-BAC includes floatation tank, O_3 contact oxidation tower, intermediate regulating tank, BAC reaction column and auxilary equipments, the processing flow is shown in Fig.1.

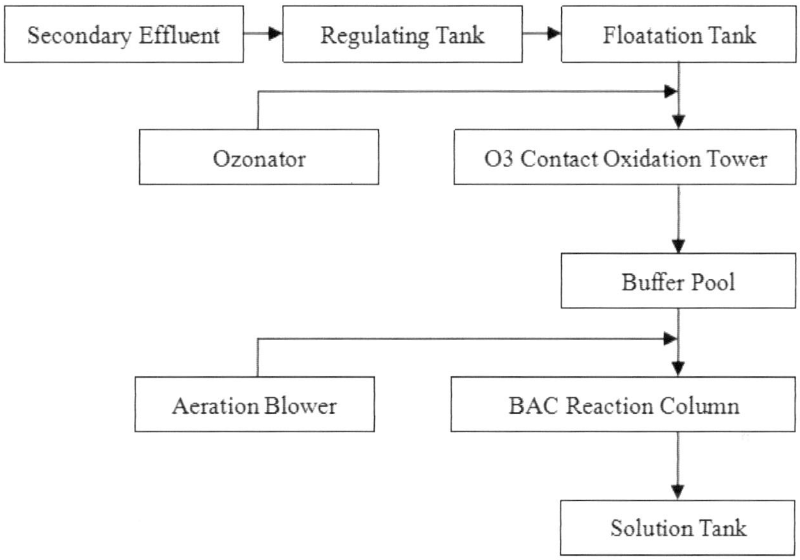

Figure 1: The processing flow of O_3-BAC

3.3. Sewage quality

The sewage comes from sewage treatment plant in Lanzhou Petrochemical Industry. The sewage quality and effluent design index are shown in Tab.1.

TABLE I. The sewage quality and effluent design inde

Items	Sewage quality	Effluent design index
COD_{Cr}/mg/L	60~70	≤30
Ammonia-nitrogen/mg/L	≤10	≤3
Petroleummg/L	≤5	≤1
SS/mg/L	≤20	≤5
PH	6~9	6~9
BOD_5/mg/L	≤10	≤3

3.4. Analytical method

COD_{Cr} is measured by the dichromate titration, and BOD_5 is measured by tacheometer, then the sewage chroma is measured by a dulution multiple method, besides Ammonia-nitrogen is measured by reagent spectrophotometer, at last, the biological phase is photographed by microscope.

3.5. Processing parameters
The processing parameters of O_3-BAC is shown in Tab.2.

TABLE II. The processing parameters of O_3-BAC

Items	Parameters
Sewage loads/kgCOD/m^{3}d	1.3~1.5
Aeration	Gas-water ratio is (2~3):1
Agent addition/mg/L	10~15
O_3 concentration/mg/L	4.2~4.5
Temperature/°C	18~25
PH	7~9

4. BAC Hanging and Domestication
Cultivation of biological membrane: the sewage temperature is about 30°C, which is beneficial to microbe cultivation. Before the operation of BAC column, it is essential to conduct the microbiological inoculation. After a week of cultivation and domestication, efficient microbe, selected from ordinary microbe, is full of BAC column with 72h aeration. Afterwards, 1/3 sewage in the column is discharged and new sewage is added to conduct the aeration again. At the beginning of aeration, the sewage is muddy. After 3-week treatment,

5. Results Analysis

5.1. The analysis of CODCr treatment efficiency
The influent COD_{Cr} is not stable and fluctuates between 50~80mg/L. The treatment efficiency of COD_{Cr} by O_3-BAC is shown in Fig.2.

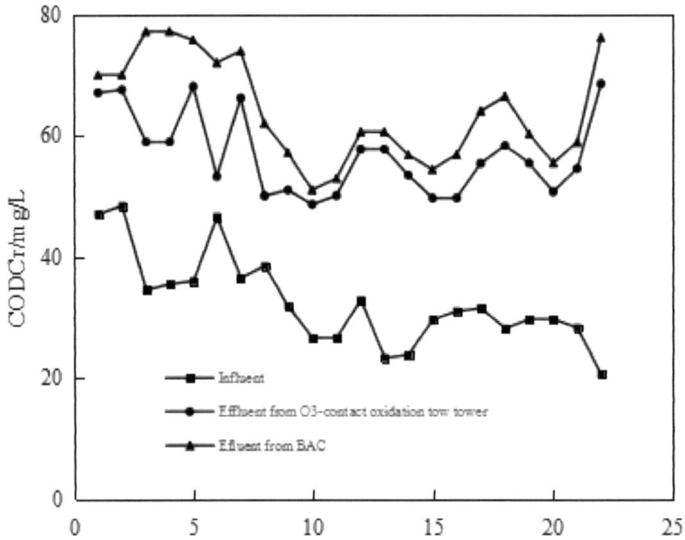

Figure 2. The CODCr removal efficiency by O3-BAC

As shown in Fig.2, when the process runs continuously, the average COD_{Cr} removal is above 50%. When the influent pollutant concentration changes, the concentration of pollutants in the effluent is slightly increased and is below 30mg/L, indicating that the O_3-BAC has a strong impact load resistance. Comparing the contribution of each unit to the COD_{Cr} removal, the contribution of O_3 contact oxidation tower is about 10% and average effluent COD_{Cr} concentration is about 56mg/L. it is seen that O_3 oxidation is not complete oxidation, which reduces the cost of sewage treatment. In addition, the BAC is main unit for degrading COD_{Cr} and the average COD_{Cr} concentration and removal respectively are 26.7mg/L and 56%, and then the BAC could continuously operate 3~4 months without washing. If the total amount of COD_{Cr} removal in each washing cycle is about 60 times more that the theory removal, it shows that the biodegradation of microbe contributes to the

COD_{Cr} removal and the BAC plays a key role during the whole process. At last, it takes about 3~4 weeks from the beginning of BAC hanging to the stable operation.

5.2. *The analysis of Ammonia-nitrogen treatment efficiency*

The fluctuation of influent Ammonia-nitrogen concentration is between 0.2mg/L and 1.0mg/L, but its total content is low. After the O_3-BAC, the Ammonia-nitrogen concentration is 0.18mg/L and removal is 75%, which could be detected in Fig.3.

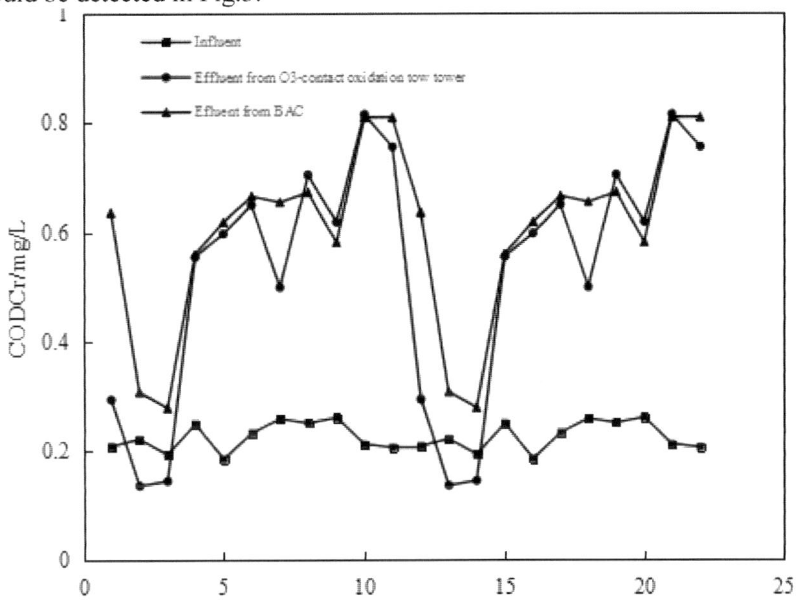

Figure 3. The Ammonia-nitrogen removal efficiency by O_3-BAC

As shown in Fig.3, The Ammonia-nitrogen removal by O_3 contact oxidation tower could be divided into two steps. At the first stage, the removal efficiency is transparent, but the Ammonia-nitrogen concentration in the second stage is almost unchanged and even higher than the influent Ammonia-nitrogen concentration. This maybe due to the organic nitrogen in sewage is oxidized by O_3, releasing and increasing the Ammonia-nitrogen concentration.

The BAC unit is main place to remove Ammonia-nitrogen. In the BAC bed, microbe grow in the pores of the carbon particles and remove Ammonia-nitrogen. At last, Ammonia-nitrogen is eventually removed and the BAC regenerated. As shown in Fig.3, although the influent Ammonia-nitrogen fluctuates dramatically, the BAC effluent is relatively stable, the effluent Ammonia-nitrogen concentration is 0.2mg/L and average removal is about 75%.

5.3. *The analysis of chroma treatment efficiency*

The influent chroma changes between 12~16 times. The effluent chroma is about 5 times. The chroma removal by O_3-BAC is shown in Fig.4.

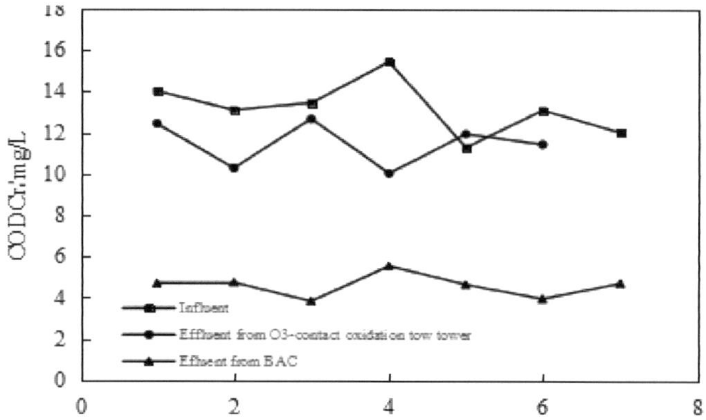

Figure 4. The chroma removal efficiency by O_3-BAC

As shown in Fig.4, BAC is the main unit of chroma removal, which accounting fore more than 90% of the total removal. However, the contribution of O_3 is relatively low and about 5%~10%, due to the chromophoric groups in sewage is hard to remove by O_3.

5.4. *The analysis of biodegradability of sewage*

The biodegradability of effluent from Lanzhou Petroleum Industry is low. Before entering the O_3 contact oxidation tower, the sewage BOD_5/COD_{Cr} ratio is about 0.08, which is difficult to degrade. In the pilot test, the O_3 addictive is about 4.0~4.5mg/L, and the BOD_5/COD_{Cr} ratio of effluent is about 0.21 due to incomplete oxidation between O_3 and organic pollutants in sewage.

5.5. *The analysis of microbe in BAC tower*

Biodegradation of BAC degradation are the main mechanism of activated carbon adsorption and biofilm, and microbial degradation of pollutants test plays a direct role in the process of test. Through the examination, the operating state of system could be marked in accordance with the species and activity of microbe. During the pilot test, the biological observation of BAC was conducted irregularly so as to better understand the operation of the system and make appropriate adjustment.

The observation of the biological phase was carried out by 100 times optical microscope. Before the observation, the carbon samples in the BAC reaction tower were placed in the beaker, and then the carbon particles were broken with tweezers. Through the observation, didinium, rotifers, nematodes, insects roaming, dun pellionella, caterpillar and other protozoa and metazoan daphnia are detected, which means there are lots of species on the microbial membrane in BAC tower. At the same time, The emergence of protozoa and metazoan showed good effluent quality and stable system operation.

6. Conclusion

The O_3-BAC process can be used for a stable treatment of the two grade biochemical effluent of Lanzhou Petrochemical Industry wastewater treatment plant. The influent COD_{Cr} concentration is fluctuated between 50mg/L and 80mg/L, and the average effluent COD_{Cr} concentration is 26.7mg/L. The influent Ammonia-nitrogen concentration is 0.2~1.0mg/L, and the average Ammonia-nitrogen concentration is 0.18mg/L. The influent chroma is about 12~16 times, and the effluent chroma is nearly 5 times. About analysis shows that O_3-BAC has a strong impact load resistance.

O_3 contact oxidation tower can not only degrade some pollutants, but also improve the biodegradability of sewage. The BOD_5/COD_{Cr} ratio of sewage before entering O_3 contact oxidation tower is about 0.08, and the effluent BOD_5/COD_{Cr} ratio is about 0.21. Through the observation, the biological composition of the activated carbon filler is complete and the microbial population is abundant, indicating the stable operation of the system.

7. References

[1] Benitez F J, Acero J L, Real F J, " Comparison of different chemical oxidation treatments for the removal of selected pharmaceuticals in water matrices," Chemical Engineering Journal, vol.168, Mar. 2011, pp. 1149-1156.

[2] D. He, N. Gao, W. Zeng, "Biofilm Colonization in Advanced Treatment of Drinking Water using Biological Activated Carbon Process," Industrial Water & Wastewater, vol.37, May. 2006, pp. 16-19.

[3] H. Yao, F. Ma, G. Li, "Treatment of Petrochemical Wastewater by O3-BAC," China Water & Wastewater, vol. 19, 2003, pp. 39-41.

[4] J. Zhang. "Study on the Removal of Micro Pollution by O3-BAC," Water & Wastewater Engineering, vol. 22, Apr. 1996, pp. 57.

[5] Li L, Zhang P, Zhu W, "Comparison of O 3 -BAC, UV/O 3 -BAC and TiO 2 /UV/O 3 -BAC processes for removing organic pollutants in secondary effluents," Journal of Photochemistry & Photobiology A Chemistry, vol.171, Feb. 2005, pp. 145-151.

[6] Li L, Zhu W, Zhang P, "Comparison of AC/O3-BAC and O3-BAC processes for removing organic pollutants in secondary effluent," Chemosphere, vol. 62, Sep. 2006, pp. 1514-1522.

[7] Zhu J P, Peng Y Z, Li X L, et al., State Environmental Protection Administration of China, Monitoring and analysis method of water and wastewater. Beijing: China Environmental Science Press, 1997.

IOP Conf. Series: Earth and Environmental Science **178** (2018) 012023 doi:10.1088/1755-1315/178/1/012023

Ecological Livability Evaluation Based on Remote Sensing Technology - A Case Study of Shijiazhuang

An Hongfei[1] and Zhao Shuqin[2*]

1 Master, Hebei GEO University, Shijiazhuang, China
2 Professor, Hebei GEO University, Shijiazhuang, China
E-mail: shuqinzhao999@163.com

Abstract: Water pollution and traffic jams make the city's living environment a great test. The public demand for creating a livable urban environment has become more and more intense with the rise of circulation, green and low-carbon development trends. Based on the statistical data and LANDSAT-8 OLI remote sensing of Shijiazhuang as the main data source, this paper extracts 19 indicators according to the principle of establishing the ecological livability evaluation index, and the practical circumstance and data of the study district based on remote sensing technology. This paper affirms the achievements of the construction of the ecological livable city, but there are some aspects to be improved, including economic development, infrastructure, science and technology culture, ecological environment and social security.

1. Introduction

With the deepening of urbanization and industrialization, the problems of urban environmental quality and ecosystem imbalance have been constantly increasing. Shijiazhuang is not only the important support area of Beijing-Tianjin-Hebei Region "third pole" and the construction of Xiong'an New Area, but also one of the most polluted areas in North China. Many scholars in our country combine economic indicators and ecological indicators from the perspective of sustainable development[1]. Relevant scholars think that livable cities not only requires high level of economic development, but also should have strong potential for development, and only improving economic level can provide material basis and social security for ecological livable construction[2]. There are many cases by applying GIS and RS technology in ecological suitability evaluation process. Fu Bo evaluated the overall livability of Changchun by using the mean variance and multi-objective linear weighting function method in combination with GIS and remote sensing technology[3]. Based on the natural features and data accessibility of the lake, Ding Leilong constructed the livable evaluation index system by using remote sensing images and other data in the study area[4]. Because of the scale problem of remote sensing image, instantaneous and imaging methods, the index precision is not accurate enough to meet the requirements, and single extraction from remote sensing image index cannot be used for one type of study, which must be combined with other indicators to analyze the comprehensive evaluation. This paper analyzes the ecological livable level in different regions, which is beneficial to improve the overall ecological livable level of shijiazhuang.

This article is among research results of Hebei Technological Department Program (Program No: 16K55407D).
An Hongfei, 1994, master, she mainly engages in study and research of land resources management and geographic information system. Email:15103313551@163.com

2. Establishment of Index System and Data Processing

2.1. Establishment of Index System

The city is a comprehensive system, which not only affects many factors, but also influences each other[5]-[6]. This paper extracts five groups of 19 indicators according to the principle of establishing the ecological livability evaluation index (Table 1).

2.2. Data sources and Processing

According to the above index system, this article selects the statistical data and remote sensing image in Shijiazhuang in 2016. Combined with the "Land Use Status Classification Standard" and the requirements of this article, the land use type of Shijiazhuang can be divided into: cultivated land, water, construction land, forest and unutilized land[7]-[8].

3. Main Research Methods

3.1. The Index Value

This paper mainly extract the ecological environmental indicator. The other indicators are derived from statistical yearbook data and economic and social development statistical bulletin.

3.2. Ascertainment of Weights

Because the data level of each index is relatively large, the selected indexes cannot be directly applied to the model in the back, and the data should be normalized[9]. The calculation model is:

$$R_{ij}=\frac{x_{ij}-\min(x_{ij})}{\max(x_{ij})-\min(x_{ij})} \quad (1) \qquad R_{ij}=\frac{\max(x_{ij})-x_{ij}}{\max(x_{ij})-\min(x_{ij})} \quad (2)$$

In this paper, analytic hierarchy process (AHP) method is used to determine the weight of the primary factors, and the weight of the secondary factors is calculated according to the principal component analysis (PCA)[10]-[11]. The final result is shown in Table 1.

Table 1. Index system and its weight.

Target layer	Primary factors	Weight	Secondary factors	Weight
Ecological Livability A	Economic development B1	0.09	Per capita revenue C1	0.074
			Per capita of GDP GDP C2	0.077
			Total investment in fixed assets C3	0.022
			Per capita living expenses of urban and rural residents C4	0.074
	Science and technology culture B2	0.09	The proportion of science and technology expenditure accounted for Fiscal expenditure C5	0.052
			The number of enterprises with R&D activities C6	0.076
			The number of primary and secondary schools C7	0.076
			The number of patent applications for industrial enterprises C8	0.073
	Infrastructure B3	0.24	Electricity consumption situation C9	0.076
			The proportion of general public service expenditure accounted for financial expenditure C10	0.034
			Annual water supply C11	0.009

Ecological environment B4	0.52	F C12	0.070
		ESV C13	0.049
		Bright temperature index C14	0.010
		HAI C15	0.060
Social security B5	0.09	The proportion of urban low-income People accounted for the total people C16	0.075
		The number of health institutions C17	0.072
		The number of registered marriages C18	0.008
		The proportion of registered unemployment accounted for the total population C19	0.013

4. Ecological Livability Evaluation of Shijiazhuang

After normalization processing on the original data, this paper calculates the ecological livable index, economic development, science and technology culture, infrastructure, ecological environment and social security according to the weight of each index by using the comprehensive weighted method, as shown in Table 2.

Table 2. Results of ecological livability evaluation.

District	A	B1	B2	B3	B4	B5
Urban district	0.034	0.2297	0.2770	0.1101	0.0033	0.1597
Luquan	0.057	0.0429	0.0944	0.0218	0.0973	0.0360
Luancheng	0.059	0.0470	0.0864	0.0214	0.1002	0.0341
Gaocheng	0.067	0.0403	0.0760	0.0106	0.1216	0.0482
Zhengding	0.054	0.0305	0.0703	0.0216	0.0910	0.0459
Xinle	0.059	0.0155	0.0598	0.0072	0.1073	0.0362
Jingxing	0.085	0.0237	0.0534	0.0122	0.1551	0.0335
Jinzhou	0.077	0.0149	0.0465	0.0265	0.1334	0.0358
Yuanshi	0.074	0.0122	0.0446	0.0107	0.1366	0.0364
Pingshan	0.085	0.0248	0.0399	0.0131	0.1546	0.0479
Xingtang	0.063	0.0058	0.0395	0.0210	0.1099	0.0352
Wuji	0.077	0.0101	0.0376	0.0395	0.1288	0.0353
Lingshou	0.072	0.0016	0.0361	0.0053	0.1348	0.0308
Shenze	0.074	0.0068	0.0358	0.0105	0.1357	0.0202
Gaoyi	0.074	0.0086	0.0328	0.0150	0.1336	0.0160
Zhaoxian	0.089	0.0121	0.0326	0.0099	0.1646	0.0388
Xinji	0.094	0.0490	0.0311	0.0252	0.1676	0.0462
Jingxing mining area	0.060	0.0147	0.0304	0.0308	0.0994	0.0080
Zanhuang	0.088	0.0077	0.0212	0.0036	0.1659	0.0270

4.1. Ecological Livable Index

The most ecological livable level in Shijiazhuang is Xinji, as shown in Figure 1 and 2, followed by Zhaoxian and Zanhuang, which has a value of 0.094. Under the premise of vigorous developing the leather city, it strives to create an eco-tourism city with distinctive features, good ecology, perfect facilities and first-class service. The urban ecological livable level is low, which is 0.034. Zhengding and Luquan are better than urban district, which is respectively 0.054 and 0.057. The reason is that these cities have poor ecological environment.

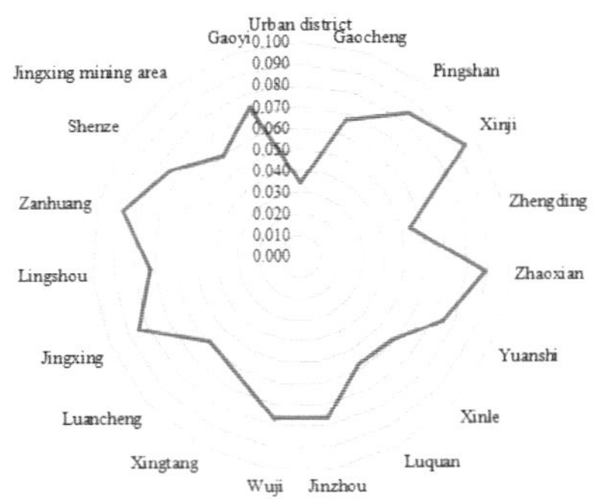

Figure 1. The radar chart of ecological livable value in Shijiazhuang.

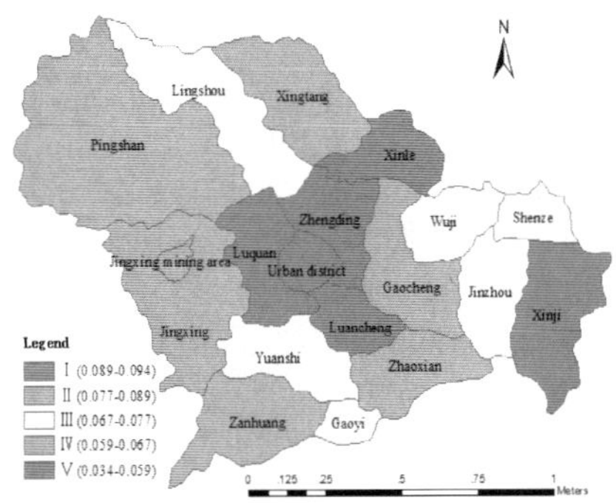

Figure 2. The map of ecological livable value in Shijiazhuang.

4.1.1. Economic development. The highest index of economic development is urban district, which has a value of 0.2297. Its per capita of GDP, total investment in fixed assets, and per capita living expenses of urban and rural residents are higher. The lowest index of economic development is Lingshou, which has a value of 0.0016. Its per capita of GDP and per capita revenue are low, which ranks at the bottom of the city.

4.1.2. Science and technology culture. The highest index of science and technology culture is urban district, which has a value of 0.277. Its the proportion of science and technology expenditure accounted for fiscal expenditure, the number of patent applications for industrial enterprises, the number of enterprises with R&D activities, and the number of primary and secondary schools in urban district are higher. The lowest index of science and technology culture is Zanhuang, which has a value of 0.0212. Its the proportion of science and technology expenditure accounted for fiscal expenditure, and the number of primary and secondary schools are lower.

4.1.3. Infrastructure. The highest index of infrastructure is urban district, which has a value of 0.1101, and followed by Wuji. The lowest index of infrastructure is Zanhuang, which has a value of 0.0036. Its the proportion of general public service expenditure accounted for financial expenditure and annual water supply is lower.

4.1.4. Ecological environment. The highest index of ecological environment is Xinji, which has a value of 0.1676. Its ESV is higher, and L_λ is lower. The lowest index of ecological environment is urban district. Its f and ESV are higher, and L_λ is higher.

4.1.5. Social security. The highest index of social security is urban district, which has a value of 0.1597. Its the number of health institutions and the proportion of urban low-income people accounted for the total people are higher. The lowest index of social security is Jingxing mining area, which has a value of 0.008. Its the proportion of registered unemployment accounted for the total population is higher, and the number of health institutions is lower.

5. References

[1] Wen Xue 2015 *Evaluation Index System of Ecological City - Shijiazhuang Comparative Analysis based on Ten Cities* (Hebei University of Economics and Business)
[2] Rui Yuan 2005 *On the Discriminant Standard of Livable City* vol(4), Economic Science pp 126-128
[3] Bo Fu 2011 *Enviornmental Evaluation of Urban Livability of Changchun Based on GIS and Remote Sensing* (Jilin University)
[4] Leilong Ding 2010 *Livability Evaluation of Weishan Lake Region Based on UAV Images* (China University of Geosciences (Beijing))

[5] Huixia Zhu 2008 *The Research of Shijiazhuang Livable City Building* (Hebei Normal University)

[6] Jijun Meng and Chunhong Zhao *Research Progress on Index System of Regional Ecological Risk Assessment* vol(20), Chinese Journal of Applied Ecology pp 983-990

[7] Baiming Chen and Xiaoping Zhou 2007 *Explanation of Current Land Use Condition Classification for National Standard of the People's Republic of China* vol(22), Journal of Natural Resources pp 994-1003

[8] Pengfeng Xiao, Shunxi Liu, Xuezhi Feng and Guangfa Lin 2006 *A Land Use/Cover Classification System Based on Medium Resolution Remote Sensing Data* vol(20), China Land Science pp 33-38

[9] Zhijuan Qi 2015 *Small Town Residential Environment Livability Evaluation Under the Background of New Urbanization - A Case Study in Handan City Linzhang* (Hebei Normal University)

[10] Xiaoshuang Wang, Xuehua Zhang and Ji Lei 2013 *Tianjin Ecological Livable City Construction Index and Evaluation Research* vol(23), China Population, Resources and Environment pp 19-22

[11] Fang Gong and Xiuming Chen 2016 *Comprehensive Evaluation of Livable City Based on the Analytic Hierarchy Process* vol(35), Journal of Changchun Normal University pp 88-94

The Establish of Liaohe Estuarine Wetland Ecological Compensation Standard Based on Muti-Source Remote Sensing Data

QIAN CHENG[1], TIE-LIANG WANG*[2], LIN-FEI ZHOU[3]

Shenyang Agricultural University, Shenyang 110866, P. R. China
Email: chengqian1979@163.com

Abstract. This article choosed LiaoHe Estuarine Wetlands National Nature Reserve as the research object. On the basis of analyzing the present situation of wetland ecological compensation, combining wetland nature reserve, economic development demand and land use requirements of the surrounding area. Combined with the 3S technology, cooperated with field reconnaissance and survey, conducted wetland ecological compensation standard analysis from space and global perspective, gave full consideration to ecological service function value of different Market value. This paper established ecological compensation standard to provide a scientific basis to establish the standard of Liaoning province.

1. Introduction

Liaohe estuarine wetland is located in southern coast of Linghai city, Liaoning Province, northeast China. Composited of large areas of reed marshes, tidal flats, wet meadows, sea water area and small rivers, the wetland has great value for scientific research and economy development. While in recent years, due to excessive reclamation and human activities, reed marshland area had experienced substantial contraction, retreat sea shoals area kept growing and the wetland environment became increasingly fragile, which seriously destroyed local ecological environment and caused enormous economic losses and resource waste [1]- [2].In this situation, how to establish a long-term mechanism of rational ecological compensation is important .This paper established ecological compensation standard to provide a scientific basis to establish the standard of Liaoning province.

2. Data Source

2.1. Overview of Study Area

Liaohe estuarine wetland national nature reserve is located at Panjin city, northern Liaodong Bay of Liaoning province, northeast China. The geographical coordinates is longitude 121°30′~ 122°00′, latitude 40°45′~ 41°10′. With the boundary of 60 km in north to south direction and 35 km in east to west direction, the wetland has a total area of $12.8\times10^4 hm^2$. This area belongs to the semi-humid warm temperate continental monsoon climate, the annual average temperature is 8.4°C, annual average rainfall is 623.2mm, annual average evaporation is 1669.6mm. There are various vegetations and rich habitations of large number of animal species. Location of Liaohe estuarine wetland national nature reserve is shown in Figure 1.

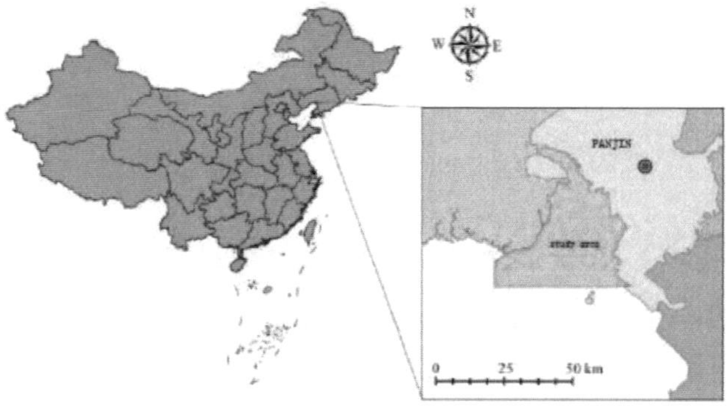

Figure 1. Location of Liaohe estuarine wetland National Nature Reserve

2.2. Data Source

Based on the wetland's landscape data in TM RS image of 1988 and 2015 and combination on text and graphic materials such as land use maps and functional zone maps in 2000, vegetation maps and field survey data, an information database of the study area was established. Image interpretation was conducted in ArcGIS platform with combination interpretation method of direct interpretation and comprehensive analysis [3]-[4]. A classification system of Liaohe estuarine wetland estuary wetland was established with consideration of image spectral characteristics and spatial features. To validate the accuracy of interpretation achievements, we conducted field GPS validation and projection conversion to ensure the data consistent with its corresponding RS image coordination. According to "Liaoning Province land use classification code" standard to establish wetland property classification code, as shown in Table 1. Changes of wetland types in Liaohe estuarine Nature Reserve from 1988 to 2015 are shown in Figure 2.

Table 1. Category interpretation of Liaohe estuarine wetland

Items	ID	1988 Area(km^2)	2015 Area(km^2)	Items	ID	1988 Area(km^2)	2015 Area(km^2)
Paddy Field	11	112.67	110.66	Reservoir	53	29.80	35.08
Dry Land	12	10.37	13.16	Tidal flat	61	99.08	74.63
Residential Area	42	74.70	102.24	Reed	63	484.73	491.69
River	52	469.68	453.58				

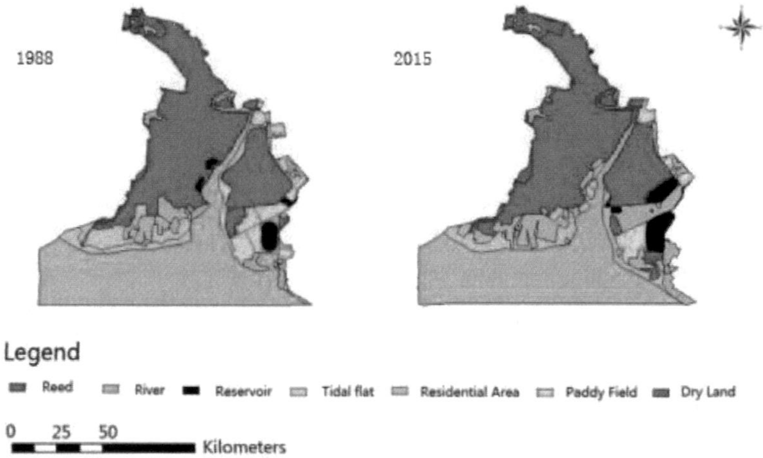

Figure 2. Changes of wetland types in Liaohe estuarine Nature Reserve (1988-2015)

3. Ecological Compensation Calculation

3.1. Ecological Compensation Based on the Value of Ecological Functions

3.1.1. Ecological service function value calculation. There were complete ecological functions and huge economic, ecological and social benefits in the Liaohe Estuarine Wetland Reserve. There were different types of wetlands and different environments in this area. The leading service functions include material production, climate regulation, water purification, bank protection, biodiversity conservation, tourism and leisure functions, education and scientific research, and historical and cultural functions[4]-[5]. Based on the data interpreted in 2015, we could estimate the total value of material production functions is 23.32×10^8 yuan,the total value of climate regulation function is 0.28×10^8 yuan, the value of purifying water quality function is 0.09×10^8 yuan, the ecosystem protection value is 2.63×10^8 yuan, the value of maintaining biodiversity is 0.48×10^8,the value of scientific research service is 4.65×10^8 yuan, the value of historical and cultural functions is 49.35×10^8 yuan, and the total value of ecological service of Liaohe estuarine Wetland Reserve is 81.09×10^8 yuan which based on the market value method, carbon tax law, substitution method, expert evaluation method, ecological value method, cost-expenditure method and so on [6]-[7].

3.1.2. Result verification. The value of ecological services per unit area of wetlands in the world was 55,420 yuan / hm^2, and the ecological service value per unit area of wetlands in China was 55,489 yuan / hm^2. The area of liaohe estuarine wetland was 128000 hm^2, the total value of ecological service was 81.09×10^8 yuan, unit ecological service value of 63328 yuan / hm^2.We could see the value of ecological service function per unit area of wetland is close to the average level of the whole country and the whole country.

3.1.3. Ecological compensation standards. The total value of ecological functions of Liaohe estuarine wetland was 81.09×10^8 yuan based on the calculation results and the unit ecological service function value was 63328 yuan / hm^2. Compared with 1988, it could be seen that the loss of ecological function value (total compensation fund) was 1,920,0496 yuan, in which the ecological compensation fun was 1548369.6 yuan of Tidal flat,the ecological compensation fun was 1019580.8 yuan of river, the ecological compensation fun was 127,289.28 yuan of Paddy Field.

3.2. The Calculation of Ecological Compensation Based on Market Value Loss

3.2.1. Calculation of market value loss. We could calculate the loss of market value of wetlands in terms of both ecological damage and environmental pollution losses.

$$EC = EC_1 + EC_2 \tag{1}$$

EC: market value loss;
EC_1: ecological damage loss;
EC_2: environmental pollution loss.

3.2.2. Market value loss caused by ecological damage. According to the research results paddy fields,Tidal flat, rivers had reduced to varying degrees.Tidal flat reduced mostly.The formula for calculating the value of ecological damage to the market value was as follows:

$$EC_1 = \sum_{i=1}^{n} E_i L_i + EC_r = E_1 L_1 + E_2 L_2 + E_3 L_3 + EC_r \tag{2}$$

EC_1 :Market value loss ;

EC_r :Indirect loss;

E_1L_1 :Tidal flat loss;

E_2L_2 :River loss;

E_3L_3 :Paddy Field loss

In a few years, the loss of paddy field area was 201hm^2, we could Specify average yield is 650kg per month.The price rice was 4660 yuan / t in 2015, so the market value loss of paddy field was 0.09 $\times 10^8$ yuan.An alternative approach was used to assess the value of canal losses, which was replaced by the investment in man-made rivers of the same construction scale. According to statistics, the cost of excavation of artificial river with a depth of 3 meters and an area of 1hm^2 was about 150,000 yuan.So we could caculate the loss of river is 1610hm^2 and the value loss of Tidal flat is 2445hm^2,so the value loss of river is 2.41 $\times 10^8$ yuan and the value loss of Tidal flat is 0.7×10^8 yuan by the analysis of the interpretation result.

Access to relevant literature,the total wetland ecological benefits compensation and returning farmland was 0.35 $\times 10^8$ yuan.the formula of the loss of market value of wetland ecological damage as follows:

$$EC_1 = \sum_{i=1}^{n} E_iL_i + EC_r \qquad (3)$$

The loss of market value of wetland ecological damage was 3.55×10^8 yuan.

3.2.3. The market value loss caused by wetland environmental pollution. The pollution types of the Liaohe estuarine wetland environment mainly included air pollution, water pollution, noise pollution and solid waste pollution, and the losses caused by them were usually measured by the cost of the project. Therefore, the loss of the total value of the wetland market was 4.36 $\times 10^8$ yuan.

3.2.4. Wetland ecological compensation standard based on market value loss. The total area of wetlands decreased by 3032 hm^2 and the compensation standard of market value loss was 6,536 yuan / hm2.a. during the period from 1988 to 2015.

4. Conclusions

The ecological compensation value per unit area was 63328 yuan / hm^2.a. according to the calculation method of ecological compensation value of ecological function value.The compensation standard per unit area was 6,536 yuan / hm^2.a. according to the calculation method of ecological compensation for the loss of market value.Therefore， the ecological compensation standard based on the loss of market value was easier to operate which was more intuitive and easier to be accepted by the masses and more in line with the local economic development level.According to this standard for ecological compensation,the compensation of river was 105,229.6 yuan, the compensation of Tidal flat was 159,805.2 yuan, and compensation of the paddy field was 13,137.36 yuan.

5. References

[1] Paul, A. K.2000. Wetland Ecology: Principles and Conservation. Cambridge: Cambridge University Press.p 51.

[2] Costanza, R. , Perez-Maqueo, O., Martinez, M. L., Sutton, P., Anderson,S. J. , and Mulder, K. 2008. The Value of coastal wetlands for hurricane protection. *Ambio： A Journal of the Human Environment*, 37(4), 241-248.

[3] Henry, P.2001. High-Resolution Wetland Prospection, using GPS and GIS: Landscape Studies at Sutton Common (South Yorkshire), and Meare Village East (Somerset). *Journal of Archaeological Science*, (28): 365-375.

[4] Cui B S, Yang Z F. 2002.Water consumption for eco-environmental aspect on wetlands. *Acta Scientiae Circum Stantiae*, 22 (2): 219-224.

[5] Liu H Y, Zhang S K, Lu X G. 2004.Wetland Landscape Structure and the Spatial-temporal Changes in 50 Years in the Sanjiang Plain. *Acta Geographica Sinica*, 59(3):391-400.

[6] Rainer Walz. 2000. Development of environmental indicator systems:experiences from germany[J]. *Environmental Management*, 25(6): 613-623.
[7] Miranda M, Porras I T, Moreno M L, et al.2003.The social impacts of payments for environmental services in Costa Rica. A quantitative field survey and analysis of the Virilla watershed[J].*the Virilla watershed."Markets for Environmental Services* Paper No.1.London:IIED.25(2):25-32.

Acknowledgement
This study is supported by the Specialized Research Fund for Cultivation Plan for Youth Agricultural Science and Technology Innovative Talents of Liaoning Province, China（No.2014038）and Provincial Department of Education research project of Liaoning, China（No. L2014250）

Synergistic Emission Reduction of SO_2 and CO_2 Based on Clean Energy Replacement in ports

Zhu Li and Qin Cuihong

China Waterborne Transport Research Institute, #8, Xitucheng Road, Beijing, 100088, China
E-mail: 545936121@qq.com

Abstract. China's energy-saving emission reduction policy calls for improved air quality while reducing greenhouse gas emissions. This paper takes the container terminal as the research object, the energy saving and emission reduction of clean energy alternative fuels such as solar energy, wind energy, geothermal energy, LNG and electricity are studied through the combination of energy substitution and emission factors. In particular, SO_2 and CO_2 Synergistic mitigation effect. The results show that the reduction SO_2 and CO_2 is positively correlated with the usage of clean energy amount. Among them, LNG has the greatest contribution to the coordinated emission reduction of SO_2, and solar energy contributes the most to CO_2 reduction. Our conclusion is that the use of LNG and shore power in ports can achieve better synergistic emission reductions for SO_2 and CO_2. And through the use of air sources, solar energy, wind energy and light energy, CO_2 emission reduction effect is better than SO_2 emission reduction. The Chinese government can use this method as a policy tool for energy conservation and emission reduction, and vigorously promoted the application of LNG and shore power in ports.

1. Introduction

There are many important ports in China. After the 20th century, China's economy has been in a high-speed development stage, at the same time, the scale of port production increased rapidly. Port cargo throughput in 2015 increased by 3.2 times to 93.48 billion tons on 22.07 billion tons in 2000. With the port construction and production scale continues to increase, the port energy consumption pressure is also increasingly prominent. The Chinese government has issued a new ambient air quality standard and issued a new carbon emission reduction task. These measures require the reduction of greenhouse gases (GHG) emissions by regulating energy consumption while improving air quality in the port area.

Ministry of Transport of the People's Republic of China in June 2014 issued 'on the promotion of port transformation and upgrading of the views of guidance' [1]. This paper puts forward the main task of optimizing the utilization of port energy and encouraging the application of liquid natural gas, wind energy and solar energy in port enterprises to improve the proportion of clean energy and renewable energy in ports. As a way to promote the port transformation and upgrading and realize the sustainable development of terminals, particularly container terminals as the main practitioners.

China's agencies have carried out the following aspects of the study: the transport sector air pollutant emissions[2] [3], forecasting and cutting, energy planning, environmental performance assessment [4], energy use and air pollution correlation analysis [5] [6], Beijing energy conservation and carbon reduction [7], and so on. In these studies, suggestions were made on optimizing the energy consumption structure and improving the quality of the air. However, for the port energy consumption, atmospheric environment improvement and greenhouse gas emission reduction of the comprehensive study is relatively small. This paper takes the container terminal as an example, uses the energy substitution method and the emission factor method to analyze the SO_2 and CO_2 emission reduction effect after using the clean energy alternative fuel in the 2013-2015 period. We hope that this study can provide a reference for the port to promote the use of clean energy.

2. Analysis method

2.1. energy use compare method.

It is the comparison of energy consumption and pollutant emissions before and after cleaning alternative fuel or on-shored power.

2.2. standard coal equivalent coefficient method.

The standard coal equivalent coefficient method is used to convert the physical quantity of energy by standard coal equivalent. It is easy to compare energy consumption. Standard coal equivalent coefficient is illustrated in Table 1.

Table 1. Standard coal equivalent coefficient[8].

	Diesel (t)	Gasonline (t)	Electricity (10,000kWh [b])	LNG (t)	Solar energy (MkJ)
tce [a]	1.4571	1.4714	3.15	1.862	29.3

[a] tce（ton of standard coal equivalent）is 1 ton of standard coal equivalent. It is conversion target based on the calorific value of standard coal to calculate the conversion of various energy .

[b]10,000kWh = 3.15tce，3.15tce is average calorific value of coal consumption of 10,000 kWh electricity in China's 2015.

2.3. Emission factor calculation method.

Emission factor calculation method is used to calculate emissions of CO_2 and SO_2 after using fossil energy. Emissions of CO_2 and SO_2 are illustrated in equation (1).

$$E{co_2, so_2} = EC \times EF \quad (1)$$

$E_{CO2,SO2}$ is emission of CO_2 (t) or SO_2(kg). *EC* is energy consumption (t , kWh). *EF* is emission factor of CO_2 (t/t , t/ kWh) or SO_2 (kg/t , kg/ kWh). $E_{CO2,SO2}$, *EC*, *EF* are illustrated in Table 2.

Table 2. Emissions factor of CO_2, SO_2 [9].

	Diesel(t)	Gasonline(t)	LNG [a] (t)	Electricity(kWh)
EF_{CO2}(t)	3.1605	2.9848	2.8376	0.0008
EF_{SO2}(kg)	0.7	0.3	—	0.0028

[a] LNG is shorthand of liquefied natural gas.

3. Analysis of Synergistic Emission Reduction of SO_2 and CO_2

"Synergistic emission reduction" refers to reducing greenhouse gas emissions while reducing local air pollutants[10].Using Ningbo port Co. Ltd. Beilun second container terminal branch (NBSCT) as an example, this paper analyzes the synergistic emission reduction of SO_2 and CO_2 in greenhouse gas before and after the application of clean energy.

3.1. Terminal background

The terminal is one of the best container terminals for super-sized container ships over 10,000 twenty-feet equivalent unit (TEU) capacities. It is one of the most advanced terminal in energy saving and emissions reduction in China. It is the first container terminal company to power tire cranes with electricity instead of diesel fuel. It is also one of the first companies in the country to spread the use of LNG container trucks. The terminal was first evaluated as one of the 'China green port' in 2016. It has used solar energy, wind energy, geothermal energy, LNG, on-shored electricity, it is the most advanced company in application of green energy, energy-saving emission reduction effect is significant.

3.2. Analysis of the energy saving and emission reduction

In NBSCT', the use of clean energy can be summarized as three aspects. First, the application of renewable energy, ground source heat pump and solar water heater alternative fuel or coal-fired boiler heating, wind and solar lighting and other alternative shore electricity, reduce energy consumption and pollutant emissions. Second, the application of LNG, mainly to replace the level of transport vehicles fuel, reduce pollutant emissions. Third, the application of shore power, mainly to replace diesel generator power generation, reduce fuel and pollutant emissions. The following is the result.

3.2.1. Air Source Heat Pumps . Currently, NBSCT has installed 5 air source heat pumps in the replacement of oil boilers to provide 35 tons hot water for the staff canteen and shower rooms each day, fuel oil was totally replaced[11]. As a result, it used 8.1 10,000 kWh electricity to replace 281.83 ton fuel oil, saving energy 377.92 tce, and reduced 823.73 ton CO_2 and 0.20 ton SO_2.

3.2.2. Solar Water Heater. NBSCT replaced oil boilers with solar heaters in staff laundries, the solar collection area of which was 5.62 m^2[11]. As a result, in 2015, NBSCT consumed 407.47 MkJ hot energy, replaced 20.25 ton fuel oil, saving 13.9 tce energy, and reduced 1296 ton CO_2 and 0.13 ton SO_2.

3.2.3. Wind and Light Hybrid Street Lights. NBSCT installed 8 wind and light hybrid street lights, saving 4380 kWh of electricity annually[11], saving 1.38 tce energy, and reduced 3.51 ton CO_2 and 0.01 ton SO_2.

3.2.4. Use of LNG. Currently, NBSCT has a total of 130 horizontal transport container trucks, including 110 LNG container trucks, accounted of 84.6%[11]. In 2015, NBSCT consumed 2628.9 ton LNG, replaced 2542.65 ton diesel, increasing energy consumption 1190 tce, reducing 576.28 ton CO_2 and 17.80 ton SO_2. According to China's current calculation method, container trucks using LNG will increase the energy consumption more 32% than that of diesel.

3.2.5. Shore power. NBSCT has installed shore power facilities supplied with low voltage and normal frequency electricity, mainly used for feeder vessels for domestic trade the power of whose auxiliary engines is no more than 120 kW[11]. In 2015, 343 ships stopped at the port, consuming 39600 kWh, replacing 11.87 ton diesel, saving 12.47 tce energy, and reduced 4.97 ton CO_2 and 0.50 ton SO_2.
The effect of energy conservation and emissions reduction results from using shore power for ocean going vessels at berth instead of power generated by auxiliary engines on board has a close relationship with level of power generation and power source. NBSCT 's electricity comes from thermal power, electricity is coal-fired electricity which is regarded as dirty electricity. However, in order to improve air quality, the Chinese government has called for ultra-low emissions of power plants[12], Which means that thermal power plants need to use a variety of pollutants efficient synergistic emission reduction technology, so that the concentration of air pollutants in line with air pollutant emission limits, Dust, sulfur dioxide, nitrogen oxide emission concentration (baseline oxygen content of 6%) were not more than 10 mg / m^3, 35 mg / m^3, 50 mg / m^3. According to the data provided by the China Electricity Council, energy resource consumed per unit electricity generated by 6MW and above power plants in China is 315gce/kWh in 2015, according to 2012 Guidelines on the Method of Calculation of the Attained Energy Efficiency Design Index (EEDI) for New Ships issued by the International Maritime Organization (IMO)[13],fuel consumption rate of

IOP Conf. Series: Earth and Environmental Science **178** (2018) 012025 doi:10.1088/1755-1315/178/1/012025

power generation for auxiliary engine of ocean going vessels is 300g/kWh, We calculated the reduction of pollutants by shore electricity instead of ship auxiliary power generation. The results show that using shore power for ocean going vessels at berth instead of power generated by auxiliary engines on board in China can save energy while reducing carbon dioxide and sulfur dioxide emissions, the results are shown in Table 3.

Table 3. Effect of application of shore power for ocean going vessels at berth in China.

	Energy consumption [gce/kWh]	CO_2 [g/kWh]	SO_2 [g/kWh]
Generated by auxiliary engines on board	428.4	934.9	15.49
Shore power	315	801.86	2.86
Emission reduction	-113.4	-133.02	-12.63

By the application of above clean energy, including air source heat pumps, solar water heater, wind and light hybrid street lights, shore power, NBSCT reduced the energy consumption significantly, reduced much CO_2 and SO_2. And the use of LNG container trucks will increase energy consumption, but it will significantly reduce CO_2 and SO_2. The results are shown in table 4.

Table 4. The reduction of energy consumption, CO_2 and SO_2.

	Air Source	Solar Energy	LNG	Shore Power	Wind and Light Energy
Energy Reduction (tce)	377.92	13.90	-1190	12.47	29.3
CO_2 Reduction (t)	889.30	1296.00	576.28	5.27	3.51
SO_2 Reduction (t)	0.20	0.13	17.80	0.50	0.01

3.3. Analysis of Clean Energy Contribution of Synergistic Emission Reduction

Assuming that L represents a certain gas (CO_2 or SO_2), $r_{i,L}$ represents the proportion of total emission reductions after the use of clean energy to the total emission reduction rate. We analyze and select clean energy that can reduce both CO_2 and SO_2 emissions. The formula is as following.

$$r_{i,L} = \Delta R_{i,L} / \Delta R_L \tag{2}$$

$r_{i,L}$ is defined as the contribution rate of clean energy to synergistic reductions,%. $\Delta R_{i,L}$ is the relative emission reduction of the clean energy i after the application of the gas L ((CO_2 or SO_2), ΔR_L is the total emission reduction of gas L after cleaning energy applications.

The relative emission reductions and total emission reductions of CO_2 or SO_2 are shown in Table 3. Various types of clean energy emission reduction contribution rate of the results are shown in Table 5.

Table 5. Synergistic Emission Reduction contribution rate %.

	Air Source	Solar Energy	LNG	Shore Power	Wind and Light Energy
CO_2	32.10	46.79	20.80	0.19	0. 13
SO_2	1.17	0.76	95.49	2.68	0.07

The results show that LNG has the greatest contribution to SO_2, and the contribution rate is 95.49% in 2015. The second is the shore power, air source, solar energy, wind energy and light energy, the contribution rate is 2.68%, 1.17%, 0.76%. The contribution rate of air energy, LNG, shore power, wind energy and light energy to CO_2 in 2015 is 32.10%, 20.80%, 0.19% and 0.13% respectively. The contribution rate of solar energy to CO_2 is the largest and the contribution rate is 46.79%. In contrast, wind energy and light energy contribute the least to synergistic reductions, mainly because wind and light can be used for outdoor lighting in ports, and the proportion of electricity consumption is relatively small, so the effect of emission reduction is relatively insignificant.

3.4. *Factor Analysis of Cooperative Emission Reduction*

On the basis of the above analysis, in order to further evaluate the synergistic effect of SO_2 and CO_2 emission reductions of various clean energy sources, this paper calculates the synergistic coefficients $k_{i,L}$ of SO_2 relative to CO_2 emission reduction. The formula is shown in (3).

$$k_{i,L} = r_{i,so_2} / r_{i,co_2} \tag{3}$$

$r_{i,So2}$ and $r_{i,co2}$ is the synergistic emission reduction contribution rate of SO_2 and CO_2 for clean energy i, as shown in Table 3. If $k_{i,L} = 1$, it indicates that the factor SO_2 and CO_2 are in a higher degree of co-ordination. If $0 < k_{i,L} < 1$, it indicates that the factor SO_2 is lower than the CO_2 emission reduction level. If $k_{i,L} > 1$, higher than CO_2 reduction. Calculation results of $k_{i,L}$ are shown in Table 6.

Table 6. Analysis of Co - ordination Coefficient of SO_2 and CO_2

	Air Source	Solar Energy	LNG	Shore Power	Wind and Light Energy
$k_{SO2/CO2}$	0.0364	0.0149	4.5907	14.1010	0.4234

By calculating and analyzing SO_2 and CO_2 emission reduction synergistic coefficients, it can be seen that LNG trucks and shore power are relatively large in terms of port energy use, so they are relatively large in fuel substitution. With the improvement of energy consumption and energy efficiency in ports, $k_{SO2/CO2}$ is greater than 1, SO_2 and CO_2 can achieve a higher degree of mitigation, SO_2 emission reduction is greater than the degree of CO_2 emission reduction. Air source, solar energy, wind energy and light energy $k_{SO2/CO2}$ less than 1, indicating that the degree of emission reduction of CO_2 is greater than the degree of emission reduction of SO_2.

4. Application Prospect of Clean Energy in Port

Due to the use of clean energy technologies facing increased costs and energy security and other issues, only the scale of production in the forefront of the industry, environmental pressure on the port terminals gradually began to demonstrate the application, such as Shenzhen Yantian Port, Ningbo Zhoushan Port, Shanghai Port, Qingdao Port, etc. It has not yet been widely used in Chinese ports.

According to the port's energy consumption structure, usually the production operations accounted for more than 98% of total energy consumption, auxiliary production operations accounted for about 2%. Solar energy, geothermal energy, wind energy and light energy and other renewable energy sources are usually used in the auxiliary production operations, energy-saving emission reduction effect is limited, but in the production operations, you can use LNG and electricity to replace fuel, energy-saving emission reduction effect is obvious. This is consistent with the conclusion of this paper. As a result, both theory and practice have proven that energy use is promoted at the port to promote clean energy while reducing SO_2 and CO_2 emissions.

During the 12th five-year period, the Ministry of Transport of the People's Republic of China issued a series of relevant documents , in order to guide the port to use LNG and shore electricity, to promote the port to optimize the energy structure, to achieve energy-saving emission reduction targets.

Such as the 'arrangement of work for energy conservation and emission reduction for waterway transport sector during the national 12th Five-Year plan' [14], there are two major jobs mentioned ,one is applications of LNG driven transport vehicle technology, the other is to promote the use of shore power technology in container ships and bulk carriers. Since 2011, supporting by the Special Fund for Energy Conservation and Emissions Reduction for Transport Sector, China Ministry of Transport gave awards to terminals which construct facilities for supplying shore power for ocean going vessels at

berth, and China shipping companies which add facilities on board in their vessels to receive shore power. Port of Lianyungang, Port of Shenzhen and Hebei ocean shipping group Co. Ltd. received more than 10 million RMB in total. Since 2014, China's Ministry of Transport announced two batches of LNG pilot and demonstration projects for water transport applications, Port of Lianyungang, Port of Shenzhen and Port of Ningbo won the first batch of demonstration project of LNG container truck.
Therefore, on the one hand is the actual needs of energy-saving and emission reduction, on the other hand is the government policy support, both aspects are conducive to promoting the use of clean energy in ports.

5. Conclusions

Through the above analysis, we can draw the following conclusions.
First, the results show that the reduction SO_2 and CO_2 is positively correlated with the usage of clean energy amount. Among them, LNG has the greatest contribution to the coordinated emission reduction of SO_2, and solar energy contributes the most to CO_2 reduction. Wind energy and light energy in the clean energy alternative to the smallest proportion, so its contribution to the smallest reduction.
Second, the use of LNG and shore power in ports can achieve better synergistic emission reductions SO_2 and CO_2. And through the use of air sources, solar energy, wind energy and light energy, CO_2 emission reduction effect is better than SO_2 emission reduction.
Finally, By the promotion of LNG and shore electricity in ports, We can bring significant effect of air pollutant emission reduction and greenhouse gas emission reduction. Therefore, this paper suggests that the transportation authorities can use this method as a policy selection tool for energy conservation and emission reduction.

6. References

[1] 2014 Promotion of Port Transformation and Upgrading of the Views of Guidance *China's Ministry of Transport*
[2] 2012 Lang Jian and Cheng Si A Study on the Trends of Vehicular Emissions in the Beijing-Tianjin-Hebei (BTH) region *Atmospheric Environment* 41 605-614.
[3] Zhou Ya, Wu Yue and Yang L 2010 The impact of transportation control measures on emission reductions during the 2008 Olympic Games in Beijing *Atmospheric Environment* 44 285-293.
[4] Wang Li, Xu Luyu and Song H M 2011 Environmental performance evaluation of Beijing's energy use planning *Energy Policy* 39 3483-3495.
[5] Pan XiCun, Yang Wei and He Kun 2007 Health benefit evaluation of the energy use scenarios in Beijing *Science of the Total Environment* 374 242-251.
[6] Hao Jiming, Wang Litao and Li Ling 2006 Contribution Rate and Control Strategy Analysis of Energy - related Air Pollution Source in Beijing *Earth Science* 35 115-122.
[7] Feng Yiyi and Zhang Luxia 2012 Scenario analysis of urban energy saving and carbon abatement policies: A case study of Beijing city *Procedia Environmental Sciences* **13** 632-644.
[8] 2015 Norm of Energy Consumption Per Unit Throughout of Container Terminal *China Standard Press* version 1 pp 4
[9] 2013 Standard for Green Port Grade Evaluation *China Communications Press* version 1 pp 17
[10] Mao Xianqiang, Xin Youkai and HuTao 2012 Analysis on Environmental Economic Path of Synergistic Emission Reduction of Sulfur, Nitrogen and Carbon in China 's Electric Power Industry Chinese Journal of Environmental Science *32* 748-756.
[11] Zhu Li and Cai Oucheng 2016 GPAS Application Form-Ningbo Port CO.LTD. BeiLun Second Container Terminal Branch pp 36-40
[12] 2015 Fully implement the work plan of ultra low emission and energy saving of coal - fired power plant *China Ministry of Environmental Protection*
[13] 2012 Guidelines on the Method of Calculation of the Attained Energy Efficiency Design Index (EEDI) for New Ships *Marine Environment Protection Committee.*

[14] 2011 Arrangement of Work for Energy Conservation and Emission Reduction for Waterway Transport Sector During the National 12th Five-Year Plan *China's Ministry of Transport.*

Research on Influence of Beijing-VI Diesel on Consumption and Emission Characteristics of Euro VI Diesel Bus

Chen lin, Wang Zhong, Qu lei, Liu Shuai
1 School of Automotive and Traffic Engineering, Jiangsu University, Zhenjiang, 212013

Abstract. The characteristics of fuel consumption and emission from a euro Ⅵ diesel bus burning Beijing-Ⅵ diesel and China-Ⅴ diesel were tested on heavy chassis dynamometer under C-WTVC testing cycle. The results indicated that the fuel consumption of Beijing-Ⅵ diesel is 1.12L/100km lower than China-Ⅴ diesel and has the advantage of saving energy. A major emission factors were decreased, NOx emission reduced by 18%, PM emission reduced by 20%, CO and HC slightly decreased under C-WTVC testing cycle. Under city conditions, road condition and highway conditions, the NOx emission reduced by 25%, 18%, 16%.

1. Introduction

The possessing capacity of heavy-duty diesel vehicles in China accounts for 5% of gross of motor vehicles, yet they contribute 78% of nitrogen oxide emission and 82% of emission of particulate matters (PM), and the emission of exhaust fume has emerged as an important urban atmosphere pollutant source in China to directly undermine citizens' health ([1]). In the face of increasingly severe pollution, on the one hand, the manufacturers attempt to reduce emission of vehicle pollutants by adopting advanced post-processing technologies such as diesel particulate filter (DPF), selective catalyzing reduction (SCR), continuously regenerating diesel particulate filter (DOC+CDPF)([2]). On the other hand, the quality of diesel also has a important effect on emission of pollutants, so emission of exhaust fume by diesel vehicle can be reduced by improving fuel quality ([3]). On January 1, 2017, Beijing implements the fuel standard of stage fume by t fume by padjusts such indices as density, viscocity, flash point, etc. according to changes in refining process of diesel, and adds clearing agent in diesel to exercise further stricter control over the major environmental indices such as sulphur content and PAH in diesel, so it of far reaching importance for energy conservation and emission reduction.

The domestic and foreign scholars researched extensively the influence of diesel quality on the performance of diesel engine and diesel cars, in the hope of reducing emission ([4]). LU studied the effect of fuel characteristics on diesel engine burning at low temperature to find that the influence of cetane value of fuel on ignition delay period is large to influence emission of PM, while other characteristics of fuel do not have a large effect on combustion ([5]). Zhao Lei et al. experimented on China-Ⅲ and China-Ⅳ diesel as well as Beijing-V diesel to find that the quality of diesel has a bearing upon fuel economy of diesel engine, and has an unobvious influence on emission of CO and HC, while has an obvious influence on emission of NO and PM([6]). Hu Zhi yuan et al. researched emission performance of China-Ⅳ buses burning diesel and B5 biological diesel on heavy-duty chassis dynamometer under CCBC cycle to find that the emission of THC and NOx for this bus burning B5 biological diesel increased, while emission of CO and PM reduced([7]). Wang Yan jun et al. adopted PEMS method and method of entire car chassis dynamometer to research the influence of different loads and testing conditions on measurement of emission factors for heavy-duty vehicle, finding that the emission factors of gaseous pollutants measured by the two methods had good correlation, while the gap between emission factors for PM was up to about 50%([8]). Current research is mostly focused on the influence of fuel quality on emission of China-III and China IV diesel engine, while little has been reported on emission of Euro-eldiesel vehicle with advanced post processing technology applied.

Heavy-duty chassis dynamometer can test on diesel vehicle in specific environment and avoid the drawbacks of poor repetitiveness and weak representativeness of engine bed test in actual road tests, so it has become an important method for domestic and foreign scholars to research performance and emission of diesel vehicle. The paper takes buses meeting Euro emission standard as test vehicle to research the oil consumption and pollutant emission of Beijing-ivdiesel and China-andiesel in condition of C-WTVC cycle on heavy-duty chassis dynamometer, in an effort to provide a certain reference frame for learning the influence of quality of Beijing-VI diesel on energy dissipation and emission of diesel vehicle and further popularization across the whole China.

2. Test Equipment and Scheme

2.1. Test equipment

This test is carried out on the four chassis dynamometer, the chassis dynamometer is produced by Germany MAHA Group, which can meet the test requirements of 4×4, 6×6, 8×8 and other driving types. It has sensitive control system and electronic inertia simulation device, and can simulate up to 35t truck weight and 2500kg to 60000kg of vehicle inertia. Measurements of pollutant emissions adopt CVS i60, AMA i60 and PSS i60 devices complying with the requirements of 2005/55 / EC. CVS i60 controls the temperature of the CVS sampling system, the acquisition capacity of the dilution tunnel, and achieve constant volume sampling. AMA i60 exhaust measurement system can be carried out during the operation of the vehicle pollutant transient measurement, with non-spectral infrared analysis of CO, measurement of hydrogen flame ion for THC, measurement of non-spectral UV chemiluminescence for NOx. PSSi60 mainly controls the particulate matter collection system so that the sampling conditions of the particles are guaranteed to be within the limits prescribed by the regulations.

2.2. Test prototype and test program

The test sample is a heavy-duty diesel city bus, its emission standard meets the Euro VI emission standard, and its main technical parameters are shown in table 1.

Table 1. The main technical parameters of test prototype

Subjects	Parameters
Vehicle type	City Bus
Total quality	11800kg
Engine type	Turbocharged Intercooler Diesel Engine
Displacement	8.4L
Rated speed	2200r/mim
Maximum power	210kW
Maximum torque	1200N·m
Emission grade	Euro VI
Oil supply system	high-pressure common-rail
Post processing system	SCR+DPF+DOC

The measurement is in accordance with GB27840-2011 (Fuel consumption test methods for heavy-duty commercial vehicles) and the test conditions is C-WTVC cycle. C-WTVC cycle are based on the WTVC cycle, according to China's actual road conditions to adjust the acceleration and deceleration of the formation of the test cycle. The whole test condition consists of urban conditions, road conditions and high speed conditions. Figure 1 is the test site.

Figure 1. Test site

2.3. Fuel characteristics

The test used 0# Beijing-Ⅵ car-use diesel and 0# China-Ⅴ car-use diesel, whose major physical and chemical indices are as shown in table 2. The two kinds of diesel fuel samples are tested by the all items test of Shanghai SGS test center.

Table 2. Major physical and chemical characteristics of diesel

Subjects	0# Beijing-Ⅵ diesel	0# China-Ⅴ diesel
Total insolubles mg/100ml	0.7	2.5
ash content %	0.001	0.01
10% steam excess carbon %	0.1	0.3
PAHs %	0.8	11
Cetane number	54.2	52.2
Sulfide mg/kg	2.5	3.6
20℃ density g/cm^3	0.831	0.834
20℃kinematic viscosity mm^2/s	4.2	4.1
flash point ℃	68	69

The table shows that the major physical and chemical characteristics of the two kinds of diesel all meet relevant standards. The main differences were as follows. The total insolubles content was 2.5 mg/100ml in the China V dieseland0.7 mg/100ml in the Beijing VI diesel. The ash was 0.01%and 0.001%, respectively.The10% steam excess carbon was 0.3%m/m and0.1% m/m, respectively. The PAH content was 11%m/m and0.8%m/m, respectively.

3. Test result and analysis

3.1. Driving curve

Figure 2 compares set velocity of test vehicle in standard condition and actual driving velocity in burning the two kinds of diesel, finding that the two testing velocities are well repetitive, highly keep in line with set velocity, and well follow velocity in condition of C-WTVC cycle.

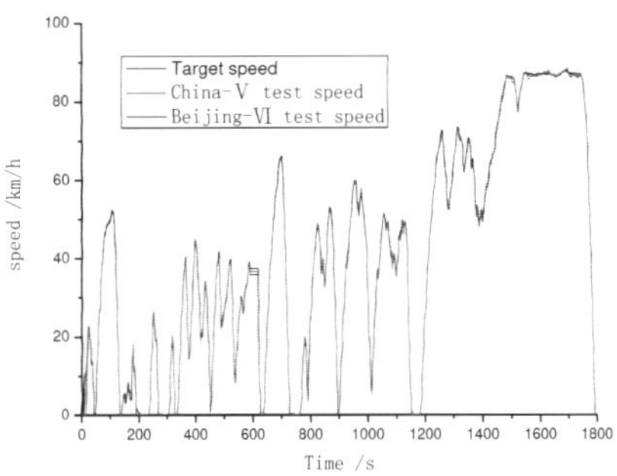

Figure 2. Testing velocity and target velocity

The running time of the whole test is 1800 s, including 900 s for urban working condition, 468 s for road condition, 432 s for high speed condition, and the running distance is 20.51 km。Under C-WTVC condition, the idling time is 186s, idling proportion is 10.3%, maximal speed is 87.8km/h, average velocity is 40.1km/h, maximal acceleration is 0.92m/s^2, maximal deceleration is 1.03 m/s^2.

3.2. Emission factor of pollutant

Able to reflect actual emission level of vehicle, emission factor is an important reference for researching pollutants. Figure 3 and figure 4 are emission factors of pollutants for the test vehicle burning China-Ⅴ diesel and Beijing- ediesel under condition of C-WTVC.

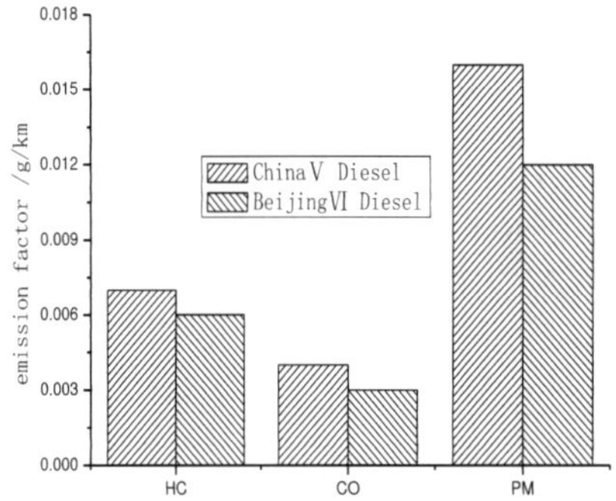

Figure 3. Emission factors of different fuel

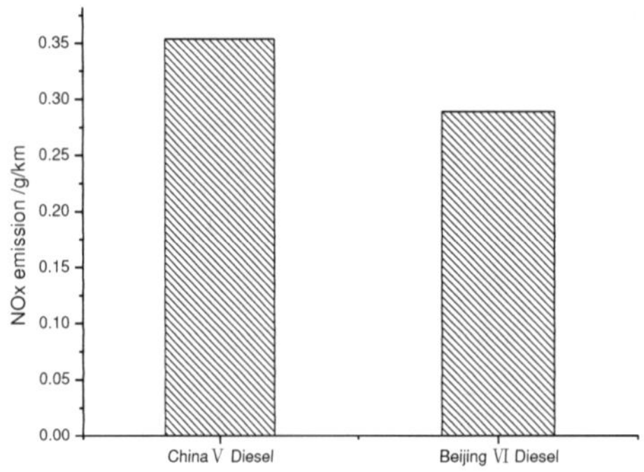

Figure 4. NOx emission factors of different fuel

It can be seen from the test results of figs. 3 and 4 that compared with the burning of China-d diesel, the emission factors of four main pollutants of the diesel engine fueled with Beijing- diesel is decreased. Among them, HC emission factor, CO emission factor, NOx emission factor and PM emission factor are respectively reduced by 0.001 g / km, 0.001 g / km, 0. 065 g / km and 0.004 g / km, and the decreasing ranges of NOx emission factor and PM emission factor are separately about 18% and 25%. The reduction of NOx and PM is mainly affected by the physical and chemical properties of fuel, the cetane number is the most important parameter of the exothermic moment, which has a significant effect on ignition delay period, it is also conducive to improve cetane value and improve fuel fire, so that the combustion is relatively full using and the incomplete combustion products is reduced([9], [10]). Diesel particulate matter is mainly composed of carbon soot particles, sulphate, incomplete combustion fuel, oil and intermediate products; and diesel fuel sulfate and gaseous SO_2 emissions are increased linearly.

To further analyze the differences in emission factors of pollutants for different fuels in different characteristic mileages, the emission factors for different stages in whole C-WTVC condition were researched. Table 3 shows emission factors for diesel vehicle burning China-l diesel and Beijing- diesel in C-WTVC cycling under conditions of town, highway and high speed.

Table 3. Emission factors in different conditions

emission	urban condition		road condition		high speed condition	
	China-Ⅴ	Beijing-Ⅵ	China-Ⅴ	Beijing-Ⅵ	China-Ⅴ	Beijing-Ⅵ
$HC/10^{-6}$	4.21	4.07	3.98	3.65	4.07	3.82
$CO/10^{-6}$	0.56	0.40	0.64	0.53	0.55	0.42
$NOx/10^{-6}$	6.70	4.97	3.32	2.71	1.51	1.26

The result shows that when the diesel vehicle drove in different conditions, the emission of HC and CO was not largely different, with burning Beijing- diesel emitting slightly less HC and CO than burning China-Ⅴ diesel. The test condition has a larger effect on NOx emission of vehicle. The emission of diesel vehicle burning China-Ⅴ diesel is 6.7×10^{-6}, 3.32×10^{-6} in highway condition and 1.51×10^{-6} in high speed condition. The emission of diesel vehicle burning Beijing- diesel is 4.97×10^{-6}, 2.71×10^{-6} in highway condition, 1.26×10^{-6} in high speed condition, showing a decrease of about 25% in town condition, a decrease of 18% in highway condition, and a decrease of 16% in high speed condition. In town condition, due to frequent accelerating and decelerating of vehicle, the diesel engine is in the state of low rotation rate and high load for long, under which state, the average air-fuel ratio of combustible mixture is on the small side, and the highest combustion temperature rises. High temperature, oxygen enrichment and endurance period of reaction are key factors influencing generation of nitrogen oxide, so emission of NOx is far higher than in that in highway condition and high speed condition. Although the temperature in air cylinder is low in high speed condition, yet the reaction time to generate nitrogen oxide provided by higher rotation rate decreases, leading to reduction of NOx emission factor ([11]).

3.3. Transient emission of pollutants

Figures 5-7 are real time emission curve of HC, CO and NOx for two kinds of diesel burnt by diesel vehicle. It is observed from figure 5 that the instantaneous emission of CO for Euro diesel vehicle has dropped to a low level, that for China- fdiesel fluctuates around 0.6×10^{-6}, and Beijing-Ⅵ diesel fluctuates around 0.2×10^{-6}. The CO emission of Beijing- diesel reduces to some extent relative to China-na-nd NOx

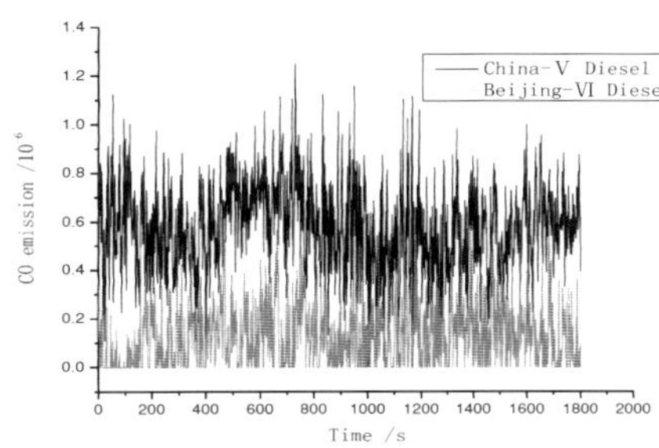

Figure 5. Instantaneous emission of CO

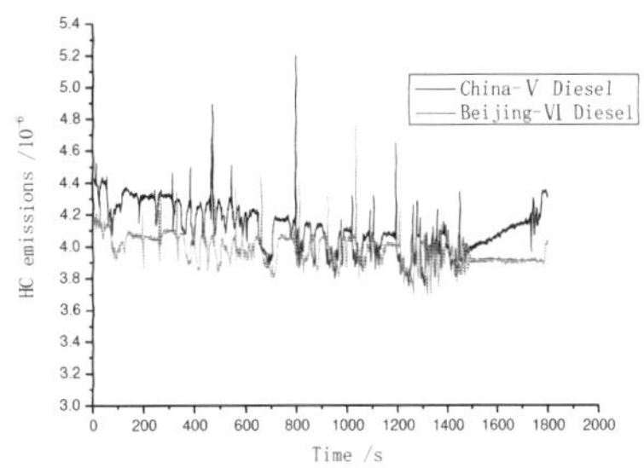

Figure 6. Instantaneous emission of HC

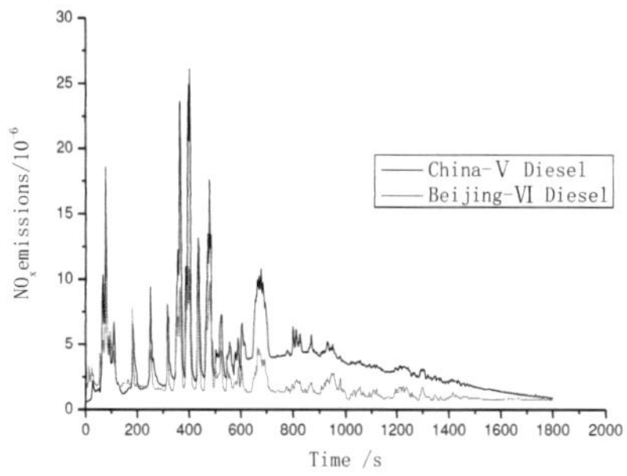

Figure 7. Instantaneous emission of NOx

It is observed from figure 6 that the emission trends of different fuels burnt by diesel vehicle are the same. The HC of China-Ⅴ diesel fluctuates around 4.1×10-6, and the HC of Beijing-Ⅵ diesel fluctuates around 4×10-6, which is a small decrease. As general control level of emission by Euro-Ⅵ diesel vehicle is high, plus the influence of interior calibration in engine and the ingredients of background air in the environment, the change in CO and HC is negligible.

It is observed from figure 7 that the NOx emission diesel vehicle burning Beijing-Ⅵ diesel reduces significantly. The highest emission concentration for diesel vehicle burning China-Ⅴ diesel is up to 26.1x10-6, and the highest emission in burning Beijing-Ⅵ diesel is 11.1x 10-6. The average velocity in high speed condition is far higher than that in low speed condition. The instantaneous emission rates of NOx are not largely different, with emission far lower than that in city condition and highway condition. Instantaneous emission concentration of NOx is remarkably related to service condition of vehicle. Within the first 900s of the cycle, the vehicle is in town condition, with average velocity of 22.9km/h, maximal acceleration of 0.917m/s2, and maximal deceleration of 1.033m/s2. In city condition, the vehicle often accelerates or accelerates rapidly, so rotation rate of engine is low and load is large. When the vehicle accelerates, the augmentation fuel charge reduces air-fuel ratio of combustible mixture, increases combustion temperature and combustion pressure, and raises emission rate of NOx.

When vehicle decelerates or idles, the fuel charge of diesel engine is reduced, thereby decreasing emission rate of NOx. So the city condition sees large fluctuation in emission of NOx. Compared with Ⅴ diesel, the emission of NOx in burning Beijing-Ⅵ diesel all reduces to a certain extent in the three conditions. This is mainly because China-Ⅴ diesel has lower cetane value than Beijing-Ⅵ diesel, leading to increase of ignition delay period, increase of heat release rate, increase of premixed combustion capacity, decrease of diffuse combustion capacity, and increase of NOx emission.

3.4. Fuel consumption

The fuel consumption for combustion-use China-Ⅴ diesel and Beijing- diesel by diesel vehicle was measured based on C-WTVC condition, with measurement result as shown in table 4.

Table 4. Fuel consumption rate

	urban conditions	road conditions	high speed conditions	Comprehensive
China Ⅴ diesel L/100km	36.83	27.55	20.26	27.52
Beijing-Ⅵ diesel L/100km	35.76	26.21	19.48	26.40

The table above shows that the fuel consumption in town condition is the largest, which is caused by constant accelerating and decelerating of vehicle. In town condition, the fuel consumption of China-b is 36.83L/100km, and that for Beijing-Ⅵ is 35.76 L/100km. In highway condition, the fuel consumption of China-3 is the 27.55 L/100km, and fuel consumption of Beijing-Ⅵ is 26.21 L/100km; in highway condition and high speed condition, the fuel consumption of China-h is 20.26L/100km, and that for Beijing-Ⅵ is 19.48; in the entire condition, the fuel consumption of China-Ⅴ is 27.52 L/100km, and fuel consumption of Beijing-Ⅵ is 26.40 L/100km, reducing by 1.12 L/100km, which is mainly largely related to cetane value of diesel. The higher the cetane value is, the higher the thermal efficiency is, the better the ignition performance, the more adequate the combustion is, hence lowering fuel consumption rate. With fuel consumption slightly lower than China-Ⅴ oil product, the Beijing-Ⅵ oil product has certain advantages in energy conservation.

4. Conclusions

This study tested the characteristics of fuel consumption and emission from a euro Ⅵ diesel bus burning Beijing-Ⅵ diesel and China-ing Beijing-Ⅵ diesel andnsumption and emission from a euro Ⅵ diesel bbetter the ignition performance,

Compared with burning China-e diesel, the emission factor of pollutants for Euro-Ⅵ heavy-duty diesel vehicle burning Beijing-Ⅵ diesel is lower, NOx reduced by about 18%, PM reduced by about 20%, while CO and HC did not reduce significantly.

In the course of instantaneous emission in C-WTVC condition, CO emission in burning China-diesel was about 0.6×10^{-6}, HC emission was about 4.1×10^{-6}. The CO emission in burning Beijing-diesel was about 0.2×10^{-6}, and HC emission was about 4×10^{-6}.

The highest emission concentration for burning China-g diesel in city condition was up to 26.1×10^{-6}, and the highest emission of combustion-use Beijing-Ⅵ diesel was up to 11.1ppm. The average velocity in high speed condition was far higher than that in low speed condition. The instantaneous emission rates of NOx were not largely different, with emission far lower than that in city condition and highway condition.

Compared with burning China-b diesel, the consumption of Beijing-n diesel by diesel vehicle reduced by 1.12 L/100km, pointing to satisfactory economy.

5. References

[1] QJ Zhang, L Wu, MY Liu, XZ Fang, et al. 2016, Pollution characteristics and health risk assessment of VOCs from vehicle exhaust in Nanjing, China. China Environmental Science, 36(10), pp.3118-3125.

[2] DM Lou, KX Zhao, PQ Tan, et al. 2014, Gaseous emission characteristics of diesel bus equipped with different after-treatments based on driving distance. China Environmental Science, 32(10), pp.79-84.

[3] F Dong, SM Hou, XL Liu et al. 2016, Experiment on the Effect of Diesel Fuel Properties on Combustion and Emissions in a Heavy-Duty Diesel Engine. Transactions of CSICE, (4), pp:289-295.

[4] T Wu, Z Huang, W Zhang, et al. 2007, Physical and Chemical Properties of GTL−Diesel Fuel Blends and Their Effects on Performance and Emissions of a Multicylinder DI Compression Ignition Engine. Energy & Fuels, 21(4), pp.1908-1914.

[5] T Lu, Z Huang, CS Cheung, et al. 2012, Size distribution of EC, OC and particle-phase PAHs emissions from a diesel engine fueled with three fuels. 438C (3), pp.3-41.

[6] L Zhao, LI Kai, LI Bo Shi, et al. 2013, Effect of Diesel Fuel Quality on Fuel Economy and Emission Characteristics of a Diesel Engine. Journal of Environmental Engineering Technology, 3(4), pp.337-341.

[7] Z Hu, Y Qin, P Tan, et al. 2015, On-road NOx and Ultrafine Particle Emission Characteristics of China IV Diesel Emission in Shanghai. JOURNAI. OF TONGJI UNIVERSITY (NATURAL SCIENCE), 43(2), pp.0286-0292.

[8] YJ Wang, Z Ji, H Yin, et al. 2014, Study of parameters influencing measurement on heavy duty diesel vehicle's emission factors. Research of Environmental Sciences, 27(3), pp.232-238.

[9] WM Chen, JX Wang, SJ Shuai, et al. 2008, Effect of Sulfur Content in Diesel Fuel on Engine Performance. Chinese Internal Combustion Engine Engineering, 29(4), pp.41-45.

[10] Z Wang, Y Zhao, M Li, et al. 2014, Kinetic model and simulation of PAHs formation pathway of methanol / diesel. JOURNAL OF JIANGSU UNIVERSITY (Natural Science Edition), 35(1), pp.1-6.

[11] JD Guo, GE Yun-Shan, JW Tan, et al. 2014, Real-world emission characteristics of China Ⅳ buses. Research of Environmental Sciences, 27(5), pp.477-483.

Analysis of Development of Hazardous Waste Disposal Technology in China

Meitong Li[1], Jin Xu[2] and Binke Li[2,*]

1 School of Chemical Engineering and Technology, Tianjin University, Tianjin 300072, China
2 School of Management, Tianjin University of Technology, Tianjin 300384, China
Email: limeitong@tjut.edu.cn, xujinbox@163.com, 15737129420@163.com

Abstract. With the rapid development of science and technology and economy, the rapid increase of hazardous waste has become the common environmental problems facing the whole world. China is the world's manufacturing power, with the accelerating process of urbanization, the resulting number of hazardous waste more and more. However, hazardous materials such as toxic, corrosive, flammable, reactive and infectious, and other dangerous characteristics if not properly handled, will cause serious harm to the environment, ecology and human health. Therefore, hazardous waste disposal research has extremely important practical significance. In this paper, the basic summary of hazardous waste, while the status of hazardous waste disposal in China to make an analysis of hazardous waste disposal in China to provide value for reference.

1. Introduction

In recent years, with the rapid development of science and technology and the continuous improvement of living standards, the problem of hazardous waste is more and more serious, and its production more and more complex composition. As early as the middle of last century, due to a large number of hazardous waste generation and improper disposal, causing a series of environmental pollution incidents. The love canal in the United States in the middle of the twentieth century as a waste dump, due to wanton and large accumulation of hazardous waste, causing groundwater pollution and soil pollution, the local ecological environment and the health of the residents seriously affected, direct economic losses of 30 Billion dollars[1]. Severe mercury pollution in the Gulf of Japan, resulting in thousands of people poisoning and more than 100 deaths in the disaster. The 1970s, the United States Missouri mixed with 2,3,7,8-TCDD sludge sludge asphalt pavement, resulting in serious soil pollution and a large number of livestock deaths, the ecological environment and the health of the residents affected. At the same time, there have been many hazardous waste pollution incidents in China, such as thousands of people caused by tin emissions in Yunnan sub-acute poisoning; The end of the last century, Jinzhou, Liaoning Jinzhou ferroalloy plant chromium residue caused by serious groundwater pollution incident and Guangxi Longjiang River upstream of the Cd leaching residue and pressure residue and other hazardous waste filtrate caused by river and underground drinking water pollution incidents[2]. Undoubtedly these hazardous waste pollution incidents on the ecological environment and human health posed a huge threat. Through the above analysis, the disposal of hazardous waste disposal should be sufficient attention to countries. At present, with the continuous improvement of China's comprehensive strength, China has become the world's largest manufacturing country, as the world's "factory", hazardous waste production can not be ignored, while a large number of hazardous wastes occupy a lot of land resources, and the harmful substances may also cause soil and water pollution and endanger human health. Therefore, how to effectively deal with hazardous waste disposal has become a new topic.

2. Brief Description of Hazardous Wastes

2.1. Classification of Hazardous Wastes

Hazardous Wastes are defined as hazardous wastes that are chemical, toxic, explosive or hazardous to humans, animals, plants and the environment, in addition to radioactive waste[2]. The United Nations Environment Program (UNEP) defines hazardous wastes as hazardous wastes. The United States, "Resource Conservation and Regeneration Law" will be defined as hazardous waste disposal due to improper handling of human health and ecological environment will cause actual or potential hazards of waste or waste compounds. Japan's "Waste Disposal and Sweep Act" Definition: After the government decides that the explosive, toxic, infectious and the ecological environment and personal health hazards of substances [3]. China in the "Solid Waste Pollution Prevention Law of the People's Republic of China" provides that hazardous waste is included in the national list of hazardous wastes or hazardous waste in accordance with national standards and methods of identification of hazardous characteristics of the waste. Hazardous waste can therefore be: ①The threat to human health and ecological waste; ②Hazardous characteristics of the two criteria for waste to determine. Hazardous wastes are categorized according to their characteristics, sources and composition, and they are classified into leaching toxicity, acute toxicity and corrosivity according to the hazardous degree of hazardous wastes [4].

2.2. Hazardous Waste Sources and Hazards

2.2.1. Sources

Hazardous wastes are widely used, of which industrial production is the main source of hazardous waste in China, accounting for 40% of the total amount of hazardous wastes, such as chemical waste, mining, metal smelting, machinery manufacturing and other hazardous waste accumulation and emissions. Followed by the medical process of hazardous waste generated, such as blood, drugs and colonies and other emissions will cause serious pollution of the ecological environment and a threat to human health. Finally, the agricultural production process and human life will also bring hazardous waste.

2.2.2. Hazards

Hazardous waste has a variety of hazardous characteristics of the ecological environment pollution and human health hazards are very serious, but did not cause enough attention, often in the cause of environmental pollution accidents or disasters were found, at this point, the consequences have reached the point where it is difficult to restore and restore. According to the potential and long-term characteristics of hazardous waste pollution, the main hazards are summarized as follows: ① Hazardous wastes are washed away by rainwater into rivers, lakes, and seas, causing surface water pollution or infiltration of filtrate into the soil, causing soil and groundwater pollution. ②Hazardous waste itself evaporation, sublimation and decomposition of organic release of harmful gases caused by air pollution. ③Hazardous waste dust, particles cause soil and air pollution. Therefore, the disposal of hazardous waste should be properly, try to avoid or reduce the hazardous waste in the nature of the continuous bioaccumulation, reduce its human and environmental threats.

3. Analysis on Present Situation of Hazardous Waste Disposal at Home and Abroad

3.1. The Status of Hazardous Waste Disposal Abroad

Some developed countries of foreign countries attach great importance to the management of hazardous wastes, after nearly half a century of development, the formation of the whole process of hazardous waste management philosophy, at the same time to establish a more comprehensive system of laws and regulations, developed an effective and effective management system, effectively curbing the hazardous waste pollution. The United States as one of the first countries to carry out hazardous waste management, as early as the 20th century, 70 years there is a hazardous waste related legal control. And the effective control of hazardous waste as a basic goal and one of the most important environmental protection work. Japan in order to solve the problem of environmental pollution of hazardous waste, has more than ten laws and regulations in the development of hazardous waste-related legal provisions, and put forward the hazardous waste management "3C" principle[5]. Accordingly, the Japanese hazardous waste pollution control has achieved remarkable results. In Germany, only the control of solid waste by the promulgation of laws and regulations there are more than ten. In Canada, the development of more detailed hazardous waste management laws and regulations and related management regulations, covering the disposal, transfer, technology, etc. In short, the developed countries, the main concept of hazardous waste management can be summarized as: "reduction, reuse, recycling," the 3R governance, the whole process and the classification of governance, economic stimulus, near the nearest governance, national and social co-governance, public participation and other principles[6-7].

The main management system is as follows: ①The smallest quantitative system. Reduce the source of hazardous waste emissions, to maximize the waste recycling. ②Standardized management system. Enterprises in the process of hazardous waste management, such as the occurrence of transfer of ownership, hazardous waste disposal and storage storage, need to go through the appropriate approval process and apply for proof. ③Industrial disposal system. Through the high-end science and technology and the sincere cooperation between countries to achieve the industrialization of hazardous waste disposal, improve the recycling of hazardous waste, and bring economic benefits. ④Information management system. The establishment of hazardous waste treatment and disposal of the whole process of information monitoring platform to improve the efficiency of hazardous waste disposal. At present, the world's environmental protection strategy has changed, has been transferred from the solid waste management to toxic chemicals and hazardous waste on the safe disposal and effective control. After continuous exploration, developed countries have established a relatively complete system of hazardous waste treatment and disposal.

The status of domestic hazardous waste disposal

Hazardous waste management in China started late, until the last century, 80 years before the establishment of specialized departments responsible for hazardous waste management. However, with the improvement of comprehensive strength in China, in the hazardous waste management work has made rapid progress. As one of the earliest parties to the Basel Convention[5]. Has always attached great importance to the management of hazardous wastes, and promulgated a series of laws and regulations and the introduction of the relevant standards and technical policies to form a set of practical standards in line with our hazardous waste management system. At the same time, China's construction of hazardous waste management institutions have also strengthened the degree of attention, has established in various provinces and cities nationwide hazardous waste management center, and hazardous waste comprehensive utilization, treatment and disposal technology research, publicity and education, technical training, International cooperation and so has made great progress. In addition, with the increasing emphasis on environmental issues in China, increasing investment in the state, the gradual improvement of laws and regulations, China's hazardous waste treatment capacity has been greatly improved. According to statistics, in 2010 the country's hazardous waste disposal capacity of about 23.25 million tons / year, the actual disposal of about 8.4 million tons, compared to 2006 increased capacity by 226%. However, as the world's largest developing country, due to extensive economic development in China, the level of technology constraints, there are still some problems in hazardous waste management. In recent years, although China's rapid economic development, people's living standards improved significantly, but the types and quantities of hazardous waste also increased, resulting in a very serious environmental problems. Figure 1 shows the distribution of hazardous wastes in China, it can be seen only Shandong, Xinjiang and Qinghai, hazardous waste production accounted for more than half of the national total, and relatively concentrated. In addition, China's hazardous waste management-related laws and regulations and pollution control standards policy system is not perfect, the lack of appropriate technical and economic policy support. Therefore, in order to effectively control the generation of hazardous wastes and their non-standard treatment and disposal, the economic and technical policies for the market operation and management of hazardous waste collection and disposal should be formulated, and the construction of centralized disposal facilities for hazardous wastes should be vigorously promoted and strict standardization of hazardous waste management system, to minimize its harm to the environment and human health.

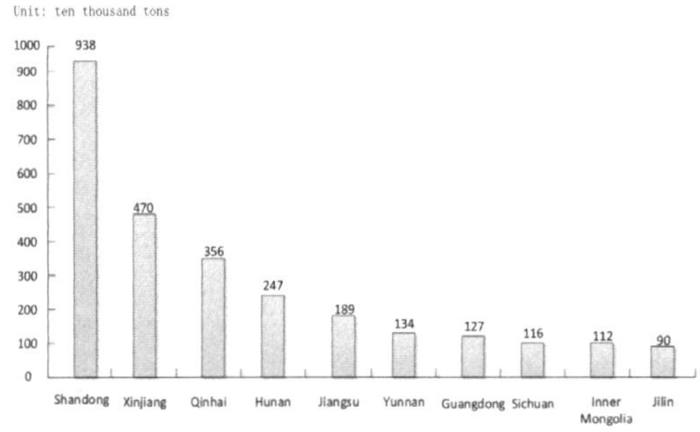

Figure 1. China's hazardous waste production distribution map

4. Analysis and Suggestion on Hazardous Waste Disposal in China

In recent years, although the hazardous waste management in China has made more positive results, but the annual production of hazardous waste is still large, and at present our country for the effective management of hazardous waste, there are still many problems. Technical immaturity and lack of economic policy support, hazardous waste disposal efficiency is low, there are environmental risks.

4.1. China's Hazardous Waste Disposal Problems

4.1.1 Hazardous Waste Generation Is Large. China as a manufacturing country, the rapid development of industry, energy and chemical industry, non-ferrous metal mining and mining, oil processing, equipment manufacturing and other industries generate hazardous waste is very large. A large number of hazardous waste can not be effectively used to deal with human health and the surrounding environment harm. With the rapid development of China's economy, the production of hazardous waste will continue to show an increasing trend. However, with the adjustment of industrial structure, the improvement of environmental management capacity, technological progress, capital investment, etc., the rise of hazardous waste production will tend to slow.

4.1.2 Hazardous waste supervision is not in place. The problem of hazardous waste pollution has not been adequately addressed for a long time, and the concept of hazardous waste in the industry is relatively vague, and the reality of hazardous waste can not be accurately and comprehensively grasped. Resulting in the chaotic state of its governance process. The medical waste management is even more chaotic, and the existence of medical waste and garbage stacked at random phenomenon, resulting in cross-infection, endangering the health of the masses. These issues require long-term cooperation in various departments, cooperate with each other and actively communicate, from the hazardous waste disposal of all aspects of collaborative governance, to achieve its "whole process of governance".

4.1.3 Electronic product waste improperly. With the rapid development of science and technology, electronic products in our lives everywhere. At the same time, the extensive use of electronic products discarded, bringing a variety of environmental problems. First of all, a large number of e-waste will seriously pollute the soil, rivers, groundwater, the atmosphere, endangering human health; Secondly, a large number of electronic products produced by the exploitation of mineral resources mining, geological disasters continue to occur, but also undermine the ecological environment, resulting in shortages and even depletion of resources and a series of problems. A large number of hoarding of electronic waste has caused a serious waste of resources, because e-waste contains a large number of precious metals and organic materials with high recycling value. With the electrical and electronic products in science and technology, society, all aspects of life penetration, the number of discarded electrical and electronic products increased year by year, but a large number of waste electrical and electronic products have not been a reasonable recycling and utilization of China's electronic waste management status, should get enough attention to the relevant departments. With the rapid economic and social development and the continuous progress of science and technology, the upgrading of electrical and electronic products is very frequent, waste electrical and electronic products will greatly increase the amount of production. Due to the lack of effective governance and supervision, as well as the industry does not have strict access system, making practitioners of the scale of operation and professional level of uneven, more chaotic market order. Therefore, China's waste electrical and electronic need for effective regulation and regulation.

4.1.4 the legal system is not perfect. Our country is still lacking of the policy of hazardous waste management, which leads to the disposal of hazardous waste can not be an effective solution to the problem. In China, hazardous waste disposal is in the initial stage, the lack of practical experience, so that the relevant legal policy is not perfect, hazardous waste management does not enter the regularization track. China's existing laws and regulations system is not perfect, the lack of hazardous waste science and flexible management and disposal, there is a standard between the fault phenomenon.

4.1.5 Hazardous waste environmental assessment is not in place. Environmental impact assessment is an important means to control environmental pollution, but some environmental assessment units are not concerned about the content of hazardous waste, the types and quantity of hazardous waste investigation is not careful enough, and the enterprise self-built hazardous waste disposal facilities treatment capacity and effectiveness of the assessment did not make an accurate assessment of the feasibility of hazardous waste pollution treatment measures at the same time vague. So that environmental protection departments to the environmental regulation difficult, but also to environmental safety and health risks.

4.1.6 Lack of advocacy. In recent years, the publicity of water pollution and air pollution has aroused the attention of the society, but the promotion of hazardous waste is not much, appears stretched, lack of propaganda. It is precisely because the propaganda is not enough and the people vigilance is not high, so most people do not know what kind of waste is dangerous, but also do not understand the hazardous waste on the environment and the human body will be what kind of impact. When the pollution accident occurs, people can not take the correct treatment and disposal methods, and even take the wrong disposal of the disposal of serious consequences.

4.1.7 Hazardous waste disposal technology is backward. At present, most of the hazardous waste disposal facilities in our country are relatively backward, and the disposal waste is single and the utilization level is low. First of all, business owners are mostly self-employed just to maximize the interests of the pursuit of investment equipment, many of which are out of technology to deal with hazardous waste for simple disposal, there is a huge second hidden environmental pollution. Secondly, a large number of non-classified e-waste and household waste into the sanitary landfill, the hazardous waste disposal delay, hazardous waste disposal program is not reported to the environmental protection department for direct landfill, resulting in secondary pollution. This shows that the lack of hazardous waste disposal facilities and technology behind the hazardous waste management to face more severe challenges.

4.2. Recommendations

4.2.1. Strengthen the supervision and management capacity of hazardous wastes. Strengthen the supervision of hazardous waste capacity should start from the following points:
①Increase the degree of government attention, strengthen management philosophy. To establish a sound regulatory body of hazardous waste and thus ensure that grass-roots environmental protection

department staffing, increase the strength of law enforcement to ensure that law enforcement requirements for professional governance.
②Establish and improve the supervision mechanism, the implementation of accountability. Give full play to the NPC and CPPCC oversight mechanism, severely punished due to the neglect of waste management seriously ignored the serious consequences of the government, the main responsible person and hazardous waste management objectives to complete the responsibility of the poor people accountability, while opening up a wide range of monitoring Responsibility channels, to expand the scope of supervision, to draw a wide range of views.
③Give full play to the role of departments, the establishment of inter-departmental collaboration mechanism. Therefore, it is necessary to give full play to the functions of the relevant functional departments, establish inter-departmental linkage mechanisms such as commerce, industry and land, and integrate industry supervision with environmental management in an effective manner.
④Accelerate the process of local laws and regulations, detailed legal constraints.

4.2.2. Improve the whole process of hazardous waste management system. First, strengthen the source of hazardous waste supervision. In the course of the project, the demonstration of clean production should be done, the approval process of environmental impact assessment should be strictly controlled, and the environmental risk assessment should be done to reduce the project environmental risk. At the same time, the establishment of hazardous waste project information transfer system, to control the production process supervision, the establishment of production reporting system to ensure the smooth development of supervision and management. Secondly, increase the intensity of law enforcement and establish a long-term environmental monitoring and monitoring mechanism. Furthermore, the listing of the illegal system. Make full use of environmental pollution deadline management system, environmental violations listing system to strengthen the supervision and management of hazardous waste, the existence of major environmental issues issued deadline for corporate governance tasks, and accountability of the relevant responsible persons. Finally, to carry out industry and regional special management of hazardous waste pollution. In order to improve the understanding of pollution prevention and control of sewage sludge, strengthen the whole process of hazardous waste management system, and strengthen the hazardous waste management of the whole process of bad environment while vigorously improve the level of control of hazardous waste control.

4.2.3. To improve pollution prevention and control measures of hazardous waste. ①Promote cleaner production and recycling economy. Through political and economic measures to encourage enterprises to promote cleaner production processes, to minimize and prevent the generation of dangerous goods. ②Hazardous waste to achieve hierarchical management. The establishment of hazardous waste classification system, hazardous waste classification system through the hazardous waste is divided into known as the very limited and hazardous waste to facilitate the disposal of hazardous waste disposal, hazardous waste management more refined and more scientific and improve management efficiency and effectiveness . ③The establishment of hazardous waste exemption / priority system, to achieve the reduction of hazardous waste, resource, sound, to achieve environmental and economic benefits of the harmonization. ④To strengthen the solid waste prevention and control of waste management technical support basis. ⑤Smooth collection and disposal of hazardous waste approach. ⑥Pntroduction and development of new technologies. And actively learn from foreign advanced technology, and constantly improve the processing technology, and actively promote the disposal of industrial, to avoid the occurrence of secondary pollution.

4.2.4. To do a good job of social supervision and publicity. With the development of the Internet, information channels more smooth, more and more obvious role of social supervision, we should give full play to the role of media supervision of public opinion to strengthen supervision and management of hazardous waste publicity. ①The use of media publicity and supervision of public opinion to carry out the environmental protection of hazardous waste warning education, and guide the masses to carry out social supervision, and promote management integrity and effectiveness of construction. ② Promoting environmental education from the basic point We shall set up environmental education courses in kindergartens, primary schools, secondary schools, universities, graduate schools, and various training and Education, and promote environmental awareness education to all levels of education. ③Strengthen training, improve quality.

5. Conclusions

Hazardous waste disposal in China is still in the initial stage, therefore, in China has potential for development. At the same time the effective disposal of hazardous waste, in line with national development needs, to protect the ecological environment and human health. At the same time, in the process of hazardous waste disposal still need to learn from the technology of developed countries and related measures to form a line with China's national conditions relative to improve the hazardous waste management system to improve the efficiency of hazardous waste disposal, bringing greater economic and social value.

6. References

[1] Lijun Zhang 2013 Current situation and Countermeasures of hazardous waste management in the region. *J. Science and Technology Information* 36 135-136
[2] Gongming Zhou 2008 Study on the Status and Countermeasures of Hazardous Waste Control in China. *J. Environmental Protection in Energy* 2 12-14
[3] Qi Wang, Qifei Huang, Huabo Duan, Feng Liu ,Yangsheng Zhou and Cheng Xu 2006 Study on technical system for characterization of hazardous wastes in China. *J. Environmental Science Research* 5 165-179
[4] Lijun Zhang 2002 Environmental management and safe disposal of hazardous wastes: Basel Convention. *J. Chemical Industry Press* 1 22-23
[5] Shenglong Sun and Hongyan Tang 2009 Analysis of hazardous waste treatment technology and management. *J. China Environmental Management* 51 72-75
[6] Cheng Bi, Keqin Ren and Ping Yu 2006 Theory of circular economy research. *J. China Environmental Science Press* 2 13-14
[7] Guangren Qian 2009 International city solid waste legislation management and practice. *J. Chemical Industry Press* 4 18-22

Acknowledgements

This research is sponsored by the Opening Foundation of Tianjin University State Key Laboratory of Hydraulic Engineering Simulation and Safety (Grant No.HESS-1701); the Project of Tianjin Science and Technology Plan (Grant No. 16ZXFWGX00090); Tianjin science and technology major projects (Grant No. 16YFXTSF00270). This research is also sponsored by the Tianjin second batch of talent development special support plan (Safety disposal and resource recycling of hazardous waste innovative team)

Present Situation and Evaluation of Contaminated Soil Disposal Technique

XU Jin[1], CHEN Xing[1] and LI Meitong[2,*]

1 China School of Management, Tianjin University of Technology, Tianjin 300384, China
2 School of chemical engineering and technology, Tianjin University, Tianjin 300072, China
Email: xujinbox@163.com, 403472703@qq.com, * limeitong@tjut.edu.cn

Abstract. In recent years, because of the abuse of pesticides and fertilizers and the random discharge of sewage, the soil pollution is becoming more and more serious. At present, China's soil environment is grim. The treatment of contaminated soil is the focus of current research in agriculture, environment and ecology. Because of the different types of soil pollution and soil properties, and the complex process of contaminated soil, many factors need to be considered. Common soil remediation techniques include: Chemical remediation, bioremediation technology. The bioremediation technologies include: animal remediation technology, phytoremediation technology and microbial remediation technology. Animal remediation technology is the use of lower animals in the soil metabolic activity degradation of organic pollutants in soil conversion. Phytoremediation technology refers to the absorption, enrichment, degradation, transformation of soil contaminants, to achieve the purpose of soil control. At present, phytoremediation technology has been widely used, which can effectively control heavy metal contaminated soil. Microbial remediation technology is the use of soil microbial metabolic activity of the soil in the degradation of pollutants into non-toxic substances.

1. Introduction

In recent years, with the acceleration of industrialization in China, environmental pollution has become increasingly serious. The abuse of chemical fertilizers and pesticides, the random discharge of sewage and the unreasonable land development have caused soil pollution, and the pollution area has been increasing year by year. Land pollution is diversified and long-term, and there are many types of land pollution, including pesticide pollution, metal pollution and organic pollution. Compared with water pollution and air pollution, soil pollution is more hidden, andThe purification ability of the soil itself is poor, these factors increase the difficulty of repairing contaminated soil. To alleviate the situation of soil pollution in China, we should adopt effective measures to actively control the soil pollution. However, in the process of soil pollution control, various factors should be taken into account to avoid the expense of some of the environment at the expense of soil management. At the same time, we need to uphold the concept of "green sustainable restoration" to reduce the secondary pollution of the environment in the process of soil remediation. This paper describes the status of contaminated soil disposal technology and evaluation.

2. Chemical Remediation

Soil remediation is a complex and tortuous process. In China, soil remediation is facing many problems. Traditional soil remediation techniques are chemical / physical remediation techniques. The so-called chemical repair technology refers to the soil by changing or destruction of the chemical properties of pollutants in order to achieve the purpose of repairing the soil. Chemical repair technology mainly rinse repair technology, redox technology.

2.1. Chemical Leaching Repair Technology

Chemical leaching remediation technology is the use of water containing flushing aids or acid / alkaline solution into the contaminated soil, washing soil contaminants, and then dealing with waste liquid containing contaminants.

The key of this technology is to find an economical and practical eluent, not only can effectively remove all forms of pollutants, but also does not destroy the basic physical and chemical properties of soil [1]. In the process of soil remediation, the nature of contaminated soils, the selection of suitable leaching remediation conditions, and the nature of contaminants in the soil are all factors influencing the remediation of chemical leaching. Soil chemical properties such as soil texture, organic matter content and soil particle size distribution affect soil chemical leaching remediation techniques. Proper chemical leaching repairing operation conditions can not only achieve the purpose of soil remediation, but also reduce economic costs and minimize secondary pollution.

Sun et al. studied the effects of different eluents on heavy metal contaminated soils under different leaching conditions and summarized the leaching effects of different eluents on heavy metals in different forms[2]. It was pointed out that the natural organic acids have high degradation and low ecological toxicity, so they are the most promising. The results show that the use of highly efficient eluent and the selection of suitable process conditions can lead to the remediation of polluted soil, increase the economic benefit and reduce the secondary pollution to the soil.

2.2. Chemical Oxidation Reduction Technology

Chemical oxidation and reduction technology is to put the soil through the chemical oxidant or reducing agent, so that the soil with the chemical reaction of pollutants in order to achieve the purpose of repairing the soil [3]. The chemical oxidation process breaks down the chemical structure of the contaminants in the soil by utilizing the oxidation function of the oxidant, so that pollutants into low-‘’pollutants sensitive to the use of chemical reduction technology to repair is the current research focus. For example, in the repair of chromium-contaminated groundwater system by the process, the hexavalent chromium is reduced to trivalent chromium, so as to achieve the purpose of repair. At present, there are some problems in the degradation of chlorine-containing organic compounds by the dechlorination property of zero-valent iron, such as passivation of iron surface activity and polymerization failure due to adsorption of soil[4]. In order to avoid these problems New catalysts and surface activation technologies are required. For groundwater and soil contaminated by contaminants at the same time can take this technology to repair.

3. Bioremediation Technology

The use of chemical / physical remediation techniques to treat large areas of soil pollution is difficult. And for large areas of soil pollution if blindly to the traditional repair technology, the resulting secondary pollution may be far more than the pollution of the soil itself hazards. In this case bioremediation can be used. Compared with the traditional method, the bioremediation technology has the advantages of simple operation, low investment, high efficiency, good effect and environmental safety. Biological methods can be divided into animal remediation, phytoremediation, microbial remediation and bioremediation[5].

3.1. Animal Repair Technology

Animal remediation technology uses low-grade animals In the soil (Such as the use of earthworms, crab, filter-feeding shellfish, etc.)and their own microbial metabolism in vivo, transformation and degradation of soil contaminants[6], thereby repairing the soil. Zhou et al found in the study by the interaction of bacterial nematodes and micro-organisms can be achieved in the degradation of contaminated soil prometryne effect[7]. But it has its own limitations. At present, there are few reports on the research of animal remediation technology, and the research on the treatment of heavy metal contaminated soil by using lower animals in soil is still in the exploratory stage.

3.2. Phytoremediation Technology
Phytoremediation uses plants' normal metabolic activities to absorb, enrich, and accumulate, and to degrade soil contaminants[8], thereby repairing contaminated soils. The remediation ability of each plant is different for polluted soils, so there are some differences in the remediation methods for different contaminated soils. In recent years, Chinese researchers have found more than 300 kinds of plants that can enrich heavy metals such as copper, arsenic and nickel. Therefore, in practical application should be fully aware of the repair capacity of each plant in order to select the appropriate plant to repair the soil, to remove soil contaminants. At present, in the process of remediation of heavy metal contaminated soil by phytoremediation, the focus is to discover the plants that can be used to repair heavy metal contaminated soils, the mechanism of plant enrichment and the factors influencing plant uptake and enrichment characteristics. In addition, plants can be used to repair organic contaminated soils.

3.3. Microbial Remediation Technology
Microbial remediation technology through the use of microbial metabolic functions, degradation of soil contaminants. So as to achieve the purpose of repairing the soil. In the natural environment, there are many microorganisms in the soil, they can automatically degrade soil contaminants, but in recent years due to human abuse of pesticides and chemical fertilizers, destroying the automatic degradation of microorganisms. At present people through artificial cultivation technology, the use of microbial remediation of contaminated soil.

3.4. Combined Bioremediation Technology
The combined bioremediation technique utilizes two or more of the above methods to repair contaminated soil. One of the most commonly used is the combination of plant and microbial bioremediation technology. It is the use of plant and microbial complex system of absorption, collation, degradation of organic matter in the soil, plant repair and microbial repair can be organically combined. Plant and microbial remediation technology mechanism is the use of the plant's own metabolic process to absorb organic pollutants in the soil and accumulate, plant-generated secretions or enzymes can soil organic pollutants degradation, soil microbial own Metabolic activities can effectively degrade organic contaminants in soil.

4. Conclusion
Chemical remediation technology developed earlier, for pollution of soil, chemical repair technology has quick, short repair time advantage. However, there are some disadvantages such as high cost, high energy consumption and secondary pollution, so it is not suitable for treating large area soil pollution.

In recent years, the rapid development of bioremediation technology, in which phytoremediation and microbial remediation development of the most rapid. Phytoremediation technology is a low cost, environmentally safe and easy to produce secondary pollution remediation of soil methods, and has become a major technology to repair soil heavy metal pollution. Soil microorganisms can use their metabolic activities to degrade organic pollutants in the soil into harmless substances. Based on this principle, the soil microorganisms can be widely used in the remediation of soil organic contaminants. In short, compared with the traditional chemical repair technology, microbial remediation technology with low consumption, high efficiency, pure ecology, easy management and other advantages, has good prospects for development.

Traditional soil remediation techniques should be taken to avoid soil treatment at the expense of certain environments. In order to remediate contaminated soil, we need to uphold the concept of "green sustainable restoration", which needs to be considered in the remediation of contaminated soils. In addition soil pollution and groundwater pollution are closely linked, so in the repair of contaminated soil at the same time should avoid secondary pollution of groundwater.

5. References
[1] Lü Qingsong, Jiang Yufeng and Yang Fan 2010 Advances in research on leaching technology of heavy metal contaminated soil. *J. Gansu Agricultural Science and Technology* 3 33-37

[2] SUN Tao, LU Kouping and WANG Hailong 2015 Advances in remediation of heavy metal contaminated soil with different eluents and leaching conditions. *J. Journal of Zhejiang Agricultural and Forestry University* 1 140-125
[3] ZHOU Qiu-sheng, ZHANG Yong-kang and QU Xue-li 2012 Study on the mechanism of sodium acetate repairing chromium-contaminated soil. *J. Environmental Pollution and Control* 5 58-62
[4] Zhang W.X 2003 Nanoscale iron particles for environmental remediation: an overview. *J. Journal of Nanoparticle Reserch* 4 323-332
[5] Zhang Chi, Gu Zhenyu and Long Yuyang 2012 Phytoremediation mechanism, genetic defects and transgenic technology of PCBs contaminated soil. *J. Journal of Nuclear Agricultural Sciences* 7 094-1099
[6] ZHANG Wei, MA Jing-jing and ZHENG Bin 2014 Effects of multiple accumulation of soil B [a] P on glutathione-S-transferase activity in earthworms. *J. Journal of Anhui Agricultural University* 4 647-653
[7] Zhou J, Li X and Jiang Y 2011 Combined effects of bacterial-feeling nematodes and prometryne on the soil microbial activity. *J. Journal of Hazardous Materials* 3 1243-1249
[8] WANG Bo, LI Kairong and CUI Bixiao 2014 Responses of antioxidant defense system of Robinia pseudoacacia seedlings to oil pollution. *J. Mizuki Retention Research* 2 251-256

Acknowledgements
This research is sponsored by the Opening Foundation of Tianjin University State Key Laboratory of Hydraulic Engineering Simulation and Safety (Grant No.HESS-1701); the Project of Tianjin Science and Technology Plan (Grant No. 16ZXFWGX00090); Tianjin science and technology major projects (Grant No. 16YFXTSF00270). This research is also sponsored by the Tianjin second batch of talent development special support plan (Safety disposal and resource recycling of hazardous waste innovative team)

IOP Conf. Series: Earth and Environmental Science **178** (2018) 012029 doi:10.1088/1755-1315/178/1/012029

Present Situation and Evaluation of Contaminated Soil Disposal Technique

Xu Jin[1], Chen Xing[1], and Zhong Wei[1]

1 School of Management,Tianjin University of Technology,Tianjin 300384,China
Email: xujinbox@163.com, 403472703@qq.com, zhongwei@tjut.edu.cn

Abstract. In recent years, because of the abuse of pesticides and fertilizers and the random discharge of sewage, the soil pollution is becoming more and more serious. At present, China's soil environment is grim. The treatment of contaminated soil is the focus of current research in agriculture, environment and ecology. Because of the different types of soil pollution and soil properties, and the complex process of contaminated soil, many factors need to be considered. Common soil remediation techniques include: Chemical remediation, bioremediation technology. The bioremediation technologies include: animal remediation technology, phytoremediation technology and microbial remediation technology. Animal remediation technology is the use of lower animals in the soil metabolic activity degradation of organic pollutants in soil conversion. Phytoremediation technology refers to the absorption, enrichment, degradation, transformation of soil contaminants, to achieve the purpose of soil control. At present, phytoremediation technology has been widely used, which can effectively control heavy metal contaminated soil. Microbial remediation technology is the use of soil microbial metabolic activity of the soil in the degradation of pollutants into non-toxic substances.

1. Introduction

In recent years, with the acceleration of industrialization in China, environmental pollution has become increasingly serious. The abuse of chemical fertilizers and pesticides, the random discharge of sewage and the unreasonable land development have caused soil pollution, and the pollution area has been increasing year by year. Land pollution is diversified and long-term, and there are many types of land pollution, including pesticide pollution, metal pollution and organic pollution. Compared with water pollution and air pollution, soil pollution is more hidden, andThe purification ability of the soil itself is poor, these factors increase the difficulty of repairing contaminated soil. To alleviate the situation of soil pollution in China, we should adopt effective measures to actively control the soil pollution. However, in the process of soil pollution control, various factors should be taken into account to avoid the expense of some of the environment at the expense of soil management. At the same time, we need to uphold the concept of "green sustainable restoration" to reduce the secondary pollution of the environment in the process of soil remediation. This paper describes the status of contaminated soil disposal technology and evaluation.

2. Chemical Remediation

Soil remediation is a complex and tortuous process. In China, soil remediation is facing many problems. Traditional soil remediation techniques are chemical / physical remediation techniques. The so-called chemical repair technology refers to the soil by changing or destruction of the chemical properties of pollutants in order to achieve the purpose of repairing the soil. Chemical repair technology mainly rinse repair technology, redox technology.

2.1. *Chemical Leaching Repair Technology*
Chemical leaching remediation technology is the use of water containing flushing aids or acid / alkaline solution into the contaminated soil, washing soil contaminants, and then dealing with waste liquid containing contaminants.

The key of this technology is to find an economical and practical eluent, not only can effectively remove all forms of pollutants, but also does not destroy the basic physical and chemical properties of soil[1]. In the process of soil remediation, the nature of contaminated soils, the selection of suitable leaching remediation conditions, and the nature of contaminants in the soil are all factors influencing the remediation of chemical leaching. Soil chemical properties such as soil texture, organic matter content and soil particle size distribution affect soil chemical leaching remediation techniques. Proper chemical leaching repairing operation conditions can not only achieve the purpose of soil remediation, but also reduce economic costs and minimize secondary pollution.

Sun et al. studied the effects of different eluents on heavy metal contaminated soils under different leaching conditions and summarized the leaching effects of different eluents on heavy metals in different forms[2]. It was pointed out that the natural organic acids have high degradation and low ecological toxicity, so they are the most promising. The results show that the use of highly efficient eluent and the selection of suitable process conditions can lead to the remediation of polluted soil, increase the economic benefit and reduce the secondary pollution to the soil.

2.2. *Chemical Oxidation Reduction Technology*
Chemical oxidation and reduction technology is to put the soil through the chemical oxidant or reducing agent, so that the soil with the chemical reaction of pollutants in order to achieve the purpose of repairing the soil[3]. The chemical oxidation process breaks down the chemical structure of the contaminants in the soil by utilizing the oxidation function of the oxidant, so that pollutants into low-''pollutants sensitive to the use of chemical reduction technology to repair is the current research focus. For example, in the repair of chromium-contaminated groundwater system by the process, the hexavalent chromium is reduced to trivalent chromium, so as to achieve the purpose of repair. At present, there are some problems in the degradation of chlorine-containing organic compounds by the dechlorination property of zero-valent iron, such as passivation of iron surface activity and polymerization failure due to adsorption of soil [4]. In order to avoid these problems New catalysts and surface activation technologies are required. For groundwater and soil contaminated by contaminants at the same time can take this technology to repair.

3. Bioremediation Technology
The use of chemical / physical remediation techniques to treat large areas of soil pollution is difficult. And for large areas of soil pollution if blindly to the traditional repair technology, the resulting secondary pollution may be far more than the pollution of the soil itself hazards. In this case bioremediation can be used. Compared with the traditional method, the bioremediation technology has the advantages of simple operation, low investment, high efficiency, good effect and environmental safety. Biological methods can be divided into animal remediation, phytoremediation, microbial remediation and bioremediation [5].

3.1. *Animal Repair Technology*
Animal remediation technology uses low-grade animals In the soil (Such as the use of earthworms, crab, filter-feeding shellfish, etc.)and their own microbial metabolism in vivo, transformation and degradation of soil contaminants, thereby repairing the soil. Zhou et al found in the study by the interaction of bacterial nematodes and micro-organisms can be achieved in the degradation of contaminated soil prometryne effect. But it has its own limitations. At present, there are few reports on the research of animal remediation technology, and the research on the treatment of heavy metal contaminated soil by using lower animals in soil is still in the exploratory stage.

3.2. *Phytoremediation Technology*
Phytoremediation uses plants' normal metabolic activities to absorb, enrich, and accumulate, and to degrade soil contaminants, thereby repairing contaminated soils. The remediation ability of each plant

is different for polluted soils, so there are some differences in the remediation methods for different contaminated soils. In recent years, Chinese researchers have found more than 300 kinds of plants that can enrich heavy metals such as copper, arsenic and nickel. Therefore, in practical application should be fully aware of the repair capacity of each plant in order to select the appropriate plant to repair the soil, to remove soil contaminants. At present, in the process of remediation of heavy metal contaminated soil by phytoremediation, the focus is to discover the plants that can be used to repair heavy metal contaminated soils, the mechanism of plant enrichment and the factors influencing plant uptake and enrichment characteristics. In addition, plants can be used to repair organic contaminated soils.

3.3. Microbial Remediation Technology

Microbial remediation technology through the use of microbial metabolic functions, degradation of soil contaminants. So as to achieve the purpose of repairing the soil. In the natural environment, there are many microorganisms in the soil, they can automatically degrade soil contaminants, but in recent years due to human abuse of pesticides and chemical fertilizers, destroying the automatic degradation of microorganisms. At present people through artificial cultivation technology, the use of microbial remediation of contaminated soil.

3.4. Combined Bioremediation Technology

The combined bioremediation technique utilizes two or more of the above methods to repair contaminated soil. One of the most commonly used is the combination of plant and microbial bioremediation technology. It is the use of plant and microbial complex system of absorption, collation, degradation of organic matter in the soil, plant repair and microbial repair can be organically combined. Plant and microbial remediation technology mechanism is the use of the plant's own metabolic process to absorb organic pollutants in the soil and accumulate, plant-generated secretions or enzymes can soil organic pollutants degradation, soil microbial own Metabolic activities can effectively degrade organic contaminants in soil.

4. Summary

Chemical remediation technology developed earlier, for pollution of soil, chemical repair technology has quick, short repair time advantage. However, there are some disadvantages such as high cost, high energy consumption and secondary pollution, so it is not suitable for treating large area soil pollution.

In recent years, the rapid development of bioremediation technology, in which phytoremediation and microbial remediation development of the most rapid. Phytoremediation technology is a low cost, environmentally safe and easy to produce secondary pollution remediation of soil methods, and has become a major technology to repair soil heavy metal pollution. Soil microorganisms can use their metabolic activities to degrade organic pollutants in the soil into harmless substances. Based on this principle, the soil microorganisms can be widely used in the remediation of soil organic contaminants.

In short, compared with the traditional chemical repair technology, microbial remediation technology with low consumption, high efficiency, pure ecology, easy management and other advantages, has good prospects for development.

Traditional soil remediation techniques should be taken to avoid soil treatment at the expense of certain environments. In order to remediate contaminated soil, we need to uphold the concept of "green sustainable restoration", which needs to be considered in the remediation of contaminated soils. In addition soil pollution and groundwater pollution are closely linked, so in the repair of contaminated soil at the same time should avoid secondary pollution of groundwater.

5. References

[1] Linbo Qian, Miaoxin Yuan and Baoliang Chen 2012 Advances of Immobilized Microbial Technology for PAHS-contaminated Soil. *J. Chinese Journal of Environmental Science* 5 1767-1776

[2] Jiachun ZHANG, Shaoxia LIN and Qinghai ZHANG 2014 Pollution characteristics of heavy metals in soil and crops around Caohai wetland in Guizhou. *J. Mizuki Retention Research* 3 273-278

[3] Bo WANG, Kairong LI and Bixiao CUI 2014 Responses of antioxidant defense system of Robinia pseudoacacia seedlings to oil pollution. *J. Mizuki Retention Research* 2 251-256
[4] Qingsong Lü, Yufeng Jiang and Fan Yang 2010 Advances in research on leaching technology of heavy metal contaminated soil. *J. Gansu Agricultural Science and Technology* 3 33-37
[5] Tao SUN, Kouping LU and Hailong WANG 2015 Advances in remediation of heavy metal contaminated soil with different eluents and leaching conditions. *J. Journal of Zhejiang Agricultural and Forestry University* 1 140-125

Acknowledgements
Supported by The Project of Tianjin Science and Technology Plan (Grant No. 15ZLZLZF00410), the Project of Tianjin Science and Technology Plan (Grant No. 16ZXFWGX00090), Humanities and Social Sciences Foundation of Ministry of Education of China(Grant No. 12YJC630248), the Opening Foundation of Tianjin University State Key Laboratory of Hydraulic Engineering Simulation and Safety, Ecological & Environmental Governance Major project (Grant No. 16YFXTSF00270).

IOP Conf. Series: Earth and Environmental Science **178** (2018) 012030 doi:10.1088/1755-1315/178/1/012030

Human Alteration of the Nitrogen Cycle and Its Impact on the Environment

Jianing Hu

Thayer Academy, Braintree, MA, USA
E-mail: jianinghu0408@gmail.com

Abstract. Nitrogen(N) is a key element that controls the functions and dynamics of Earth's various ecosystems. In recent years, however, human activities, including the use of N fertilizers and the combustion of biomass and fossil fuels, have been injecting artificially transferred nitrogen into the atmosphere. Human alteration of the nitrogen cycle poses a threat to various aspects of the environment. Organisms in terrestrial and aquatic ecosystems used to lower nitrogen levels struggle with the new conditions; plant species richness and biodiversity are reduced due to excessive use of N fertilizers; reactive nitrogen in different forms negatively alter the ozone layer and exacerbate the greenhouse effect. Although most nitrogen inputs are serving human needs, the long-term environmental consequences are serious and should not be ignored.

1. Nitrogen Facts

All life depends on nitrogen, one of the most important elements on planet Earth, to survive. Nitrogen is the major component of DNA, RNA, and organic molecules like proteins.[1] Although the Earth's surface contains 78 percent nitrogen gas, most organisms cannot use it directly, as they do with oxygen and carbon dioxide. This is because the triple bonds between nitrogen atoms in a N2 molecule make it unreactive, while living organisms utilize relatively reactive nitrogen.[2] As a result, nitrogen is considered one of the essential limiting factors in the dynamics of our environment. However, the usable nitrogen in the atmosphere has increased and the nitrogen cycle altered significantly in the recent years due to human activities such as fertilizer production and fossil fuel consumption. To secure agricultural production, availability of resources, biodiversity, and functioning of ecosystems, and to minimize nitrogen's effects on climate change, immediate actions need to be taken.

2. Nitrogen as a Useful Resource

With the fast development of technology in the recent decades, people have had increasing knowledge about the element nitrogen. We have discovered nitrogen to be a useful resource and have utilized nitrogen both in our everyday life and industries.

2.1. Uses in Pharmaceuticals

The utilization of nitrogen is a widespread method adopted by many pharmaceutical industries. For example, nitrous oxide is used as an anesthetic, and nitrogen itself can be used to preserve biological specimens such as sperm, eggs and blood.

2.2. Metal Production

Nitrogen is a key component to prevent oxidation in the process of metal productions. For example, in the production of stainless steel, nitrogen is required during the process to strength the steel and to prevent corrosion.

2.3. *Blanketing Agent*
Due to nitrogen's unique property as an inert gas, it is often used to separate unstable elements from the contact with air, to prevent possible hazardous reactions.

2.4. *Preservation of Food and Beverage*
Because of nitrogen's extremely low temperature, it is an effective agent in the process of food cooling and freezing. Other advantages also include the reduction of food damage, oxidation and worsen-flavors during long-distance transportation.

2.5. *Pressurizing Gas as Chemicals*
During chemical processes, nitrogen can act as a propeller that helps liquids get through pipelines. It can also remove unstable organic chemical from the process.

2.6. *Fertilizers*
Nitrogen is a key nutrient needed by the plants, as a result, adding nitrogen to the soil supplements nutrition deficiency. [3]

3. The Nitrogen Cycle
As with other human-caused global change, before any further evaluation of the extensive human alteration on nitrogen's increasing amount and its effects on the environment, it is necessary for us to first understand a major process involving nitrogen, and a process to secure all living organisms - the nitrogen cycle. The nitrogen cycle is a process of "fixing" nitrogen and the movement of nitrogen in various forms throughout the atmosphere. In order for plants and animals to successfully incorporate nitrogen into their cell function, nitrogen must be "fixed", in other words, to be extracted from the atmosphere and to bond with hydrogen and oxygen to form inorganic compounds.[4] During the cycle, lightning accounts for some naturally occurring nitrogen: the electrical energy during a thunderstorm separate the nitrogen atoms in the atmosphere; the atoms then falls with rainwater to the soil, and acts as a natural fertilizer. While lightning transfers only a small portion of the total nitrogen amount, nitrogen fixation carried out by bacteria in the soil is the primary contributor. Bacteria such as those in the genus Rhizobium exists in a symbiotic relationship with the host plant, and they have the ability to transform the N2 molecules in the atmosphere into biologically available forms that is necessary in order for the plants to grow. Nitrogen produced by nitrogen-fixing bacteria is estimated to be 32-53 Tg/yr.[5] Animals which consume the plants and protein then become important nitrogen sources by passing the nitrogen through their waste. In addition, when the unconsumed plants decompose, they will transfer the nitrogen to the soil for reuse by other plants.[6]

4. Fertilizers' Contribution to the Nitrogen Cycle
The largest contributor to the sharply increasing nitrogen amount in the nitrogen cycle is N fertilizer. With knowledge about how beneficial the presence of nitrogen is to the development of crops, farmers eventually started to increase the amount of nitrogen when cultivating their land. However, resources available to farmer were very limited during the nineteenth and twentieth century. At first, farmers could only rotate lands between regular crops and those nitrogen-fixing plants, or add natural nitrogen sources like animal manure.

5. Amount of Nitrogen Produced in N Fertilizers
According to research done by James Galloway and his colleagues, whose work on nitrogen has been published in BioScience, humans were creating about 15 teragrams (Tg) of nitrogen per year in the 1860s.[7] At this time, however, German scientists Fritz Haber and Carl Bosch developed the Haber-Bosch process, a method that combines hydrogen and nitrogen under high temperature and pressure to produce large amount of ammonia, which is the essential component of nitrogen fertilizers. The highly effective fertilizer dramatically impacted human's ability to grow food, and it served as the "detonator of the population explosion".[8] Although Haber-Bosch process was able to boost crop production and the world population, it releases significant amount of chemically produced nitrogen into the atmosphere. Everyday, giant fertilizer factories around the world constantly perform the Haber-Bosch process to produce hundred million tons of fertilizers per year.[9] According to research lead by Peter

M. Vitousek, a scientist from the Department of Biological Sciences in Stanford University, Nitrogen fixation used in N fertilizers has increased exponentially from approximately zero in 1940. In addition, while the use of N fertilizers in developed countries has stabilized since 1970, applications in developing countries has been increasing significantly. The extra nitrogen that's being spread on farmlands around the world and forced into the nitrogen cycle would eventually pose a threat to the balance and stability of many of Earth's ecosystems.

6. Fossil Fuel Combustion's Contribution to the Nitrogen Cycle

The combustion of fossil fuel also contributes to the global reactive nitrogen load. It has long been known that burning fossil fuel releases nitric oxides into the air and leads to smog and acid rain, but the extent to which it impacts the environment has remained unclear in the past few decades. To further evaluate this issue, researchers from Brown University and University of Washington explored a new approach. In a study published in Science, these researchers were able to trace the nitrates back to its source: nitric oxide produced by burning fossil fuels. The study also analyzed nitrogen isotopes found in nitrates from a Greenland ice core, and concluded that the ice core contains nitrates from about 1718, which corresponds to the beginning of the Industrial Revolution when fossil fuel came into view. "What we find is there has been this significant change to the nitrogen cycle over the past 300 years," said Meredith Hastings, assistant professor of geological sciences at Brown and the paper's lead author. "So, we've added this new source — and not just a little bit of it, but a lot of it." The combustion of fossil fuel transfers about 20 Tg of reactive nitrogen from natural storage to the atmosphere each year.[10]

7. Total Amount of Nitrogen Injected by Human

According to Figure 1, a graph from a technical report published in Issues in Ecology in 1997, led by professor Peter Vitousek and his colleagues, other sources generates reactive nitrogen as well. For example, burning of biomass--combustion of organic matters-- transfers more than 40 Tg each year; the turning of forests and grasslands into agricultural use converts 20 Tg per year; draining wetlands allows organic materials in the soil to oxidize, which mobilizes 10 Tg/yr or more. Although these sources seem rather insignificant, with all of them function at the same time, there can be dramatic impacts. In fact, human activities inject around 140Tg in total into the nitrogen cycle per year.[11]

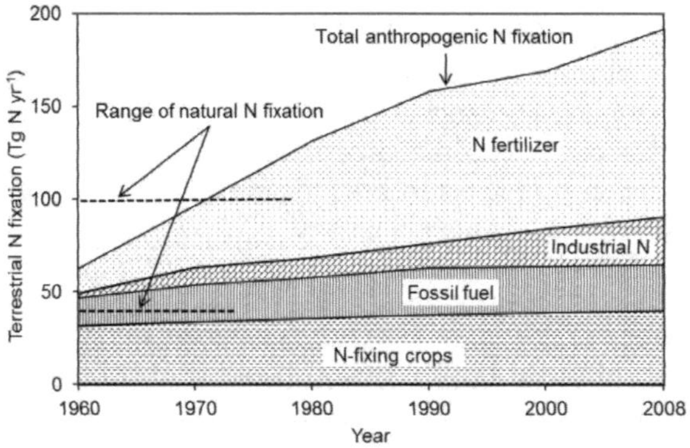

Figure 1. Amount of nitrogen injected into the nitrogen cycle by human activities [12]

8. Impacts on Plants

It is undeniable that the industrial uses of nitrogen have brought a series of advantages to our society, but as nitrogen level continue to rise, it poses a threat on agriculture, terrestrial and aquatic ecosystems, and human health. To begin with, large amount of nitrogen fertilizers can have negative effects on soil health, and can lead to decreased plant diversity. A team of scientists lead by Samuel M. Simkin studied more than 15,000 sites in the United States, and found that nitrogen deposition exceeded critical loads for loss of plant species richness in 24 percent of the sites. The result highlights that the overuse of nitrogen is affecting plant diversity across the United States.[13]

9. Impacts on Aquatic Ecosystems and Drinking Water

Furthermore, nitrogen produced by human activities is dramatically affecting the aquatic ecosystems. When nitrogen level rises in water, it often leads to algae overgrowth, which would harm water quality, food resources and habitats. When algae die and decompose in the water, the amount of organic materials increase. Oxygen is often required during this process, and the decreasing level of oxygen in water will threaten aquatic life.[14] In addition, high reactive nitrogen level can also infiltrate drinking water, as soluble nitrogen from fertilizers and animal gets into rivers, streams and underground water. Groups such as children or the elderly will be exposed to potential diseases when come in contact with polluted water or aquatic organisms. For example, high nitrate concentration can cause methemoglobinemia, also called “blue baby disease”, due to nitrate ions weakening blood’s ability to transfer oxygen. In fact, scientists have also found connections between nitrate and various cancers and reproductive problems.[15]

10. Impacts on Climate Change

Excessive nitrogen produced by human activities also has a great impact on the atmosphere and is connected closely to climate change. Different forms of nitrogen in the atmosphere include: N2O formed by nitrogen-fixing organisms, NOx produced by burning fossil fuel and biomass, and NH3 emitted by N fertilizers.[16] NOx and NHx (NH3 and NH4+) are able to accumulate in the troposphere, but they cannot return to the form of N2. As a result, nitrogen in the form of NOx and NHx in the atmosphere returns to ground level eventually. Combined with the volatile organic carbon compound, increasing NOx level will lead to increased ozone concentration, and enhance the greenhouse potential of the atmosphere.[17] In addition, N2O acts as a potent greenhouse gas as well, with each molecule absorbing outgoing radiation as much as 200 times more than carbon dioxide. Although reactive nitrogen increases ozone concentration in lower altitudes, it actually destroys ozone in higher altitudes. What happens in the stratosphere is that ultraviolet light breaks N2O apart, and forms NO molecules, which acts as a catalyst to break down the ozone layer. Unfortunately, destroying the ozone layer allows more ultraviolet light to enter the surface of the earth, and as a result, worsen the greenhouse effect.[12]

11. Conclusion

To sum up, human activities have been altering the nitrogen cycle in a significant way. Although the application of nitrogen has brought unprecedented improvements to our society, it’s overuse can have serious consequences. So, it is necessary for us to understand both side of the issue and to act wisely.

12. References

[1] [Special issue]. (n.d.). Current Issue. https://doi.org/10.1073/pnas.1515241113

[2] Blaszczak-Boxe, A. (2014, December 22). Facts about nitrogen. Retrieved August 12, 2017, from Livescience website: https://www.livescience.com/28726-nitrogen.html

[3] Charles, D. (2013, May). Fertilized world. National Geographic. Retrieved from http://ngm.nationalgeographic.com/2013/05/fertilized-world/charles-text

[4] Fields, S. (2004, July). Global nitrogen: Cycling out of control. Retrieved August 14, 2017, from National Center for Biotechnology Information website: https://www.ncbi.nlm.nih.gov/pmc/articles/PMC1247398/

[5] Five Applications of Nitrogen in Everyday Life. (2016, June 2). Retrieved August 13, 2017, from Linkedin website: https://www.linkedin.com/pulse/five-applications-nitrogen-everyday-life-industrial-oxygen

[6] Galloway, J. N. (2003, April 1). The nitrogen cascade. BioScience, 53(4), 341-356. Retrieved from https://academic.oup.com/bioscience/article/53/4/341/250178/The-Nitrogen-Cascade

[7] Glass, D. (2012, January 3). Lightning helps fertilize the soil. Retrieved August 13, 2017, from Moment of Science website: http://indianapublicmedia.org/amomentofscience/lightening-helps-fertilize-soil/

[8] Harrison, J. A. (2003). The nitrogen cycle: Of microbes and men. Retrieved August 12, 2017, from Visionlearning website: http://www.visionlearning.com/en/library/Earth-Science/6/The-Nitrogen-Cycle/98

[9] Interesting facts about nitrogen element. (n.d.). Retrieved August 13, 2017, from Worlds of Chemicals website: http://www.worldofchemicals.com/436/chemistry-articles/interesting-facts-about-nitrogen-element.html

[10] Kahl, H. (2004, September 1). Role and importance of nitrogen in your soil. Retrieved August 13, 2017, from OrganicNZ website: https://organicnz.org.nz/magazine-articles/role-importance-nitrogen-soil/
[11] Lamb, J. A., Fernandez, F. G., & Kaiser, D. E. (2014). Understanding nitrogen in soils. Retrieved August 13, 2017, from University of Minnesota Extension website: https://www.extension.umn.edu/agriculture/nutrient-management/nitrogen/understanding-nitrogen-in-soils/
[12] New proxy reveals how humans have disrupted the nitrogen cycle. (2009, June 4). Retrieved August 14, 2017, from Brown University website: https://news.brown.edu/articles/2009/06/nitrogen
[13] Nitrogen. (n.d.). Retrieved August 13, 2017, from Praxair website: http://www.praxair.co.in/gases/buy-liquid-nitrogen-or-compressed-nitrogen-gas/?tab=industries
[14] Nutrient pollution. (n.d.). Retrieved August 15, 2017, from United States Environental Protection Agency website: https://www.epa.gov/nutrientpollution/problem
[15] Schmidhuber, J. (n.d.). Most influential persons of the 20th century. Retrieved August 14, 2017, from http://people.idsia.ch/~juergen/haberbosch.html
[16] Vitousek, P. M. (1997). Human alteration of the global nitrogen cycle sources and consequences. Retrieved from https://era.library.ualberta.ca/files/9880vs143/EA_7_1997_737.pdf
[17] Vitousek, P. M. et al. Human alteration of the global nitrogen cycle: sources and consequences. Ecol. Appl. 7, 737–750 (1997).

IOP Conf. Series: Earth and Environmental Science **178** (2018) 012031 doi:10.1088/1755-1315/178/1/012031

Survey on the Effects of Frequency of Experiencing Water Environments on Moods in China

C C Peng[1], M Rooks[2] and Y Hu[1]

1 Psychological Center, Dalian University of Technology, Liaoning, China
2 Graduate School of Maritime Sciences, Kobe University, Hyogo, Japan
E-mail: pengchenchen2125@163.com

Abstract. This study describes findings of the impacts of water area on environmental factors including ocean, rivers, lakes and so on, and how these affected human mood in China. We used a short version of Sakano et al.'s Mood Inventory to measure tension and excitement, fatigue, depressive, anxious, and refreshed moods. 207 people (121 males and 86 females) were randomly selected from a Chinese university, and categorized into three groups according to different frequencies (high, medium, and low frequencies) of experiencing water area environment views. We found that a high frequency of experiencing water environment views had a more significant effect on mood than did medium and low frequencies. There were vast differences in scores on the subscales assessing depressive and refreshed moods among the three frequencies, and (2) for males, the differences in the mood inventory scores among the high, medium, and low frequencies were more pronounced and significant than they were for females. The effects of water environment on mood had large variations and substantial effects, especially on depressive and refreshed moods. These findings may provide new insights for future living environment and design guidelines for structuring more water-enriched environments.

1. Introduction

"A quarter of individuals suffer from mental or behavioural disorders both in developed and developing countries" reported in the World Health Organization in 2001 [1]. Many studies that have evaluated similar aspects have frequently shown that natural environments play an important role in mood [2]-[5]. Living in a natural environment was associated with fewer stress-related illnesses [6] [7]. On the basis of these findings, Peng et al. [8] confirmed that compared to Japanese residents who lived inland, those who lived by the seaside showed higher positive psychological effects. Furthermore, Peng et al. [9] reported that a high degree of enthusiasm for beach-going has a greater significant effect on mood and mental health than did lower degrees of enthusiasm for beach-going in Japan. Ashbully et al. [10] reported that approximately 15 families had a major improvement in health and well-being that was related to being in beach environments. The qualitative and quantitative relaxing effects of living by the ocean were confirmed by visits to the beach [11]. In general, ocean views are believed to influence humans and result in positive psychological effects such as calmness and feeling relaxed [12] [13].

According to this theory, aquatic settings showed positive impacts on individuals' well-being compared to those without exposure to such surroundings [14]. In particular, environments with water – positively referred to as natural "blue spaces" – are associated with higher preferences, greater positive effects and higher perceived restorativeness [15]. Tajima et al. [16] reported that people who

live near a waterfront area had positive, high moods. In general, waterfront areas have come to be utilized more and more by an increasing number of people for feeling the freedom and carefree.
Depression and negative moods may be outcomes of the psychosocial stress from living in adverse environments that can increase the risk for poor health outcomes and other diseases. As more and more individuals are moving to and living in cities, the mental health burden of illness is rising. This raises the question of how to prevent, mitigate or reduce the effect of urban stress on individuals living in cities. The Chinese government was making great efforts to strengthen the water environment; however, no studies had evaluated the effects of water treatment on human mood by water environment in China. The present article made an attempt to examine the correlation between water environment and mood in China. We premised our investigation on the following two hypotheses: (1) that there were mutually reinforcing linkages between water area environment and human mood in China; (2) that for both males and females, water environment had a positive influence on mood. Furthermore, in males, there was a higher significant influence than females. We aimed to evaluate three groups by university students' frequency (high frequency, medium frequency, and low frequency) of water area environment views to investigate differing effects on mood, and to examine the three frequencies of water area environment experiences on the basis of gender.

2. Methods

Subjects
207 people (121 males and 86 females) were recruited from university students in China to participate in this survey. The subjects were categorized into three groups by their frequency of water area environment views: (1) high frequency (more than ten times a year); (2) medium frequency (less than ten times a year, more than once a year); (3) low frequency (once a year or not at all). We explained the aim of the study to the participants. In this questionnaire answers, there was no right and wrong, good or bad to any of the items.

Questionnaire
In this study, the self-administered questionnaire included questions about socio-demographics (gender, age and birthplace), the frequency of water area environment views (e.g., ocean views, river views, lake views, etc.) and a mood inventory.
We used a short version of the original Mood Inventory developed by Sakano et al. [17]. The mood inventory is a self-administered questionnaire designed to measure an individual's state of mood. The questionnaire had five subscales: tension and excitement, fatigue, depressive mood, anxious mood, and refreshed mood, and each subscale comprised three items. The participants were required to respond each item on a 4-point Likert scale (1 = Strongly Disagree, 2 = Disagree, 3 = Agree, and 4 = Strongly Agree). The score on each subscale was calculated by taking the mean of the item responses.

Statistical Analysis
Questionnaires with missing or incomplete data were excluded from analysis. In our statistical analysis, the three groups of high frequency, medium frequency, and low frequency of water area environment views were examined, and the gender associated with different frequencies of water area environment views were also calculated.
The data were entered and analyzed using the Statistical Package for Social Sciences (SPSS, 2007) software. A one-way analysis of variance (ANOVA) and Tukey HSD post-hoc analysis for multiple comparisons were conducted in order to summarize and interpret the descriptive data. Statistical significance was set at $p < .05$, $p < .01$.

3. Results and Discussion

We received completed questionnaires from 207 participants, 121 males and 86 females. The proportion of frequency of water area environment views: (1) 65 participants who having water area environment views more than ten times a year were classified as the high frequency group; (2) 129 participants who having water area environment views less than ten times and more than once a year were classified as the medium frequency group; (3) 13 participants who having water area environment views once a year or not at all were classified as the low frequency group.

3.1. The effect of three frequencies of experiencing water environment views on moods

The data in Table 1 show that there were significant differences in increase ($p < .05$) in scores between the high and medium frequency groups on the depressive mood and refreshed mood subscales. There were no significant differences with regard to variations in the participants' scores between the high and medium frequency groups on the tension and excitement ($p = 0.89$), fatigue ($p = 0.25$), and anxious mood ($p = 0.61$) subscales. Between the high and low frequency groups, there were strong significant differences in crease ($p < .01$) in scores on the depressive mood and refreshed mood subscales, and significant differences in crease ($p < .05$) in scores on the fatigue and anxious mood subscales. No significant difference was found with regard to tension and excitement ($p = 0.89$). Moreover, between the medium and low frequency groups, only refreshed mood subscale had a significant difference ($p < .05$). There were no significant differences in term of the tension and excitement ($p = 0.97$), fatigue ($p = 0.81$), depressive mood ($p = 0.88$), and anxious mood ($p = 0.48$). Figure 1 show that there was a tendency for the average mood scores to improve as the frequency of experiencing water area environment increased. The mean scores on the fatigue, depressive mood and anxious mood subscales showed an ascending trend, and the refreshed mood scores showed a descending trend among the high, medium, and low frequencies groups; in other words, fatigue, depressive mood, anxious mood and refreshed mood improved for those who had a higher frequency of water environment. Obviously, the participants with high frequency of experiencing water area environment showed stronger significantly than did the lower frequency of experiencing water environment, on the depressive mood, refreshed mood, fatigue and anxious mood. Specifically, from among all the subscales, vast differences in scores were obtained on the depressive mood and refreshed mood. This suggests that a high frequency of water area environment had a greater influence on participants' moods than did a lower frequency of water environment.

Table 1. Three frequencies of having water environment views
Means, SE, and p-values from the ANOVA and Tukey HSD post-hoc analysis.

Subscales	High-Medium			High-Low			Medium-Low		
	M1	**SE1**	***p***	**M2**	**SE2**	***p***	**M3**	**SE3**	***p***
Tension and excitement	0.16	0.08	0.89	0.02	0.10	0.89	-0.05	0.01	0.97
Fatigue	-0.15	0.09	0.25	-0.21	0.10	*	0.04	0.13	0.81
Depressive mood	-0.24	0.11	*	-0.41	0.10	**	-0.17	0.11	0.88
Anxious mood	-0.13	0.11	0.61	-0.21	0.12	*	-0.19	0.21	0.48
Refreshed mood	0.26	0.12	*	0.39	0.14	**	0.26	0.17	*

* $p < .05$, ** $p < .01$ M1 = M (high) – M (medium) M2 = M (high) – M (low) M3 = M (medium) – M (low)

It is important to note that people who had higher frequency of experiencing water environment had reduced odds of depression mood and mental distress. Depression mood may be outcomes of the psychosocial stress from living in complex environments that can increase the risk for poor health outcomes and other diseases. People in a relax environment tend to be in a refreshed mood, and have better mental health. Water area environments including pleasant blue sea, calm lake, sunshine waterfront, and these views can enhance the feeling of being comfortable. Many Chinese people would like to go to the ocean, river, or lake to do some sports, such as swimming, fishing, and boating. These types of sports make their time much more sacred as well as enjoyable. In addition, many Chinese people would like to live on or near the sea, because the sea gives them some sense of openness and freedom. In this study, it has been confirmed that water area environment promoted

Chinese people's moods. This survey suggests that the more frequently Chinese people have the experiencing of water environment, the better their improving moods and mental health.

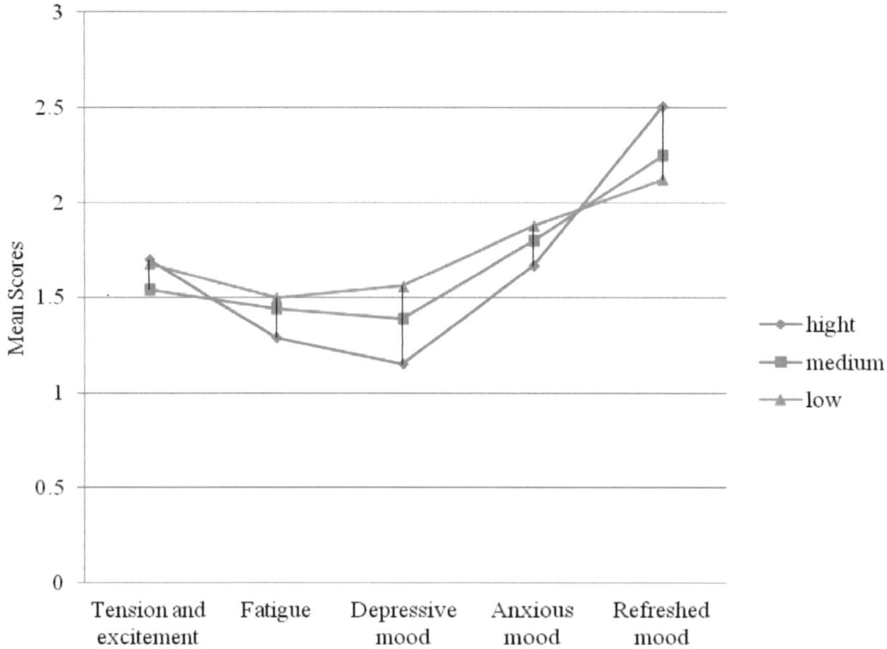

Figure 1. Mean scores for moods influenced by the three frequencies of having water environment views

Associating gender and three frequencies of experiencing water environment views on moods

The data in Table 2 present the results pertaining to the association between three groups by the frequency of water area environment views and gender. In term of males, there were significant differences ($p < .05$) on depressive mood and refreshed mood subscales between the high and medium frequency groups, and no significant differences on the tension and excitement ($p = 1.00$), fatigue ($p = 0.26$), and anxious mood ($p = 0.15$). Regarding the differences between the high and low frequency groups, there were strong significant differences in crease ($p < .01$) on the depressive mood and refreshed mood subscales, and significant differences in crease ($p < .05$) on the fatigue and refreshed mood. With regard to the mean tension and excitement ($p = 0.11$) was no significant difference. In the case of the medium and low frequency groups, tension and excitement ($p = 0.32$), fatigue ($p = 0.21$), depressive mood ($p = 0.11$), anxious mood ($p = 0.71$) and refreshed mood ($p = 0.21$) had no significant differences.

In term of females, the only significant difference ($p < .05$) was with regard to the depressive mood between the high and low frequency groups. With regard to the tension and excitement ($p = 1.00$), fatigue ($p = 0.55$), depressive mood ($p = 0.87$), anxious mood ($p = 0.15$) and refreshed mood ($p = 1.01$) were no significant differences between the high and medium frequency groups. Moreover, between the high and low frequency groups, tension and excitement ($p = 0.13$), fatigue ($p = 0.99$), anxious mood ($p = 0.16$) and refreshed mood ($p = 0.70$) were no significant differences. With regard to the tension and excitement ($p = 0.11$), fatigue ($p = 0.71$), depressive mood ($p = 0.21$), anxious mood ($p = 0.55$) and refreshed mood ($p = 0.69$) were no significant differences between the medium and low frequency groups. In particular, the mean mood scores of females were lower than males.

Figure 2 and Figure 3 show that the frequency of experiencing water area environment affected males' mood, and didn't affect females' mood. Both for males and females, among the high-medium frequencies, high-low frequencies, and medium-low frequencies of water area environment, only depressive mood had a significant effect. In term of males, the changes in scores for fatigue, depressive mood, and anxious mood showed an ascending trend, and the scores for refreshed mood showed a descending trend as the frequency of experiencing water area environment increased. However, in term of females, there was no obvious change as the frequency of experiencing water area environment increased.

Table 2. Three frequencies of having water environment views and gender Means, SE, and p-values from the ANOVA and Tukey HSD post-hoc analysis.

	Subscales	High-Medium			High-Low			Medium-Low		
		M1	SE1	*p*	M2	SE2	*p*	M3	SE3	*p*
Males	Tension and excitement	0.15	0.25	1.00	0.33	0.25	0.11	0.18	0.25	0.32
	Fatigue	-0.21	0.25	0.26	-0.44	0.16	*	0.20	0.17	0.21
	Depressive mood	-0.43	0.14	*	-0.69	0.14	**	0.32	0.18	0.11
	Anxious mood	-0.35	0.18	0.15	-0.40	0.19	*	0.19	0.28	0.71
	Refreshed mood	0.43	0.22	*	0.57	0.22	**	0.22	0.22	0.21
Females	Tension and excitement	-0.01	0.14	1.00	-0.36	0.17	0.13	-0.34	0.20	0.11
	Fatigue	0.07	0.18	0.55	-0.03	0.23	0.99	-0.13	0.13	0.71
	Depressive mood	-0.04	0.23	0.87	-0.41	0.43	*	-0.28	0.20	0.21
	Anxious mood	0.21	0.17	0.71	-0.29	0.17	0.16	-0.39	0.16	0.55
	Refreshed mood	0.28	0.16	1.01	0.25	0.37	0.70	0.23	0.26	0.69

* $p < .05$, ** $p < .01$ M1 = M (high) – M (medium) M2 = M (high) – M (low) M3 = M (medium) – M (low)

The gender-based prevalence of moods was compared among the three frequencies of water area environment. In China, as more and more individuals are moving to and living in cities, the moods and mental health burden of illness is rising. This raises the question of how to reduce the effect of urban stress on individuals living in cities as the populating continues to rise. Having water environment like ocean, rivers and lakes, to be the most relaxing of all types of space, followed by looking out of windows at natural spaces. Chinese men engaged in more sports such as fishing, swimming, and surfing, which improved their mental health and released their stress. Males who enjoyed engaging in different types of sports had better moods as well as mental health. It could also be that males with more time and money to visit the beach are happier. For females, large quantities of Chinese women need to work, raise children, and do housework, so they couldn't really concentrate on relaxation. Even though, Chinese females are on a scenic and beautiful waterfront, it is difficult to unwind and let go. Furthermore, a lot of Chinese women don't like sports, they prefer to stay at home, watch TV rather than do some sports outside. So they couldn't experience the fun of swimming, boating and fishing. In sum, the main finding revealed that the water area environments could improve males' moods much more than it could in the case of the females.

IOP Conf. Series: Earth and Environmental Science **178** (2018) 012031 doi:10.1088/1755-1315/178/1/012031

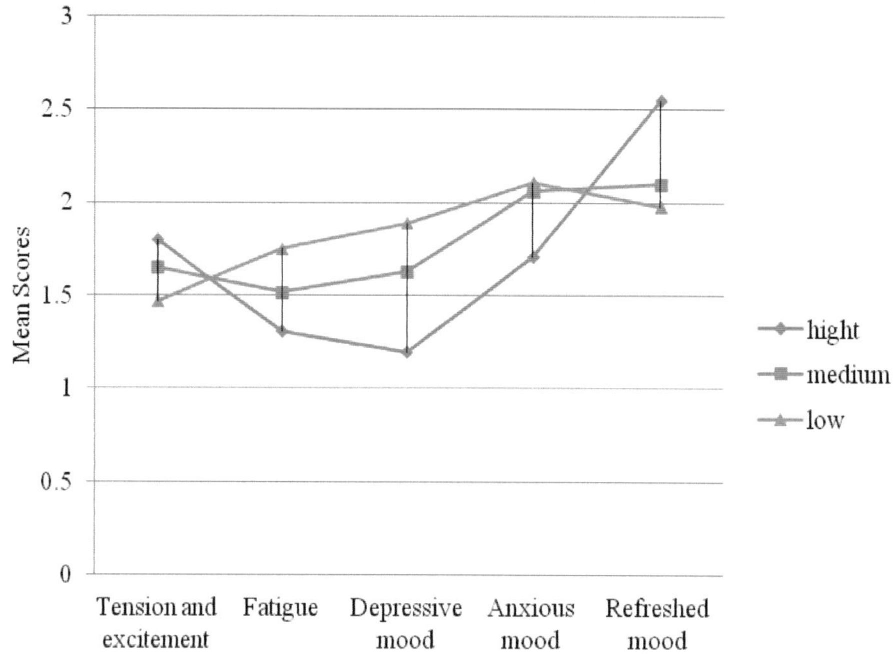

Figure 2. Mean Scores of males for moods influenced by the three frequencies of having water environment views

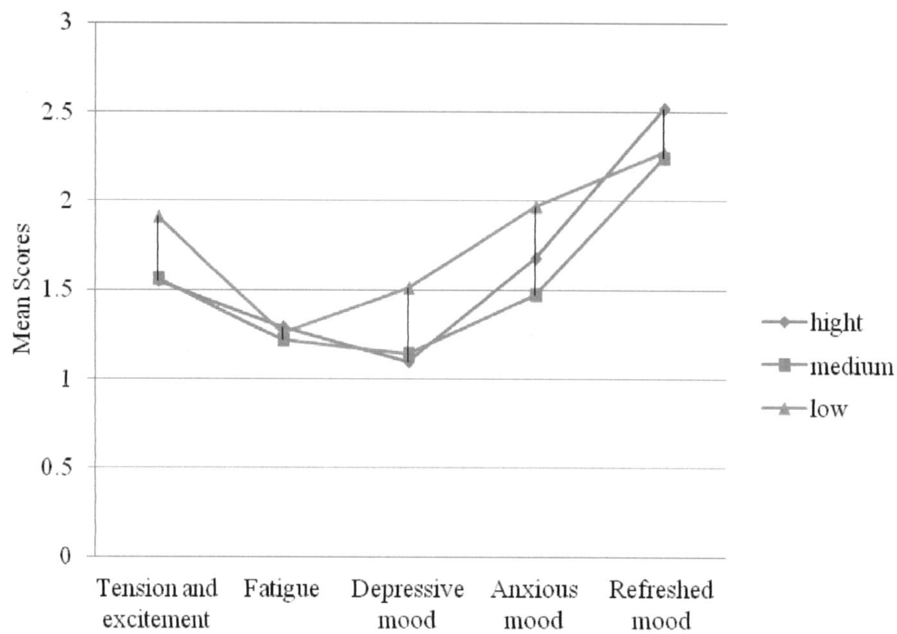

Figure 3. Mean Scores of females for moods influenced by the three frequencies of having water environment views

4. Conclusions

The results of our study suggested that water environment views improved human mood in China. In particular, the impact of the water environment on mood was more pronounced for men than for women. We found a significant gender difference with regard to the effects of the water environment on mood. Our results were as follows:

1. From among the three frequencies of having water environment views, a high frequency of experiencing water environment views had a more significant effect on mood than did medium and low frequencies. In particular, there were vast differences in scores on the subscales assessing depressive and refreshed moods the subscales measuring among the three groups with varying frequencies. As a result, it can be considered that the frequency of experiencing water environment improves the mood of Chinese people.
2. The findings also indicate that in both men and women, the water environment has a positive influence on mood. Furthermore, for males, the differences in the mood inventory scores among the high, medium, and low frequencies were more pronounced and significant than they were for females. The effects of water environment on mood had large variations and substantial effects, especially on depressive and refreshed moods. For women, the difference among the high, medium, and low frequencies with regard to mood status was slight and almost not significant.

Unsurprisingly, water environment was positively associated with human mood, and the depressive and refreshed moods had significant corresponding impacts on mood. The effects were stronger for males. Experiencing water environments led to improved mood among Chinese people. The results of this study offer compelling evidence to support the effect of water environment on mood in China. These finding may provide new insights for future living environment and design guidelines for structuring more water environments.

5. References

[1] World Health Organization WHO 2001 World Health Report Geneva *WHO*

[2] Rutter M 2005 How the environment affects mental health *The British Journal of Psychiatry* 186 4-6

[3] Yamazaki S, Inatani F and Nonaka M 2010 Comparison in tendency and frequency of reminiscence between the elderly and the middle-aged relation between quality of reminiscence and psychological well-being *Kurume University Psychological Research* 9 57-61

[4] Barton J and Pretty J 2010 What is the best dose of nature and green exercise for improving mental health? A multi-study analysis *Environmental Science & Technology* 44 3947-3955

[5] White MP, Pahl S, Ashbullby K, Herbert S and Depledge MH 2013 Feelings of restoration from recent nature visits *Journal of Environmental Psychology* 35 40-51

[6] Kawakubo A, Kasahara R and Oguchi T 2015 Effects of natural environmental picture and sound on relieving stress *Rikkyo Psychological Research* 57 11-19

[7] Thompson Coon J, Boddy K, Stein K, Whear R, Barton J and Depedge M 2011 Does participating in physical activity in outdoor natural environments have a greater effect on physical and mental wellbeing than physical activity indoors? A systematic review *Environmental science & technology* 45 1761-1772

[8] Peng CC, Yamashita K and Kobayashi E 2016 Effects of the coastal environment on well-being *J. Coast Zone Manag* 19 1-7

[9] Peng CC, Yamashita K and Kobayashi E 2015 An empirical study on the effects of the beach on mood and mental *J. Coast Zone Manag* 18 1-6

[10] Ashbullby KJ, Pahl S, Webley P and White MP 2013 The beach as a setting for families' health promotion A qualitative study with parents and children living in coastal regions in Southwest England *Health & Place* 23 138-147

[11] Miyamoto T, Shigeeda M, Sugiyama S, Oshima A and Togashi M 2001 Measurement of Relaxing Effects from a Visit to the Beach *Journal of International Society of Life Information Science* 19 38-46

[12] Wheeler BW, White M., Stahl-Timmins W and Depledge MH 2012 Does living by the coast improve health and wellbeing? *Health & Place* 18 1198-1201

[13] White MP, Alcock I, Wheeler BW and Depledge MH 2013 Coastal proximity, health and well-being results from a longitudinal panel survey *Health & Place* 23 97-103

[14] Wherrett JR 2000 Creating landscape preference models using internet survey techniques *Landscape research* 25 79-96

[15] White M, Smith A, Humphryes K, Pahl S, Snelling D, et al. 2010 Blue space The importance of water for preference, affect, and restorativeness ratings of natural and built scenes *Journal of Environmental Psychology* 30 482-493
[16] Tajima K, Watanabe H and Kuroyanagi A 1997 A study on effect of urban waterfront area in high density living environment *Journal of Architecture, Planning, and Environmental Engineering* 4 277-284
[17] Sakano Y, Fukui T, Kumano H, Horie H, Kawahara K, et al. 1994 Development and validation of a new mood inventory *Jpn J Psychosom Med* 34 629-636

Acknowledgments

This work was financially supported by the Fundamental Research Funds for Dalian University of Technology (No. DUT16RC(3)107).

Maximum and Minimum Temperature in Naiman County, China

Shuxia Yao and Chuancheng Zhao

1 Lanzhou City University, Lanzhou 730070, China
E-mail:zhao_chch1978@163.com

Abstract. The trends of mean temperature, maximum temperature (Tmax) and minimum temperature (Tmin) data from Naiman National automatic meteorological station were analyzed on annual, seasonal and monthly scales over the 1970-2013 periods. Results indicated: (1) The average annual mean temperature, Tmax and Tmin in Naiman County was 7.1°C, 13.5°C and 1.6°C, respectively, and average annual mean temperature, Tmax and Tmin all increased, at a rate of of 0.29°C/10a, 0.24°C/10a and 0.28°C/10a, respectively, which were all higher than global nearly 50 years (0.13°C/10a); (2) The average annual mean temperature of spring, summer, autumn and winter was 8.3°C, 22.6°C, 7.7°C and -10.1°C, respectively; The increasing speed of mean temperature, Tmax and Tmin were all most obviously at spring, at a rate of 0.38°C/10a, 0.3°C/10a and 0.38°C/10a, respectively; (3) The highest and lowest mean temperature was obtained at July (23.9°C) and January (-12.3°C), respectively; The highest and lowest Tmax was obtained at July (23.9°C) and January (-5.9°C), and the highest and lowest Tmin was obtained at July (19.1°C) and Dec. (-14.3°C).

1. Introduction

Air temperature as one of the most important variables in climate research and global change, controls many biological and physical processes between the hydrosphere, atmosphere and biosphere[1]-[2]. The analysis of long-term trends of air temperature is essential for the assessment of the impacts of climate variability and change in a given area[3]. The interest in assessing climate changes using air temperature resides not only in mean temperature, but also maximum and minimum temperatures (Tmax and Tmin), because changes in mean temperature are caused by changes in either Tmin or Tmax, or both[4]-[6].

Naiman County is located in the eastern part of Inner Mongolia, China. This area has become one of the most severely affected by desertification in northern China due to long-term overgrazing, reclamation, gathering of fire wood, and excessive utilization of water resources[7], and are generally thought to originate from sand and dust storms that occur frequently in the arid and semi-arid regions of northern China[8]-[9].

The main objective of this study was to provide a brief trend of mean temperature, Tmax and Tmin at the scales of monthly, seasonal and annual in Naiman County, and original data collected from Naiman National automatic meteorological station with continuous record over the period 1970-2013.

2. Materials and Methods

2.1. Study Area Description

The study was conducted in the eastern part (42°55′N, 120°42′E; 360 m elevation) of Inner Mongolia, China (Figure 1). The climate is temperate, semi-arid, continental, and monsoonal. The area receives

357 mm of precipitation annually. Mean annual potential pan-evaporation is approximately 1935 mm. The mean annual temperature is 6.8°C. The annual frost-free period is 130–150 days. The average annual wind speed is 3.4 m s^{-1}[9]. Soils are sandy, light yellow, have loose structure, and are particularly susceptible to wind erosion[7]. Moving dunes are the most typical feature in this area; depth to groundwater is between 12 and 15 m, therefore, groundwater is unable to supply surface soil water.

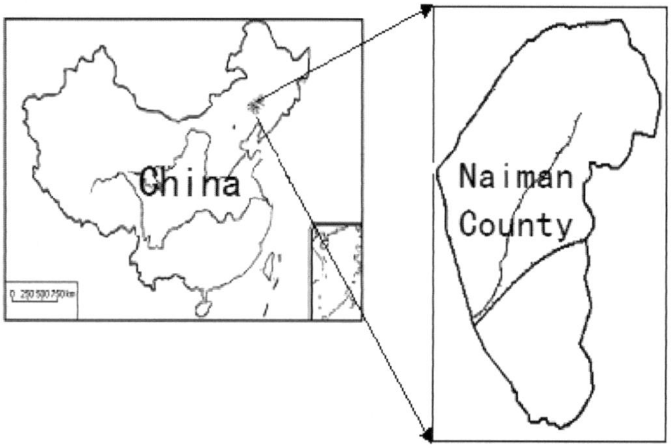

Figure 1. Locations of study area

2.2. Data

All data used hereby were extracted from the Naiman National automatic meteorological station. This station with continuous data record was involved in the analysis. The analysis was applied on the mean temperature (℃); mean maximum temperature (Tmax) (℃); mean minimum temperature (Tmin) (℃), for the 1970-2013 intervals.

2.3. Methods

Temperature (y) changed with year was regressed by the linear equation: $Y = at + b$, where: t is the year; $A=a*10$ was the trend rate of temperature. The value of A can describe the variation of temperature series, and a positive (negative) A represents linear increasing (decreasing) of the temperature in the statistical time[10]. Trends of significance test chose F-test method. The trend analysis and F-test were completed by Orgin8.0 software.

3. Results and Discussion

3.1. Annual Temperature

The average annual mean temperature, Tmax and Tmin in Naiman County was 7.1℃, 13.5℃ and 1.6℃, respectively (Table 1). The coefficient of variation of annual temperature was smallest for Tmax (CV=0.06), next was mean temperature (CV=0.1), and the coefficient of variation of Tmin was bigger (CV=0.4).

Table 1. Statistical characteristic of average annual mean temperature, Tmax and Tmin in annual, different months and different seasons.

Month	Mean temperature			Tmax			Tmin		
	Mean	SD	CV	Mean	SD	CV	Mean	SD	CV
Jan.	-12.32	2.17	-0.18	-5.90	2.34	-0.40	-17.13	2.13	-0.12
Feb.	-8.37	2.40	-0.29	-1.55	2.75	-1.78	-13.82	2.27	-0.16
Mar.	-0.91	2.07	-2.28	5.80	2.17	0.37	-6.72	2.13	-0.32
Apr.	9.07	1.89	0.21	15.83	2.36	0.15	2.70	1.48	0.55

May	16.66	1.19	0.07	23.26	1.47	0.06	10.16	1.00	0.10
Jun.	21.46	1.24	0.06	27.55	1.52	0.06	15.74	0.97	0.06
Jul.	23.89	1.11	0.05	29.28	1.38	0.05	19.06	0.89	0.05
Aug.	22.46	0.90	0.04	28.20	1.08	0.04	17.29	0.94	0.05
Sept.	16.52	0.96	0.06	23.24	1.28	0.06	10.50	0.90	0.09
Oct.	8.29	1.47	0.18	15.07	1.86	0.12	2.54	1.21	0.48
Nov.	-1.87	1.79	-0.96	4.35	2.20	0.51	-6.66	1.59	-0.24
Dec.	-9.74	1.71	-0.18	-3.68	1.95	-0.53	-14.27	1.63	-0.11
Annual	7.10	0.71	0.10	13.45	0.78	0.06	1.62	0.68	0.42
Spring	8.27	7.43	0.90	14.96	7.46	0.50	2.05	7.11	3.47
Summer	22.61	1.48	0.07	28.34	1.51	0.05	17.36	1.65	0.09
Autumn	7.65	7.69	1.00	14.22	7.97	0.56	2.12	7.15	3.36
Winter	-10.11	2.66	-0.26	-3.70	2.96	-0.80	-15.01	2.47	-0.16

The linear trends of average annual mean temperature, Tmax and Tmin along with year (between 1970 and 2013) (Figure 2) showed that they were all increased. The linear regression equation of average annual mean temperature, Tmax and Tmin (Table 2) showed that the intercept of line was between 0.024~0.029, the R^2 were all larger than 0.15. The variability of mean temperature was 0.29℃/10a, the variability of annual Tmax and Tmin was 0.24℃/10a and 0.28℃/10a, respectively, which were all higher than global nearly 50 years (0.13℃/10a)[11]. The highest and lowest mean temperatures occurred at 2007 (8.6℃) and 1974 (5.7℃), respectively.

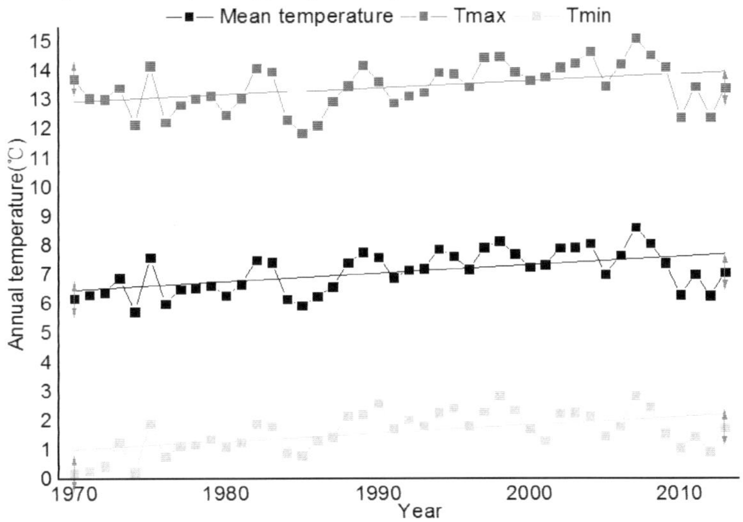

Figure 2. Linear trends of annual mean temperature, Tmax and Tmin with year.

Table 2. Linear regression equation of mean temperature, Tmax and Tmin of annual, spring, summer, autumn and winter.

Time	Temperature	Linear equation	R^2	Time	Temperature	Linear equation	R^2
	Mean	y = 0.038x + 7.419	0.168		Mean	y = 0.026x + 22.010	0.171
Spring	Tmax	y = 0.030x + 14.291	0.088	Summer	Tmax	y = 0.024x + 27.813	0.094
	Tmin	y = 0.038x + 1.183	0.190		Tmin	y = 0.022x + 16.875	0.176
	Mean	y = 0.025x + 7.081	0.116		Mean	y = 0.021x - 10.593	0.028
Autumn	Tmax	y = 0.027x + 13.623	0.085	Winter	Tmax	y = 0.015x - 4.048	0.011
	Tmin	y = 0.019x + 1.703	0.078		Tmin	y = 0.027x - 15.641	0.051
Annual	Mean	y = 0.029x + 6.453	0.269				

Tmax	y = 0.024x + 12.916	0.154
Tmin	y = 0.028x + 0.989	0.276

The increasing trend of temperature (mean temperature, Tmax and Tmin) in Naiman County was consistent with the trend in global temperatures. In addition, Tmin was increased faster than Tmax, this was consistent to the related results, e.g., Hua et al.[6] indicated that the average annual mean Tmax and Tmin over China increased by 0.147℃/10a and 0.283℃/10a respectively; Hughes showed that Australia has warmed ~0.8°C over the last century with Tmin warming faster than Tmax[12]; Uddin et al. showed that there is an increasing trend in the annual mean Tmax and Tmin in Bangladesh region between 1948 and 2013, and the annual mean Tmax increased by 0.0037°C per year, while the annual mean Tmin experienced a 0.069°C increased between 1948 and 2013[13]. The related research results also showed that there was obvious warming over the past century for many parts of the world[6],[14]-[15], and the distributions of both daily Tmax and Tmin have significantly shifted towards higher values[16]. In most regions, both Tmax and Tmin have become more positively skewed in more recent decades which indicates a change in shape towards the hotter part of the distribution [16].

The trend rate and significance test of linear regression equation of mean temperature, Tmax and Tmin showed (Table 3) that the trend rate (*A*) of annual temperature (mean, Tmax and Tmin) were 0.24~0.29, and they all increased insignificant (F>7).

Table 3. Temperature trend and its significance test of linear regression equation of mean temperature, Tmax and Tmin at annual, spring, summer, autumn and winter.

Time		Spring		Summer		Autumn		Winter		Annual	
Parameters		A	F	A	F	A	F	A	F	A	F
	Mean	0.38	8.47	0.26	8.63	0.25	5.50	0.21	1.17	0.29	15.46
Temperature	Tmax	0.30	4.03	0.24	4.37	0.27	3.92	0.15*	0.45	0.24	7.66
	Tmin	0.38	9.84	0.22	8.99	0.19	3.53	0.27	2.21	0.28	16.04

*p<0.5.

3.2. Seasonal Temperature

The division of the season in the study area as following: spring was Mar. to May, summer was Jun. to Aug., autumn was Sept. to Nov., and winter was Dec. to Feb. of next year[10].

The statistical characteristics of mean temperature, Tmax and Tmin at four seasons were also listed in Table 1. It indicated that the mean temperature of spring, summer, autumn and winter was 8.3℃, 22.6℃, 7.7℃ and -10.1℃, In addition, the mean temperature and Tmax changed with season in the order: summer >spring > autumn >winter, but for Tmin, it was higher at autumn than spring. The highest coefficient of variation of mean temperature, Tmax and Tmin was autumn, winter and spring, respectively, and the coefficient of variation of three temperatures (mean, Tmax and Tmin) were all lowest at summer. That was, the temperature of summer was relatively stable in different years.

The linear trends of mean temperature, Tmax and Tmin at four seasons indicated that (Figure 3) they all increased with year. The linear regression equation, the trend rate and significance test of mean temperature, Tmax and Tmin at spring, summer, autumn and winter were also listed in Table 2 and Table 3, respectively.

The increasing trend of mean temperature was most obviously at spring (0.38°C/10a), and weakest at winter (0.21°C/10a). The highest and lowest mean temperatures of spring were 11.1°C of 2002 and 6.1°C of 2010; The highest and lowest mean temperature of winter were -6.9°C of 2006 and -13.1°C of 2012. The increasing trend of Tmax was also most obviously at spring (0.3°C/10a), and weakest at winter (0.15°C/10a), but the increasing trend of Tmin was most obviously at spring (0.38°C/10a), and weakest at autumn (0.19°C/10a). In four seasons, for mean temperature, Tmax and Tmin , only Tmax of winter increased significant (F<0.5) (Table 3).

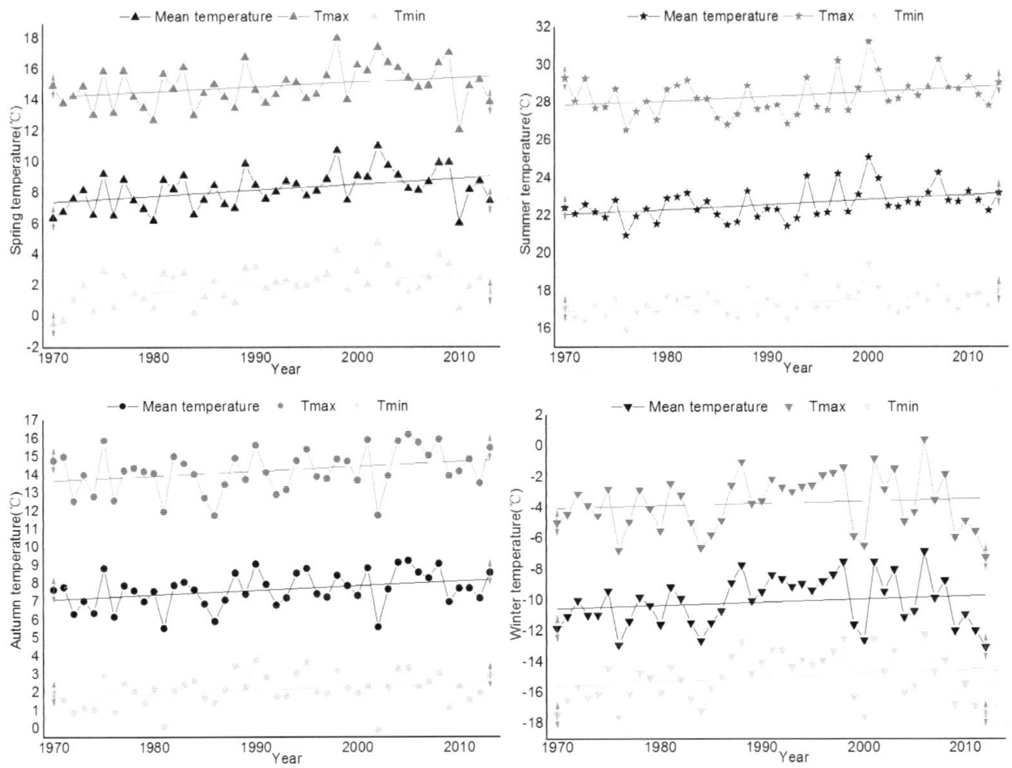

Figure 3. Linear trends of seasonal mean temperature, Tmax and Tmin with year.

3.3. Monthly Temperature

The statistical characteristics of mean temperature, Tmax and Tmin in different months were also showed in Table 1.The highest and lowest mean temperature was obtained at July (23.9℃) and January (-12.3℃), respectively. In addition, the mean temperature and Tmax of different months were decreased in the order: Jul. >Aug. >Jun. >May. >Sept. >Apr. >Oct. >Mar. >Nov. > Feb. >Dec. >Jan. However, the order of Tmin in different months was different to mean temperature and Tmax, e.g., Tmin at Sept. was higher than May, and Nov. was higher than Mar. The highest and lowest Tmin was obtained at Jul. (19.1℃) and Dec. (-14.3℃).The coefficient of variation of mean temperature, Tmax and Tmin were highest at March, Feb and Apr. And it was lower for mean temperature, Tmax and Tmin from May to Sept. The bigger coefficient of variation of temperature (mean, Tmax and Tmin) occurred between Feb. to Apr. mainly because frequent winds between November to May makes this period the major wind-erosion season, especially between Feb. to Apr.[8].paragraph text follows on from the subsubsection heading but should not be in italic.

4. Conclusions

We presented a trend analysis in mean temperature, Tmax and Tmin at scales of annual, seasonal and monthly over 1970-2013. The main conclusions are as follows.

(1) The annual mean temperature, Tmax and Tmin in Naiman County was 7.1°C, 13.5°C and 1.6°C, respectively, There is a general warming signal for mean temperature, Tmax and Tmin over Naiman County, with increasing speed of 0.29℃/10a, 0.24℃/10a and 0.28℃/10a, respectively;

(2) The mean temperature of spring, summer, autumn and winter was 8.3°C, 22.6°C, 7.7°C and -10.1°C; The increasing degree of mean temperature, Tmax and Tmin were all most obviously at spring, with increasing speed 0.38℃/10a, 0.3℃/10a and 0.38℃/10a, respectively;

(3) The highest and lowest mean temperature was obtained at July (23.9°C) and January (-12.3°C), respectively; The highest and lowest Tmax was obtained at July (23.9°C) and January (-5.9°C),

respectively; The highest and lowest Tmin was obtained at July (19.1°C) and Dec. (-14.3°C), respectively.

5. References

[1] Nieto H., Sandholt I., Aguada I., et al. 2011. Air temperature estimation with MSG-SEVIRI data: Calibration and validation of the TVX algorithm for the Iberian Peninsula. Remote Sensing of Environment, 115, 107–116.
[2] Zhu W.B., Lű A.F., Jia S.F. 2013. Estimation ofdaily maximum andminimum air temperature using MODIS landsurface temperature products. Remote Sensing of Environment, 130: 62–73.
[3] Marin L., Birsan M.V., Bojariu R., et al. 2014. An overview of annual climatic changes in Romania: Trends in air temperature, precipitation, sunshine hours, cloud cover, relative humidity and wind speed during the 1961–2013 period. Carpathian Journal of Earth and Environmental Sciences, 9(4):253–258.
[4] Vose R.S., Easterling D.R., and Gleason B. 2005: Maximum and minimum temperature trends for the globe: An update through 2004. Geophys. Res. Lett., 32, L23822, doi: 10.1029/2005GL 024379.
[5] Elguindi N., Rauscher S.A., and Giorgi F. 2012. Historical and future changes in maximum and minimum temperature records over Europe. Climate Change, 117(1-2), 415–141.
[6] Hua W., Shen S.S.P., Wang H.J. 2014. Analysis of Sampling Error Uncertainties and Trends in Maximum and Minimum Temperatures in China. Advances in Atmospheric Sciences, 31: 263–272
[7] Zhao X.Y., Zuo X.A., Huang G., et al. 2010. Desertification reversion in relation to land use change and climate in Naiman County, Inner-Mongolia, China. Sci Cold Arid Reg, 2(1):15–20
[8] Li F.R., Zhang A.S., Duan S.S., 2005. Patterns of reproductive allocation in Artemisia halodendron inhabiting two contrasting habitats. Acta Oecologica, 28:57–64.
[9] Liu X.P., He Y.H., Zhang T.H., et al. 2014. The response of infiltration depth, evaporation, and soil water replenishment to rainfall in mobile dunes in the Horqin Sandy Land, Northern China. Environ Earth Sci, DOI 10.1007/s12665-015-4125-0.
[10] Qu C.P., Guan D.X., Wang A.Z., et al. 2009. Chinese Journal of Ecology, 28(11): 2326-2332.
[11] IPCC. 2007. Climate Change 2007: The Physical Science Basis. Intergovernmental Panel on Climate Change [EB /OL]. [2007-04-08]. http: //www. ipcc. ch /ipccreports /index.
[12] Hughes L. 2003. Climate change and Australia: Trends, projections and impacts. Austral Ecology, 28: 423–443
[13] Uddin M.N., Mondal M.S.A., Nasher N.M.R. 2014. Spatio-temporal Analysis of Maximum and Minimum Temperature in Bangladesh. Sci. & Natural Resources, 7(2):73–77.
[14] Bonsal B.R., Zhang X., Vincent L.A., et al. 2001. Characteristics of daily and extreme temperatures over Canada. J. Climate, 14(9), 1959–1976.
[15] Meehl G., Tebaldi C., Walton G., Easterling D., et al. 2009. Relative increase of record high maximum temperatures compared to record low minimum temperatures in the U.S. Geophys., Res., Lett., 36, L23701, doi:10.1029/2009GL040736.
[16] Markus G.D., Lisa V.A. 2012. The shifting probability distribution of global daytime and night-time temperatures. Geophysical research letters, 39, L14707, doi: 10.1029/ 2012GL052459.

Acknowledgments

This study was supported by the National Natural Science Foundation of China (No. 41771087). We are very grateful for the Naiman Desertification Research Station of Cold and Arid Regions of Environmental and Engineering Research Institute, Chinese Academy of Sciences for providing relation data.

The Status of Existing Building Materials in Thailand to Cope with Future Climate: Wall Component

Patcharawan Bootkaew and Nachawit Tikul

Faculty of Architecture and Environmental Design, Meajo University, Nongharn, Sansai, Chiang Mai, 50290, Thailand
Email: patcharawan.bkw@gmail.com

Abstract. Climate change tolerant building materials, such as high temperature resistance and flooding resistance, affect living condition of people in an accommodation. This study focuses on durability features of building material related with climate change in Thailand. The result suggests that the property of each wall material type cannot show proper value of tolerance and climate change resistance. Therefore, the conclusion for future tolerant building material properties are that the U-value should be greater than 0.64 W/m^2k, the R-Value should be greater than 0.64 m^2k/W, and the K-Value should be less than 0.155 $W/m^{\circ}C$. Such values can cope with temperature up to approximately 49 degree Celsius. Apart from that, flood resistance should be less than or equal to 2%, water penetration should be less than or equal to 1.5% and swelling test should be less than or equal to 12%. This will be able to cope with flood that lasts about 60 days. The results of the study is a guideline for future wall material design and development.

1. Introduction

Climate change causes changes in temperature, rainfall, extreme weather condition, hurricane, storm surge and season period. Nowadays many countries face these problems. Similarly, Thailand is undergoing the rising temperature and flood[1]. According to Thailand's future climate change prediction, it is likely to have higher average temperature and more precipitation across the country[2]. The changes affect various sectors of the agricultural industry, including the housing and building sector which is relatively affected. The damage caused by climate change prompts many organizations to find ways to reduce environmental pollution and adapt themselves to cope with it. The high temperature has direct impact on roof and wall. Flooding causes the decay of wall and floor, while wall is affected by both high temperature and flooding[3]. In the past, material selection factors are price, quality, beauty and brand reliability[4]. However, wall material has to be able to tolerate with future climate change. Therefore, the high temperature tolerance and water resistance. Due to what mentioned above, this study will find the answers of the questions; what is the suitable building materials for future climate changes and how much material will be available to support that change. This study aims to find possibility of creating and developing wall material for future climate.

2. Methodology

2.1. Climate Change Information

This study use existing information of future climate to estimate wall material properties in order to cope with future climate. The RCM (regional climate model) from SEA START RC database is mainly used[5]. This data covered 100 years (2000-2099). In addition, the database from Meteorological Department, Geo-Informatics and Space Technology Development Agency (Public Organization) – GISDA[6] and Hydro and Agro Informatics Institute (HAII) are also used in this study. This study makes use of the highest temperature (degree Celsius) and highest precipitation (millimetre) related to the number of flooding period of each year.

2.2. Climate Change Effect Analysis

After collecting climate change data, the data is put in a grid format and plotting lines to find out what the prominent value of the climate change is by dividing the most obvious period into four periods of 25 years. For high temperature, information is directly used from the database. Due to too many related factors such as speed, strength, water level ,storm surge, water volume in the dam, drainage and frequency, the study investigates only flood data on rainfall and number of flood days by studying the damage that has occurred in the past. And because precipitation is the main cause of flood, estimating flood risk by using data from linear regression of the relation between precipitation and number of flood days is another method. 'Figure 1' is a figure's linear regression of flooding that demonstrates correlation (R) equal 0.597. Even if the number of correlation is not notable height, its consistency with rainfall is the most important factor creating flood.

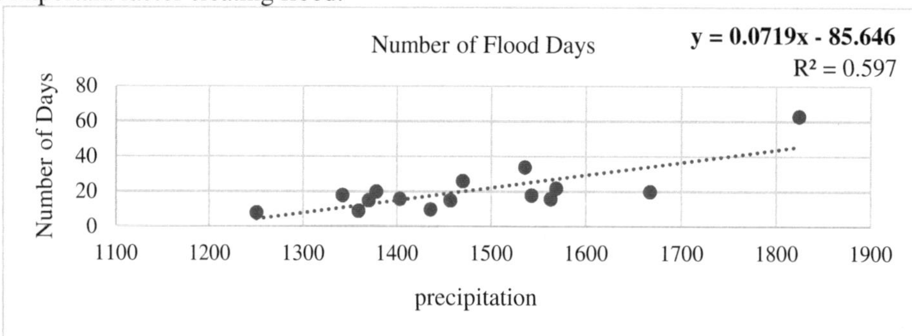

Figure 1. Linear regression of flooding

After mentioned methods, the next is selecting the range of both temperature and flood line graph in 'Figure 2' where the lines are most sloping and similar to find the characteristic of durability of wall construction material.

2.3. Material Characteristic Analysis

The factor related to building material and climate change is heat resistance which includes U-Value, R-Value and K-Value. The followings are durability parameter and durability equation.

$$Q = U \times A \times \Delta T \quad (1) \qquad U = \frac{1}{R} \quad (2) \qquad R = \frac{\Delta X}{K} \quad (3)$$

Q = heat transfer rate , U = overall heat transfer coefficient *(W/m² °C)*, A = heat transfer surface area *(m²)*, ΔT = difference in temperature, R = overall thermal resistance *(m²°C/W)*, ΔX = the wall thickness *(m)* and K = the thermal conductivity of the material *(W/(m·K)*

Water resistance properties include water absorption *(%)*, permeability *(%)* and swelling test value *(%)*.

2.4. Building Material Survey

Data is collected by creating a tool for collecting data from both building material and research. Wall material can be divided into 21 types which are 1.White brick 2.Brick or red brick 3.Light weight brick 4.Concrete block 5.Real wood 6.Artificial wood 7.PVC/plastic 8.Precast concrete 9.Metal/sheet metal 10.Iron/Steel 11.Aluminum 12.Glass 13.Gypsum 14.Fiberboard 15.Ceramic 16.Terracotta Tile 17.Granite Tile 18.Tile Lane 19.Insulation 20.Color and 21.Coating.

Primary survey comes from baan&BEYOND home department store, The Siam Cement Group Public Company Limited (SCG) and Global House Public Company Limited. Secondary survey comes from researching the information from websites of ThaiLIS –thai library integrated system and copyright information from department of intellectual property.

3. Result

3.1. Climate Change in the Future
According to the collection of highest temperature of each year. The second period (2025-2049) was chosen. By using a temperature of 49 degrees as the initial temperature to calculate the durability, the second period's graph is steeply above the other ranges and the average temperature is 44.75 degrees Celsius[2]. The flood data can be used to estimate future climate by using historical data for making linear regression. The number of the longest flood period is 50-60 days, which is the flood occurred in Thailand in 2011.

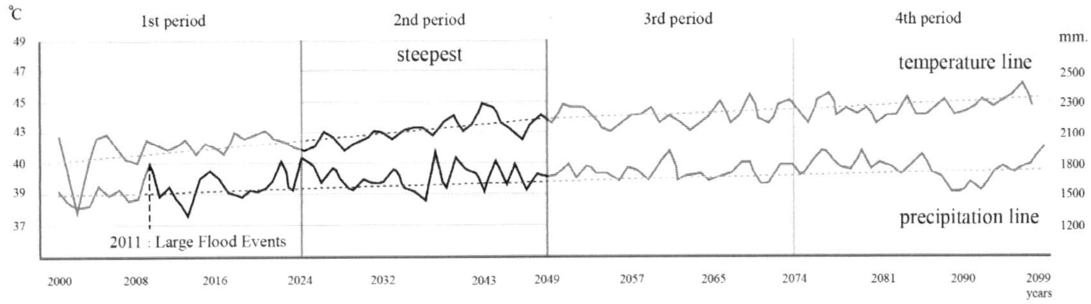

Figure 2. Average high temperature and precipitation graph

3.2. Properties of Wall Material Suitable for Future Climate Change.
For high temperature resistance, the heat transfers to the building wall shall be less than or equal to 30 watts per square meter[7]. The result is as following; Heat Resistance (U) $\geq$ 1.55 W/m2K from equation (1(, Heat Resistance (R) $\geq$ 0.64 m2K/W from equation (2) and Heat Conductivity (K) $\leq$ 0.155 W/m °C from equation (3). Flood durability of wall material should have new criteria. According to the research, the material durability varies in each type. The method are collecting information and find the median value to determine the value that all types of material can use. The results are water absorption $\leq$ 2 % from EN 12087, water penetration $\leq$ 1.5 % and swelling test $\leq$ 12 %[8].

3.3. Present Situation of the Wall Materials
The situation is analyzed via GAP ANALYSIS and statistical data (Matrix Diagram), with creating the table with the new criteria. Durability data and the most durable value are shown by using bold and italic font to indicate the value qualified durable by the criteria specified in Table 1.

Table 1. Situation of the wall materials.

No	Types	Heat resistant			Water resistant		
		$U \geq 1.55$ W/m^2K	$R \geq 0.64$ m^2K/W	$K \leq 0.155$ $W/m\,^{\circ}C$	water absorption ≤ 2 %	water penetration ≤ 1.5 %	swelling test ≤ 12 %
1	White Brick	0.75	×	×	< 18	×	×
2	Brick	×	×	×	≤ 18	×	×
3	Light weight Brick	0.98	***1.5***	***0.13***	30	×	×
4	Concrete Block	×	×	×	≤ 5	×	×
5	Real Wood	×	×	×	×	×	×
6	PVC / Plastic	***1.854***	0.539	×	×	×	×

No	Types	Heat resistant			Water resistant		
		U ≥ 1.55 W/m²K	R ≥ 0.64 m²K/W	K ≤ 0.155 W/m °C	water absorption ≤ 2 %	water penetration ≤ 1.5 %	swelling test ≤ 12 %
7	Artificial Wood	×	×	×	×	×	×
8	Precast Concrete	0.88	***1.14***	0.175	13	×	×
9	Metal / Sheet Metal	×	×	×	×	×	×
10	Steel / Metal	×	×	×	×	×	×
11	Aluminum	×	×	×	×	×	×
12	Glass	0.024	×	×	< 5	×	×
13	Gypsum	0.191	***5.592***	***0.027***	***≤2***	×	×
14	Fiber board	***1.854***	0.539	***0.083***	30	< 5	×
15	Ceramic	×	×	×	10	×	×
16	Terracotta Tile	×	×	×	×	×	×
17	Granite Tile	×	×	×	×	×	×
18	Tile Lane	×	×	×	***0.05***	×	×
19	Insulation	0.025	***2.679***	×	***0.00015***	×	×
20	Color	×	×	×	×	×	×
21	Coatings	×	×	×	×	×	×

Of 309 researches, the researches focusing on high temperature resistance are 98.38% and focusing on flood resistance are 1.61%.

4. Discussion and Conclusion

The gap of this study is flood resistant material. Material with high water resistance and water permeability is rarely found[9]. Swelling value, which is a test when the material is soaked and inflated, may pass the criteria but that is still not satisfying. Future material should have swelling value at least 12%. Because the value affects the most on material durability[10], this durability should be highly developed[11]. Flood resistant material is scarcely focused by researchers. Apart from that, despite not much satisfying, high temperature resistant wall material passes the new criteria. U-value is 0.304 W/m2K higher than new criteria. R-value with the highest thermal resistance is 5.592 m2K/W. K-value is 0.128 W/m °C. The results suggest that high temperature resistant wall material should be developed for better heat resistance.

Standard criteria of building material should always be adjusted and developed in order to support the current situation and cope with the future circumstance[12]. Besides, quality is another important factor for consumers. They should be able to distinguish the characteristics of each material in order to choose the most proper material for demanding application[13]. Because of not being well informed in the past, they lacked understanding about building material properties. Therefore, the government and the council of Engineers should take part in acknowledging consumers[14].

5. References

[1] Martin Parry OC, Jean Palutikof, Paul van der Linden, Clair Hanson. Climate Change 2007: Impacts, Adaptation and Vulnerability. 32 Avenue of the Americans, New York, NY 10012-2473, USA: Canbridge University Press; 2007.

[2] Department M. Changes in Temperature and Rainfall over Thailand under. 4353 SUKHUMVIT ROAD, BANGKOK 10260, THAILAND; 2016.

[3] Kaczmarek A. *Factors Affecting Humidity Conditions of a Face Wall Layer of a Heated Building*. Procedia Engineering. 2017;197:205-10.

[4] Doktalyong S. A Study Decision Making Styles Of Consumer Which Effect to Buying Decision the of Modern Trade Construction Material Stores in Bangkok and Suburban area. National Acandamic Conference2015.

[5] Climate Data Distribution System [Internet]. 1996. Available from: http://ccs.gms-eoc.org/climatechange/home/index.html.

[6] Thailand Flood Monitoring System [Internet]. 2000. Available from: http://flood.gistda.or.th.

[7] Suaunka K. The Evaluation of Energy Saving for Home Design And Construction Materials. 2014.

[8] Chusit P. *The Product of ceiling Plate-Bar from base bark of Physic-nut by Phenol-Cyanate*. Rajamangala University of Technology National Conference. 2014;5(RMUTP Research Journal Special Issue):155-67.

[9] Haruehansapong S. *Effect of Nanosilica Particle Size on the Water Permeability, Abrasion Resistance, Drying Shrinkage, and Repair Work Properties of Cement Mortar Containing Nano-SiO2*. Advances in Materials Science and Engineering. 2017;Volume 2017:11.

[10] Pimentel E. *Existing Methods for Swelling Tests – A Critical Review*. Energy Procedia. 2015;76:96-105.

[11] Peduzzi P. *Flooding: Prioritizing protection?* nature climate change. 2017;7(9):625-6.

[12] Tagg A. A new standard for flood resistance and resilience of buildings: new build and retrofit. European Conference on Flood Risk Management (FLOODrisk 2016); 20 October2016. p. 6.

[13] Ovichit P. Green buildings: defiding sustainable construction materials in Thailand.: Iowa State University; 2007.

[14] Commission OotCS. Proactive Senior Executive Service. Bangkok Thailand2012. p. 37-68.

Acknowledgement

Financial support from the National Research Council of Thailand (Scholarships for Graduate Students) in 2018 for this research project is gratefully acknowledged.

IOP Conf. Series: Earth and Environmental Science **178** (2018) 012034 doi:10.1088/1755-1315/178/1/012034

Spatio-Temporal Variation Characteristics of Vegetation Coverage and Its Correlation with Climate Factors in Henan Province from 2006 to 2015

Dandan Zhang[1], Zhihui Tian[2], Xiaohui He[2], Hengliang Guo[2], Haitao Wei[2]

1 College of Water Conservancy and Environment, Zhengzhou University, Zhengzhou 450001, China
2 The Smart City institute, Zhengzhou University, Zhengzhou 450001, China

Abstract. Vegetation coverage is an important parameter for comprehensively quantifying vegetation status. Using the remote sensing data of MODIS MODND1M and the meteorological data such as temperature and precipitation, on the base of ArcGIS, taking maximum value composite(MVC), vegetation index, kriging interpolation, linear regression analysis, correlation analysis and significance test methods, this study analyzed the spatio-temporal variation characteristics and the trend of changes of vegetation coverage, and its relationships with temperature and precipitation from 2006 to 2015. The results showed that: 1) From the aspect of Space, the vegetation coverage is generally higher in the eastern and western parts and lower in the middle part in Henan Province. 2) From the aspect of time, the vegetation coverage with a variation range of 85.62% -87.78% firstly increased and then decreased and then increased from 2006 to 2015. 3) The change trend of vegetation coverage is obvious regional spatial difference characteristics.4) The responses of different vegetation types to climate factors were different.

1. Introduction

Vegetation, as an important part of the earth's ecosystem, plays an irreplaceable role in the climate, hydrology and energy balance[1]. Its Spatio-temporal variation reflect the environmental impact of natural and social activities. Vegetation coverage is an important index to reflect the vegetation cover of the surface and to monitor the ecological environment[2]. With the continuous development of remote sensing technology, monitoring the vegetation cover using remote sensing images has become a hot spot of common concern in the scientific community. NDVI can accurately reflect the degree of vegetation change and growth status, and has become one of the most commonly used vegetation activity monitoring indicators. Many scholars[3-5] use NDVI to dynamically monitor the changes of vegetation on the surface and achieved good results. Henan Province is a major agricultural province in China. The dynamic changes of its ecosystem directly affect food security and affect Henan's social and economic development. Therefore, the real-time and effective monitoring of vegetation cover in Henan Province is an important guarantee for coordinating resources and environment in an all-round way, making full use of the advantages of Henan's natural and economic conditions and improving the sustainable development of economy.

2. Study Area

Henan Province (110 ° 21'-116 ° 39'E and 31 ° 23'-36 ° 22'N) is located in the central-eastern part of China. The terrain is generally high in the west and low in the east, with elevations between -60 and 2400 m. The annual average temperature is 13-16°C. The average annual precipitation is 500 ~

1100mm. Henan Province is the Central Plains Economic Zone's main business, but also China's important agricultural food base. The dynamic changes of its ecosystem can not be underestimated for the socio-economic development.

3. Material and Method

3.1. Data Source and Processing

The MODND1M data used in this study comes from the geospatial data cloud platform. With the support of MRT software, the format conversion and reprojection of MODND1M were performed and then cropped. The maximum annual NDVI value was obtained using maximum value composite (MVC). The meteorological data are from the monthly average temperature and precipitation of 20 benchmark meteorological stations provided by China Meteorological Science Data Sharing Service. The Kriging interpolation, resampling and cropping are used to obtain the raster image with the same cell size as the vegetation coverage data.

3.2. Vegetation Index Method

In this study, vegetation index method was used to extract vegetation coverage. The formula[6] is as follows. Among them, NDVI is the normalized difference vegetation index, taking NDVIV = 0.70 and NDVIS = 0.00. And, Fv takes a value of 1 when a pixel's NDVI value is greater than 0.70, and Fv takes a value of 0 when NDVI is less than 0.00.

$$\text{Fv}=(\text{NDVI}-\text{NDVIS})/(\text{NDVIV}-\text{NDVIS}) \tag{1}$$

3.3. Linear Regression

The univariate linear regression analysis can eliminate the influence of extreme climate in a particular year, so this method is used to analyze the trend of vegetation coverage in Henan Province. The rate of change is calculated as follows[7]:

$$\theta_{slope} = \frac{n \times \sum_{k=1}^{n} (k \times X_k) - \left(\sum_{k=1}^{n} k\right)\left(\sum_{k=1}^{n} X_k\right)}{n \times \sum_{k=1}^{n} k^2 - \left(\sum_{k=1}^{n} k\right)^2} \tag{2}$$

3.4. Correlation and Significance

Correlation analysis is often used to study the degree of correlation and related directions between elements. In this study, pixel-based Pearson correlation coefficient method was used to study the relationship between vegetation coverage and temperature, as well as precipitation with $P < 0.10$ and $P < 0.05$ confidence levels. Correlation and significance analysis is as follows [8]:

$$r_{x,y} = \frac{\sum_{k=1}^{n} (x_k - \bar{x})(y_k - \bar{y})}{\sqrt{\sum_{k=1}^{n} (x_i - \bar{x})^2 \sum_{k=1}^{n} (y_k - \bar{y})^2}} \tag{3}$$

$$r_{xy_1 \cdot y_2} = \frac{r_{xy_1} - r_{xy_2} r_{y_1 y_2}}{\sqrt{1 - r_{xy_2}^2}\sqrt{1 - r_{y_1 y_2}^2}} \tag{4}$$

$$R_{x \cdot y_1 y_2} = \sqrt{1 - (1 - r_{xy_1}^2)(1 - r_{xy_2 \cdot y_1}^2)} \tag{5}$$

$$T = \frac{r_{xy_1 \cdot y_2}}{\sqrt{1 - r_{xy_1 \cdot y_2}^{\ 2}}}\sqrt{n - m - 1} \tag{6}$$

$$F = \frac{R_{x \cdot y_1 y_2}^{\ 2}}{1 - R_{x \cdot y_1 y_2}^{\ 2}} \times \frac{n - k - 1}{k} \tag{7}$$

4. Results Analysis

4.1. Spatial Distribution

The average NDVI and vegetation coverage in Henan Province during 2006-2015 are shown in Figure1. The annual average vegetation coverage was mostly between 0.95 and 1, which was consistent with the spatial distribution of NDVI. Overall, the vegetation coverage was higher in the east and west and lower in the middle. The coverage of low vegetation is mainly distributed in the city and the surrounding areas, indicating that the rapid development of urbanization and industrialization bring economic benefits while reducing the quality of ecological environment. Vegetation coverage is also low due to sparse vegetation near large reservoirs and rivers. The vegetation in the southeastern part of Henan Province is dominated by broad-leaved forests and has a strong photosynthesis. Therefore, the vegetation coverage in this area is extremely high.

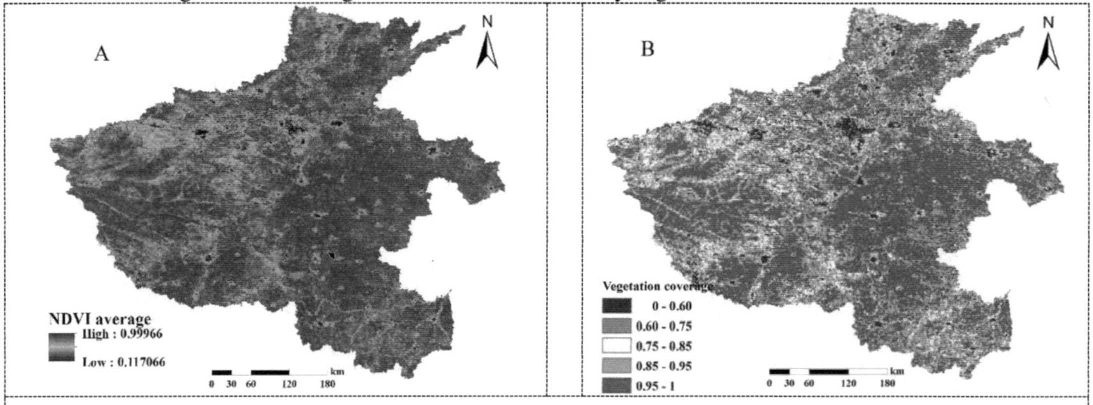

Figure 1. Distribution of mean NDVI(A) and Vegetation coverage (B) in Henan Province.

4.2. Temporal Variation

Using land cover type data in Henan Province, based on ArcGIS, the average vegetation coverage in different statistical regions such as urban and built-up areas, farmland areas and natural vegetation areas in Henan Province during 2006-2015 was statistically analyzed (Table 1). The vegetation coverage increased firstly and then decreased and then rose again in Henan Province from 2006 to 2015, with a range of 85.62% ~ 87.78%. The lowest vegetation coverage in Henan Province in 2006 was 85.62%. The highest vegetation coverage in 2015 was 87.78%.

Table 1. The composition characteristics of Vegetation coverage in Henan Province from 2006 to 2015. (%)

Area \ Year	2006	2007	2008	2009	2010	2011	2012	2013	2014	2015
Urban and Built-up	62.76	61.51	60.64	59.75	59.63	58.34	57.88	57.96	56.43	56.68
Farmland	77.35	77.04	76.57	76.38	75.72	76.25	76.37	77.16	76.28	75.41
Forest	87.57	89.51	89.68	89.89	91.24	91.64	91.88	90.85	87.63	92.69
Shrublands	81.34	85.44	86.24	86.38	87.66	87.56	87.87	88.97	83.41	89.34
Grasslands	76.49	76.21	75.68	76.10	75.01	75.02	76.42	76.29	75.98	76.39
Henan Province	85.62	86.01	86.23	86.41	87.26	87.45	87.61	86.54	86.21	87.78

4.3. Trend Change

Based on pixel size, a linear regression analysis was conducted on the interannual variability of vegetation coverage in Henan Province from 2006 to 2015. The trend of vegetation coverage of each grid in the study area is obtained, and the results are classified by standard deviation (Figure 2). The change trend of vegetation coverage is obvious regional spatial difference characteristics. The areas with the increasing vegetation coverage are mainly located in southern Xinyang, northeastern Nanyang, southwest of Kaifeng and other regions. The areas with the declining vegetation coverage are mainly located in the vicinity of Taihang Mountain, along the Yellow River ecological conservation zone, Funiu Mountain, Xiong'ershan area, parts of the Danjiang River basin and surrounding areas of cities and counties. The interannual variation of vegetation coverage in Henan Province from 2006 to 2015 was between -0.08 and 0.09. The rate of change in most areas was between 0.0009 and 0.0071 (Table 2). The area which is basically unchanged trend accounts for 48% of the total area in Henan Province. The area which is upward trend accounts for 28.09%. The area which is downtrend accounts for 28.09%.

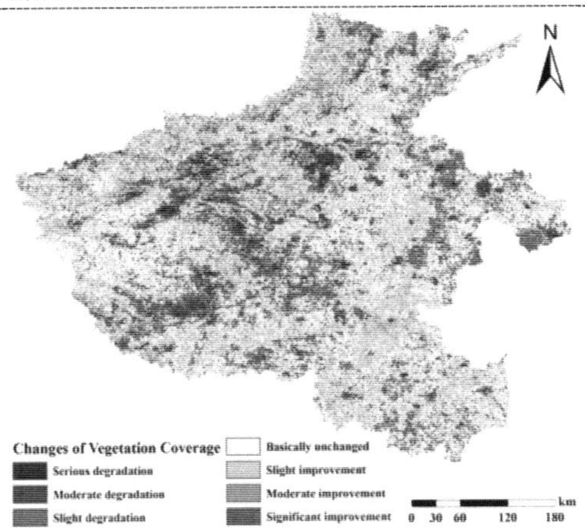

Figure 2. The spatial distribution of Vegetation Coverage slope in Henan Province from 2006 to 2015.

Table 2. Statistical result of the change trend of Vegetation Coverage from 2006 to 2015

Change Trend of Vegetation Coverage	Degree of change	Area/10^4Km^2	Area%
Slope≤-0.0116	Serious degradation	0.28	1.73
-0.0116≤Slope≤-0.0054	Moderate degradation	0.68	4.21
-0.0054≤Slope≤0.0009	Slight degradation	3.02	18.69
0.0009≤Slope≤0.0071	Basically unchanged	7.64	47.28
0.0071≤Slope≤0.0133	Slight improvement	3.91	24.20
0.0133≤Slope≤0.0196	Moderate improvement	0.53	3.28
Slope≥0.0196	Significant improvement	0.10	0.61

4.4. Correlation Analysis

The correlation coefficient between vegetation coverage and average temperature is -0.90 ~ 0.93(Figure 3). The positive and negative correlations accounted for 59.82% and 40.18% of the study area respectively. 63.1%, 41.4% of the pixels passed the 0.10 level, 0.05 level of T-significance test, respectively. The correlation coefficient between vegetation coverage and rainfall ranged from -0.91 to 0.98(Figure 3). The positive and negative correlations accounted for 74.22% and 25.78% of the study area respectively. 76.3%, 53.6% of the pixels passed the 0.10 level, 0.05 level of T-significance test, respectively. Vegetation coverage changes of different types vegetation have different responses to climatic factors (Table 3). The correlation coefficients between vegetation coverage of forest, shrublands, grassland, other vegetation and climatic factors showed positive and negative coexistence.

IOP Conf. Series: Earth and Environmental Science **178** (2018) 012034 doi:10.1088/1755-1315/178/1/012034

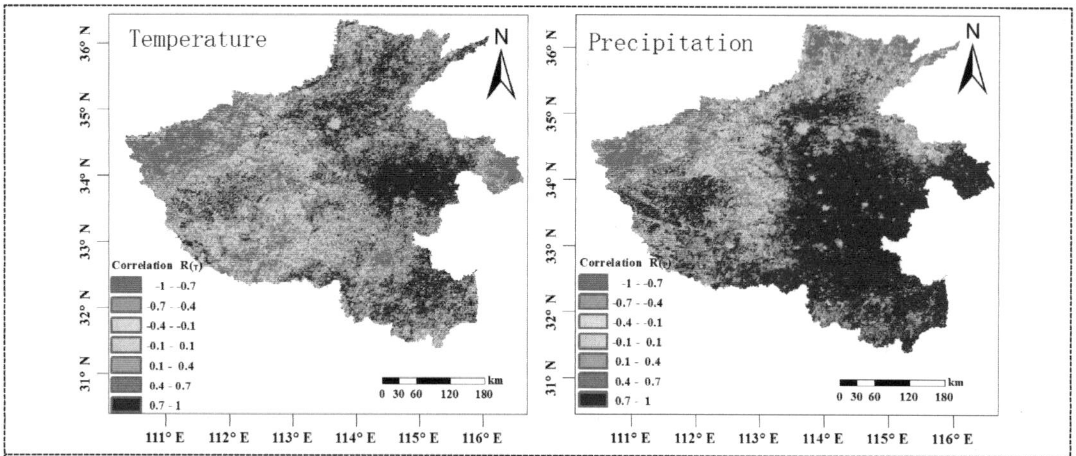

Figure 3. Spatial distribution of correlations between Vegetation Coverage and temperature, as well as precipitation in Henan Province during 2006—2015.

Table 3. Coefficients of Vegetation Coverage in different vegetation types with temperature and precipitation

Correlation coefficients	Forest	Shrublands	Grassland	Farmland	Others
R_{NT}	-0.23	0.41	-0.22	0.28	0.15
R_{NP}	0.52	-0.23	0.49	0.24	-0.24
R_{NT-P}	-0.31	0.57	-0.32	0.58	0.14
R_{NP-T}	0.73	-0.34	0.52	0.54	-0.23
R_{N-TP}	0.63	0.48	0.61	0.43	0.35

5. Conclusion

(1) From the aspect of Space, 70% of the plain is farmland in Henan Province. The annual average vegetation coverage was mostly between 0.95 and 1. Overall, the vegetation coverage was higher in the east and west and lower in the middle.

(2) From the aspect of time, the vegetation coverage with a variation range of 85.62% -87.78% firstly increased and then decreased and then increased in Henan Province from 2006 to 2015. In different types of land cover area, the highest vegetation coverage is the forest area and the lowest value is the urban and built-up area in each period

(3) The change trend of vegetation coverage is obvious regional spatial difference characteristics. The interannual variation of vegetation coverage in Henan Province from 2006 to 2015 was between -0.08 and 0.09. The rate of change in most areas was between 0.0009 and 0.0071.

(4) The correlation between vegetation coverage and temperature as well as precipitation showed obvious regional differences. Vegetation coverage changes of different types vegetation have different responses to climatic factors.

6. References

[1] Dengke L, Jianzhong F and Juan W 2010 Characteristics and causes of vegetation coverage in Shaanxi province *J. Chinese Journal of Applied Ecology* 21(11)2896-903.

[2] Gang W and Dongsheng G 2012 Effects of vegetation coverage and normalized humidity index on thermal landscape pattern-A case study of Guangzhou *J. Chinese Journal of Applied Ecology* 23(09)2429-36.

[3] Weijun W, Xueyan Z, Wenyu W, Hua L and Bing X 2016 Changes of vegetation coverage and its responses to climatic changes in Gannan plateau from 2000 to 2014 *J. Ecology Journal* 35(09)2494-504.

[4] Wenji Z, Zhaoning G, Min Z and Yunbao F 2017 Temporal and spatial variation of climate in the north China plain and its impact on vegetation coverage *J. Acta Ecologica Sinica* 37(02)576 -92.

[5] Minjie G, Tingting Z, Jianjun Z, Lili C and Xiaoping Z 2014 Response of vegetation coverage to climate change in the Loess Plateau from 1982 to 2006 *J. Chinese Journal of Soil and Water Conservation* 21(05)35-40.

[6] Zhenghong M, Zhiming L, Zongming W, Kaishan S, Chunying R, Jia D and Lihong Z 2010 Dynamic remote sensing monitoring of vegetation coverage in Jilin province based on MODIS NDVI *J. Remote Sensing Technology and Application* 25(03)387-93.

[7] Chuanhua L and Jun Z 2013 Spatial and temporal variations and driving factors of NPP in Shiyanghe river basin from 2000 to2010 *J. Ecology Journal* 32(03)712-8.

[8] Shaojie M, Jianlong L, Yizhao C, Chengcheng G, Wei Z and Weimin J 2012 Temporal and spatial variations of vegetation coverage in Inner Mongolia from 2001 to 2010 *J. Journal of Geography* 67(09)1255-68.

Acknowledgments
This work was supported by grants from the China National Critical Project for Science and Technology on Water Pollution Prevention and Control (2015ZX07210-002-005)

Analysis of Land Use Efficiency in Provincial-Level Development Zones—Case Study of Wanjiang City Belt, China

R H He[1] R Wang[1] Y N Ren[1] C Che[1] and J He[2]

1 School of Economy and Management, Anhui Agricultural University, Hefei, China
2 School of Transportation, Southeast University, Nanjing, China
E-mail:1. heruhai88@163.com; 2. hejie@seu.edu.cn

Abstract. Analyses of land use efficiency in development zones can provide insight into promoting effective land use and improving development efficiency. However, few studies have systematically evaluated land use efficiency in development zones at the provincial level. This study aims to analyze the economic efficiency of land use in 40 provincial-level development zones in the Wanjiang City Belt. By conducting a data envelopment analysis (DEA), we derive the following results. First, there remains considerable room for efficiency improvement. Second, the zones located on the development axis along the Yangtze River demonstrate high economic efficiency in land use. Third, the land use inefficiency of development zones located in a county is worse than that in a prefecture. In conclusion, as for efficiency improvements at the provincial level, development zones in county areas, especially in the north area of the Yangtze River, need attention.

1. Introduction

A land development zone is a multi-input and output system. Analyzing the economic efficiency of land use can be underst ood as studying the effects of developing park land, capital, labour, technology, and other input factor on economic output[1-2]. It reflects a reasonable degree of land resources and various factors included in configuring a comprehensive development zone, also known as the level of economic land use. A quantitative evaluation of the economic land use efficiency includes the ascertaining of the utilization efficiency of resources and analyzing land situations, problems, and obstacles. In addition, it contributes to the timely adjustment and improvement in land management practices and industrial policies. However, several Chinese scholars have expressed their concern regarding the efficiency of various land evaluations.

Many existing studies have proposed static and dynamic measures to evaluate the overall efficiency of decision-making units (DMUs) in terms of land use efficiency[1, 3-11]. Although studies have adopted research methods that are based on DEA[1][4-6][8-11], their results tend to differ[3][7][12]. Related research focuses on the national and regional levels as well as the economic and technological development of high-tech land analyses; however, few studies analyze land at a provincial level. Recent studies, nevertheless, are gradually extending to different sectors[12], thus offering diverse insights into land use at the micro level[7][12].

The Wanjiang City Belt is the first to relocate its industries to the national demonstration area in China. Land analyses at the provincial level include industrial transfers, although few studies have systematically evaluated efficiency at the provincial level. This study attempts to bridge this gap by conducting a provincial-level analysis using 2012 land data for 40 provinces in the Wanjiang City Belt.

To do so, it performs a DEA that categorizes land, capital, labour, science and technology, and other economic factors under four output efficiency levels to evaluate economic land use efficiency. The analysis provides us with an overview of the land area situation at the provincial level, land use efficiency in development zones, and other characteristics and problems. The results serve as a foundation to adjust industrial policy, improve land management zones, and promote the transformation and upgrading of quality and efficiency development.

2. Study Area

The Demonstration Zone for Industrial Transfer of the Wanjiang River City Belt (hereinafter Wanjiang City Belt) was established in 2010 and is located in Anhui of the Yangtze River basin (Fig. 1). Its regional land area is 7.79 x 104 km^2 and was the first and currently remains the only region to relocate its industries to China's national demonstration area. The Demonstration Zone is the first area to demonstrate collaborative scientific development. The central region is important to the growth of the country and is an advanced manufacturing and modern service industry base that has undertaken the industrial transfer process from within the country to abroad[13]. The demonstration area includes Hefei, Wuhu, Maanshan, Tongling, Anqing, Chizhou, Chuzhou, Xuancheng, Jinan, and Shucheng of Lu'an city. The area's industrial development structure lies on the development axis along the Yangtze River. Its two central regions are Hefei and Wuhu and Chuzhou and Xuancheng show an industrial distribution pattern along the axis.

As can be seen, important regional industries have been relocated and the present economic demonstration zones (technology), development zones, industrial parks, and processing zones have transferred from coastal regions to the eastern Anhui Province to accelerate the upgrading and restructuring of the regional industrial structure and promote industrialization in these regions. By the end of 2013, the demonstration area reported 108 types of development parks, covering a total area of 2,350 km^2 and accounting for 61.71% and 55.62% of the province's development zones. As shown in Fig. 2, as of 2013, the region has 47,884 companies and 231.51 million practitioners. The total fixed assets was 590.677 billion yuan, the total imports and exports was \$22.053 billion, operational (sales) income was 2.31345 trillion yuan, total industrial output value was 1.683716 trillion yuan, and state tax revenue was 66.601 billion yuan. The provincial-level development zones occupy half of the demonstration area and the transfer of industry is indispensable.

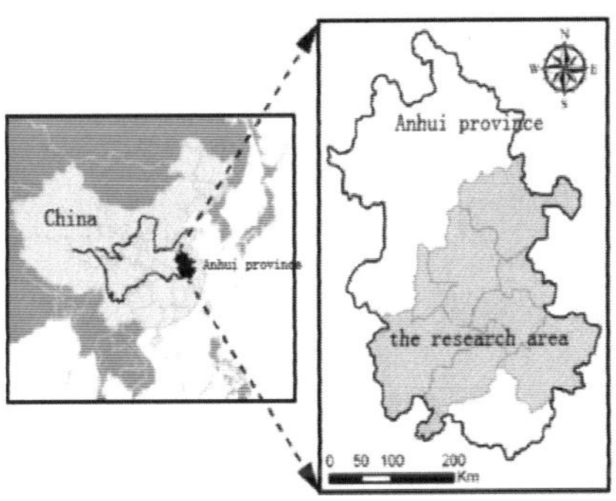

Figure 1. Location diagram of the Wanjiang City Belt[14]

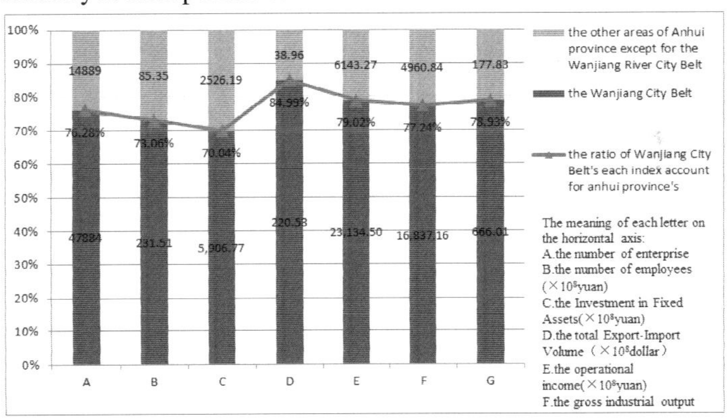

Figure 2. Development of various development parks in the Wanjiang City Belt

3. Data and Methods

3.1. Evaluation model

DEA is a relative efficiency evaluation model. It uses an optimal method to determine the weight of input elements to evaluate the relative efficiency for several DMUs with the same type of multi-input and output system[15,16]. This non-parametric statics method does not require non-dimensionalization or an estimated index and thus, easily provides information to improve DMUs that are not efficient [4], rendering it an ideal multi-objective decision method. At present, DEA is a universally applied method to estimate land use efficiency, including but not limited to urban areas[17-19], development zones [9-11], small towns[20], and counties[21-22]. The land of development zones is a special multi-input and output system. This research uses the DEA method and CCR model to evaluate land use efficiency in development zones.

3.2. Construction of index system and data sources

The present evaluation index system to evaluate the economic efficiency of land use in development zones is based on DEA. It is divides factors into two types, inputs and outputs. Inputs include land, capital, labour, and technology and output are economic benefits. X1 is the supplied construction land area (ha), X2 is the total amount of industrial enterprises' fixed assets investment (million), X3 is the number of practitioners (people), and X4 is the annual expenditure on scientific and technological activities (million). Explanatory variables are land, capital, labour, and technology. Industry total production value (Y1; million), total income of industrial enterprises (Y2; million), and total annual tax (Y3; million) are explanatory variables.

Assuming a time-lag effect of actual land supply and incorporating production factors[23-24], we set the lag period to one year. For the index X1, we use refer to the 2012 dataset of intensive-use assessments for development zones in the Anhui Province. Other indicators are from cross-section data for the end of 2013 and are taken from the development report for the Anhui Province. All of the above are positive variables. We select 40 provincial-level development zones with complete data. The sample accounts for 89% of the total study area, of which 26 counties are located within the area and 14 are in the city (prefecture). To facilitate the identification and simplify the analysis, we number the regions as shown in Table 1.

Table 1. Provincial-level development zones in the Wanjiang City Belt

1 Hefei Yaohai Economic development zone	11 Wuhu Jiujiang Economic development zone	21 Wangjiang Economic development zone	31 Hanshan Industrial zone
2 Hefei Luyang Industrial zone	12 Xinwu Economic development zone	22 Tongcheng Economic development zone	32 Hexian Economic development zone
3 Hefei Shushan Economic development zone	13 Nanling Industrial zone	23 Lu'an Economic development zone	33 Quanjiao Economic development zone
4 Hefei Baohe Industrial zone	14 Wuwei Economic development zone	24 Shucheng Economic development zone	34 Dingyuan Economic development zone
5 Changfeng Shuangfeng Economic development zone	15 Anqing Changjiang Bridge Economic development zone	25 Xuancheng Economic development zone	35 Fengyang Industrial zone
6 Feidong Economic development zone	16 Huaining Economic development zone	26 Xuanzhou Economic development zone	36 Mingguang Industrial zone
7 Feixi Taohua Industrial zone	17 Zongyang Economic development zone	27 Guangde Economic development zone	37 Tongling Lion Rock Economic development zone
8 Lujiang Economic development zone	18 Qianshan Economic development zone	28 Ningguo Economic development zone	38 Tongling Bridge Economic development zone
9 Chaohu Economic development zone	19 Taihu Economic development zone	29 Maanshan Cihu Advanced technology development zone	39 Tongling Jinqiao Economic development zone
10 Chaohu Fuhuang Industrial zone	20 Susong	30 Dangtu	40 Dongzhi Xiangyu Chemical engineering industrial zone

	Economic development zone	Economic development zone

Note: Every development zone is numbered as follows: Hefei 1–10, Wuhu 11–14, Anqing 15–22, Lu'an 23–24, Xuancheng 25–28, Maanshan 29–32, Chuzhou 33–36, Tongling 37–39, Chizhou 40.

4. Results

We consider each development zone as an independent DMU. The total amount of investment is two times greater than the number of variables of output units, which is in line with the requirements of the rule of thumb. Using SPSS (20.0), we conducted a Pearson correlation test for the input and output variables (Table 2). In addition to Y1 and X2 being statistically significant at the 5% level, other input and output variables are significant at the 1%, level, indicating a significantly positive correlation between the amplitude expansion, which is in line with the requirements for input and output variables[25]. We use DEAP 2.1 package and a CCR model and set a variable status to guide investments and variable returns on the basis of a relative comparison of the effectiveness of DMUs with the the same level of output. The evaluation results are presented in Tables 3 and 4.

Table 2. Pearson correlation test results for output and input indicators

	X1	X2	X3	X4
Y1	0.628**	0.374*	0.842**	0.598**
Y2	0.774**	0.983**	0.432**	0.471**
Y3	0.575**	0.469**	0.780**	0.449**

Note: * and ** denote significance at the 5% and 1% levels.

Table 3. Non-DEA efficiency evaluation results for effective DMUs

DMU	Overall efficiency CE	Pure technical efficiency VE	Scale efficiency SE	Stage	DMU	Overall efficiency CE	Pure technical efficiency VE	Scale efficiency SE	stage
1	0.811	1	0.811	-	19	0.839	1	0.839	+
5	0.789	1	0.789	-	20	0.715	1	0.715	+
6	0.955	1	0.955	-	24	0.723	0.764	0.946	+
7	0.895	1	0.895	-	27	0.885	0.902	0.981	-
8	0.812	0.942	0.862	+	30	0.986	1	0.986	-
9	0.828	0.835	0.992	-	31	0.898	1	0.898	+
10	0.936	1	0.936	+	32	0.830	1	0.830	+
13	0.868	0.881	0.984	-	33	0.774	0.806	0.960	-
15	0.894	0.897	0.997	+	34	0.706	1	0.706	+
16	0.938	1	0.938	-	36	0.870	0.884	0.984	-
17	0.838	0.840	0.998	-	37	0.931	1	0.931	+

Note: "+" and "–" indicate increasing and decreasing returns to scale.

4.1. *Comprehensive efficiency*

Overall, the comprehensive efficiency of development zones in the Wanjiang City Belt at the provincial level is high, that is, above 0.7 with an average of 0.918. The industrial park in Hefei Luyang and 17 other development areas meet the CE = 1 and S− = S+ = 0 conditions and achieve the technical and scale efficiency. This suggests that under the present economic and technical levels, input and output of land utilization for the development zones are at its best. The economic efficiency of land utilization is also optimal. The other 22 development zones do not fully achieve DEA efficiency (Table 3). Among them, Changfeng, Susong, Shucheng, Quanjiao, and Dingyuan are the lowest with less than 0.8.

To evaluate comprehensive efficiency by regional distribution, we first calculate the comprehensive efficiency mean of the seven areas and obtain a value greater than two. The average efficiency in ascending order are Tongling (0.977), Xuancheng (0.971), Wuhu (0.967), Maanshan (0.929), Anqing (industry group; 0.9164), and Hefei (industry group; 0.8965). The value for "north core" (Hefei) is less than that for "south core" (Wuhu), the value for "north wing" (Chuzhou) is less than that for "south wing" (Xuancheng), and those for "north core" and "north wing" are lower and along the river development axis (Maanshan, Wuhu, Tongling, and Anqing).

It is noteworthy that the land comprehensive efficiency of the "north core" or the Hefei Provincial Development Zone is not ideal. The main obstacle is that the efficiency of the development zone of Hefei's industrial group is low, while most of the DMUs are not DEA-effective and show diminishing returns to scale. This may be because Hefei's economic development level is higher than that in other cities. In addition, the higher level and superior technical strength contribute to the industrial transfer process, although its collaboration with advanced local enterprises in the same industry is low. However, the technology advantage is not obvious and thus, the technology spillover is smaller[26]. From a regional development perspective, in the present study area[27] , in both zones[14] , the industrial transfer formed Hefei forms a single pattern. The demonstration area (dual core)[13] does not match the current pattern and is not conducive to industries in a city layout for coordination and collaborative development[27]. Moreover, some drawbacks were visible in the land use efficiency of the provincial-level development zone. In addition, comprehensive efficiency of spatial distribution in the provincial development zone reflects the industrial belt along the Yangtze River and the process of industrial transfer may be relative to the proper handling of the relationship between efficiency and benefit, which is helpful to future rapid and efficient growth.

In addition, the evaluation results show that in the absence of an effective DEA, of the 22 development zones, the three zones located in the city (district) and 19 in counties (cities) account for 21.43% and 73.08% of all types of development zones. In contrast, the land utilization efficiency at the county level is not optimistic. This can be attributed to the following aspects. First, most of the cities and counties under the jurisdiction of the development zone are located downtown and in the suburbs. In addition, the development zone and central city show an integrated trend for planning and construction and the land utilization efficiency for the county (city) and development zone is lower than that in the city and has a certain rationality. Second, combined with the development of the study area, most of the counties are located in the provincial development zones and the economic volume and scale are too small. The industrial clustering effect is weak and the industry characteristics are not obvious. Further, the industry chain is incomplete and competitiveness relatively low. In particular, during the period of an economic decline, the larger the county development zone, the bigger the total pressure assessment. Some counties (cities) in existing development zones are yet to form a sufficiently large scale and ignore quality in an attempt to achieve quantity.

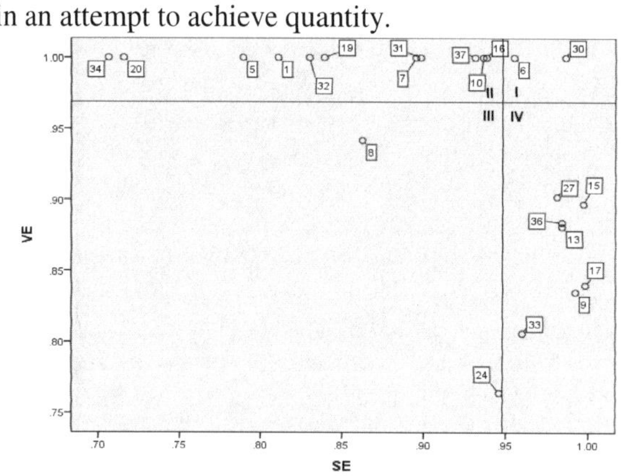

Figure 3. Non-DEA effective VE-SE DMU map

4.2. Technical and scale efficiency and stage of scale returns

Comprehensive efficiency alone cannot reflect the relationship between the comprehensive efficiency value of each DMU and the technical and scale efficiencies. The above results of each efficiency values are generally higher. Referencing an existing research[28], we drew a scatter plot based on technical efficiency (VE) and scale efficiency (SE) using SPSS 20.0. We adopt two efficiency means (0.969, 0.948) for all DMUs as the division standard. We were able to draw a VE-SE spatial map for non-DEAs and completely effective DMUs (Fig. 3). Combined with its stage of scale returns (Table 3), the research indicated the following.

First, five economic development zones, Quanjiao (−), Chaohu (−), Zongyang (−), Guangde (−), and Anqing Yangtze River Bridge (+), and two industrial parks, Nanling (−) and Mingguang (−), are located in the fourth quadrant; their comprehensive efficiency is low due to the low technical efficiency. In other words, there are heavy scale and light technology issues in these development zones, mainly owing to scale efficiency needed to maintain a certain level of development. In addition, apart from the economic development zone of the Anqing Yangtze River, all zones show a decreasing returns to scale. Therefore, there is a need to control the scale of land use, improve the management of industrial parks, and promote technological progress to improve the industrial structure and achieve transformation and upgrading.

Second, eight economic development zones, Dingyuan (+); Susong (+); Huaining (−); Taihu (+); He county (+); Shuangfeng, Changfeng Country (−); Yaohai district, Hefei city (−); and Shizishan district, Tongling city (+), and three industrial parks, Hanshan; Taohua Town, Feixi Country (−); and Fu huang (+), are located in the second quadrant. Their comprehensive efficiency is low mainly due to the low scale efficiency. Among the 11 development zones, the returns to scale of Huaining, Yaohai district, Shuangfeng, and Taohua Town continue to decrease and thus, continuous attention is needed towards controlling the scale of land use. As for the other seven development zones, while promoting technological progress a certain degree of integration and expansion should be considered and the scale properly expanded if conditions and policies permit, thus taking its scale efficiency to a new level.

The two development zones, Lujiang (+) and Shucheng (+), are in the third quadrant. Both zones report lower than average technical and scale efficiency. The technical efficiency of Shucheng's economic development zones is at the lowest among all development zones.

We found 20.0% of input redundancy and 22.5% of output deficiency in the provincial development zones at different levels in the study area (Table 4).

Table 4. input redundancy S- and insufficient output S+

DMU	X1	X2	X3	X4	Y1	Y2	Y3
8	11.90	0	0	0	407313.08	0	11951.36
9	0.21	0	519.02	0	418144.90	0	11753.99
13	3.42	0	158.55	0	0	77880.21	59777.41
15	432.51	41181.96	0	0	0	0	17755.60
17	1.49	0	4731.03	0	0	0	9162.60
24	0	0	0	0	128837.18	0	34828.17
27	0	0	238.08	74318.96	0	0	62596.95
33	16.42	0	2176.72	103.07	101335.56	0	35271.15
36	0	0	4267.75	0	477596.71	0	35460.81

Under the precondition of maintaining the existing output level, 40 development zones needed input and structural adjustments in the Wan Jiang City Belt: construction land (465.95 ha), industrial enterprise investment in fixed assets (412 million yuan), employment population (12,100 people), and science and technology activities (744 million yuan) accounted for 2.97%, 0.13%, 1.32%, and 0.13%

of total inputs. From the viewpoint of insufficient output, under the precondition of maintaining the existing investment level, improving the efficiency of production inputs can increase industrial output by 15.332 billion yuan, industrial enterprise income by 779 million yuan, tax by 2.786 billion yuan that, respectively, account for 2.76%, 0.20%, and 13.84% of actual output. Some development zones did not most positively affect tax in leveraging the reversed efficiency of transitioning enterprises and actively adjusting industrial structure. In general, the degree and range of input redundancy and output deficiency are small.

5. Conclusions

The main conclusions from this study are as follows.

First, the comprehensive efficiency of the Wan Jiang City Belt's provincial development zone is high, but there remains room for improvement. Increased capital investment and average land investment in science and technology can improve the land utilization and economic efficiency of the studied area.

Second, the comprehensive efficiency of regional distribution shows two characteristics: The comprehensive land use efficiency in Hefei's "north core" is not ideal, but and the development axis along the river shows good efficiency. Second, the efficiency of county development zone is a focus within provincial development zones in the study area, especially in the north of development zones along the Yangtze River.

Third, improving development efficiency should be kept separate from improving technical and scale efficiency. The technical–scale efficiency map is helpfull for identifing reasons underlying the low efficiency of the 22 development zones.

6. Policy suggestions

From the above analysis, we propose the following suggestions.

First, it is important to manifest a dominant position to undertake industrial transfers of industries along the upper reaches of the Yangtze River, promote coordinated development of regional industrial. Moreover, coastal industrial transfers should be of high quality. In addition, the investment scale should be appropriately increased and land and industrial policies should aim at the development axis along the river to promote the transfer of industries. Finally, the regions must continue to nurture high-tech industries and strategically emerging industries; lead the development of regional clusters of industries[26]; and facilitate technological advances in production efficiency.

Second, focus should be on promoting land use efficiency in the development zones of counties. It should encourage high-tech development zones and municipal and county jurisdiction development zones should cooperate to build gardens and other zones and have a certain capacity for independent innovation for multi-level, multi-field cooperation and interactions with other development zones.

Third, development zones stocks should be controlled, optimized, and integrated. In addition, a project access system should be established and improved upon and be strictly enforced in industrial land to control land use, monitor multi-story factory buildings, and improve the level of intensive land conservation zones. Development zones short of land and needing expansion should be prioritized through integrated approaches. Efforts should also be made to distribute idle land among development zones, regularly conduct clean-ups, and explore secondary development of inefficient land use in the zones.

Finally, taxation should serve as an incentive to advance quality and efficiency. On the one hand, it strengthens tax administrations and can be implemented in every tax system, and on the other hand, it establishes a sound system for differential taxation to encourage and support the strategic emergence of high-tech industries through, for example, tax rebates.

7. Reference

[1] Wang,H.F.,Shi,Y.S.,&Yin,C.Y.(2014). Land use efficiencies and their changes of Shanghai's development zones employing DEA model and Malmquist productivity. Geographical Research,33(9),1636-1646.

[2] Zhai,W.X., Huang,X.J., & Zhang,Q. (2006). Analysis on land intensive use of economic developing zones-a case for Jiangsu province. Areal Research & Development, 25(1),101-105.

[3] Wu,Y.L.,Qu,F.T., & Feng,Z.L.(2006). Analysis on region diversity of land resource allocation efficiency in development zone in China. China Population Resources & Environment,16(5),112-116.

[4] Fu,L.P.,Gu,Y.J.(2008). Evaluation of land use efficiency based on data envelopment analysis. Journal of Xidian University,18(3),103-107

[5] Song,D.,Zhong,S.M.,Li,S.J. (2009). Efficiency analysis of land use in state-level development zones of china based on DEA model. Scientific & Technological Management of Land & Resources.,26(6),94-98.

[6] Zhuang,H.W., & Li,H. (2011). The evaluation research of industrial land use efficiency of different regions development zone in Hunan. Economic Geography, 31(12),2100-2104.

[7] Zhang,L., & Wang,Y.H. (2014). A study on influencing factors of industrial land output efficiency from the perspective of micro-enterprises—an empirical analysis based on 2088 samples of industrial enterprises. East China Economic Management, 28(9),43-48.

[8] Bai,X.J., Jiang,K.,& Pang,R.Z.(2008). Operation efficiency and upgrading models of main national-level development zones in china from foreign capital and land utilization. China Industrial Economics, 22(8), 26-35.

[9] Jiang,L.W., & Zhao,X.M.(2010).Application of land intensive use scale efficiency evaluation in development zone based on data envelopment analysis method. China Population Resources & Environment, 20(3),115-118.

[10] Zheng,Y.(2012). Research on the efficiency of resource allocation for shanghai development zones based on data envelopment analysis. Chinese Journal of Management Science, 20(11),198-203.

[11] Cai,S.Z., & Lu,L. (2014). Efficiency measurement and spatio-temporal differentiation of economic and technological development zones in China. Scientia Geographica Sinica, 34(7),794-802.

[12] Zou,W., Zhao,J., & Wu,Q.(2014). Impact of property tax on land use behavior of industries: evidences from typical industrial enterprises in Jiangsu province. China Land Sciences. 28(1),60-67.

[13] National Development and Reform Commission. Urban Zone Undertaking Industrial Transferring Demonstration Area Program along Yangtze River in Anhui.2010-01-21

[14] Qi, X.Y, Yang, X.L, & Li, G. (2014). Spatial variations of land use intensity in state-and-province level developing zones in wanjiang urban belt. Chinese Agricultural Science Bulletin, 30(15), 157-162.

[15] Li,S.J.,Song,D.,Liu,Z.S.,&Dou,S.,(2012). Regional effect analysis of intensive land-use of development zones:a case study in the middle development zones of Jilin province. China Population Resources & Environment, 01,311-319.

[16] Guo,G.C.,&Xiong,Q.(2014) Study on the Urban Industrial Land Use Efficiency and Its Influencing Factors in China. China Land Science. 28(4),45-52.

[17] Wu.D.W., Mao,H.Y., & Zhang,X.L., (2011). Assessment of urban land use efficiency in China. Acta Geographica Sinica, 66(8), 1111-1121.

[18] Zhang,Z.H.(204).Study of Chinese Urban Land Use Efficiency.The Journal of Quantitative & Technical Economics,134-149.

[19] Xu,J.W., Xu,X.Y., & Zhu, M. X. (2013). Urban land use efficiency and its change of the Yangtze River delta based on data envelopment analysis. World Regional Studies, 22(3),121-129.

[20] Song,G.J., Peng,Y.K.(2005).An empirical analysis based on the land use efficiency of small towns in Chengdu.Statistics and Decision,40-41.

[21] Shen,C.L., Li,M.C., & Li,X.F.(2011). Assessment of rural land use efficiency in Fenshui town based on DEA. China Land Science. 25(1),16-21.

[22] Zhang,Y.W, Zhang,F.G, & Chen,Y.F. (2010). An assessment of agricultural production efficiency using the dea model in a transect area along the national highway 204 in Jiangsu province. Resources Science, 32(2), 353-358.

[23] Dong,G.X., & Zou,J. (2006). The positive analysis of contribution of arable land loss to economic development——a study of Zhejiang. Journal of South China Agricultural University.5(3),41-47.

[24] Zhang,W.B.,Chen,Y.,& Zhang,R.Z. (2013). Study on the relationship between cultivated land occupation and economic growth based on the decoupling method —taking Kangle county of Gansu province as an example. Journal of Natural Resources, 28(4),560-570.
[25] Xie, H.J, & Zhang, H. (2015). Productivity change and convergence of hi-tech industry in Yangtze Economic zone. Science & Technology Progress & Policy, 24:34-39.
[26] Zhu, P. F., Zhang, H., & Zhang, H. M. (2014). The efficiency research on the wan-jiang urban belt receiving industrial transfer from the Yangtze River delta region under the constraints of undesired output. China Soft Science, 07,105-114.
[27] Ding,J.Y., Feng,D.L.,& Fu,X.J.(2012). Research Report on the economic development of Wanjiang. Beijing: Economic Science Press, 139-146.
[28] Xi,M.Q.(2012). Emprical research on the relationship between city efficiency and city size—— based on city panel data from 2001 to 2009 in China. On Economic Problems, 37-41.

Acknowledgments
The authors would like to thank the Project of the National Natural Science of China (Grant No. 41371155), Ministry of land and resources of China (supported by "public welfare scientific research funding", Grant No. 201111010-01), Ministry of Agriculture of China (Soft science project, Grant No. 20140604), and Department of land and resources of Anhui Province (Soft science project, Grant No. 2012010). Their assistance is gratefully acknowledged.

IOP Conf. Series: Earth and Environmental Science **178** (2018) 012036 doi:10.1088/1755-1315/178/1/012036

Analysis on Updated Results of Benchmark Land Price in Small Towns — A Case Study of Downtown in Xichong County, Nanchong City

Fachao Qin[1], Zhen Xu[2] and Ting Jiang[3]
1 Profess ,China West Normal University, Nanchong, China
2 Student, China West Normal University, Nanchong, China
3 Student, China West Normal University, Nanchong, China
Email: larcqfc@163.com

Abstract. As an important constituent of Chinese urban land price system, benchmark land price is the basis of land price evaluation in China. Taking downtown of Xichong County, Nanchong City as an example, this paper calculates the benchmark land price in the downtown of Xichong County by combining the land classification ranking and benchmark land price. According to the analysis of the updated results, the benchmark land price in Xichong County at each level is as follows: commercial land>residential land>industrial land, and 2017>2011. Based on the comparison of the standard land price of some counties in Nanchong, commercial land: Langzhong>Nanbu>Xichong; Residential land: Langzhong>Xichong>Nanbu; Industrial land: Langzhong> Xichong>Nanbu. The updated benchmark land price results of Xichong County are basically consistent with the local social and economic development level, and the standard price of industrial land meets the requirements of relevant documents. Therefore, the updated results of benchmark land price in Xichong County are scientific and reasonable.

1. Introduction

Benchmark land price refers to the basic standard land price of state-owned land in cities and towns. That is, in accordance with the scientific assessment methods, according to various land types, land transactions and real income in towns, the average price of commercial land, industrial land, and housing land in land range at different levels or homogeneous areas is assessed in a certain period[1]. It reflects the general level and spatial distribution of urban land price. According to benchmark land price, the country implements macro control and management of land prices in the real estate market. Meantime, it is also the basic standard for the state to levy the tax on land use and participate in the distribution of urban land revenue[2] [3]. The rapid economic development in small towns accelerates the pace of urbanization in China. In recent years, with the promotion of the new rural construction, urban and rural integration, new industrialization and major construction projects, the economic level in the county has been continuously improved. Particularly, land transactions have become increasingly active, and rural construction land circulation has been increased. Therefore, it is particularly important and urgent to determine the scientific and reasonable benchmark land price of small towns[4] [5], which is a unified criterion and key work to coordinate the price of all kinds of land parcels[6].

2. Regional survey

Xichong County is located in the center by north of Sichuan Basin, the watershed between the Jialing River and the Fujiang River. The geographic position is in 30°52′04″~31°15′07″N and 105°36′04″~

106°24′07″E[7]. It is in the back zone of the Jialing River and the Fujiang River, and belongs to shallow hill topography. The terrain is high in the northwest and low in the southeast. The mountains range from north to south, and the altitude is 256~889 meters, belonging to the middle subtropical humid climate zone of Sichuan Basin. The update area of urban standard land price in Xichong County is 17.79km², including the built-up area and the planning area. Xichong County is under the administration of Nanchong city, Sichuan Province. The abundant labor force and low land price create a good development condition for the industrial diffusion in Nanchong City and the industrialization development in Xichong County.

3. The methods and purpose of urban benchmark land price updating in Xichong County

3.1. Methods

According to Regulation for Gradation of Urban Land (GB/T 18507-2014), the land grading factors that affect land types (commercial, residential, and industrial) are selected to calculate the factor action score[8]. According to the weight determined after grading various factors by several experts, the factor function scores are superimposed in the coordinate network of the entire grading range. According to the total score after the superposition, the total score frequency curve method is adopted to preliminarily divide the land grade based on the score frequency histogram, and the land grade is verified by the differential income of land. According to Regulation for Valuation of Urban Land (GB/T 18507-2014), in the benchmark land price assessment, according to the land price data of the real estate market in Xichong County, the principle and method of land price calculation are employed to calculate the sample land price of the land use type in each grade and the benchmark land prices at all levels with the classified land level as the unit.

3.2. Purpose

With the establishment of the objective model of the socialist market economic system reform in China, the system reform of the paid use system of urban land has been continuously deepened, which not only realizes the land ownership and the rural collective land ownership economically, but also promotes the clarification of the land property right and forms market circulation by continuous decomposition. This effectively enhances the effective management of land resources and the efficient use of land assets. Land value and land price management should be based on scientific urban land grading and evaluation. The important significance and function of urban land grading and benchmark land price evaluation are as follows: (1) Conducive to maintaining the normalization, institutionalization and healthy development of land market management; (2) Promote the implementation of the overall land use planning; (3) Enhance the implementation of the national macro-control policy.

Urban land price is the inevitable outcome of urbanization and economic development, which is determined by the scarcity and non-mobility of land. As a fixed production factor, the land has formed a dynamic market. Strengthening land price management and control is not only an important prerequisite to ensure the rational and orderly development of urban land, but also a vital means to regulate land transaction and manage the land market. In addition, scientific and reasonable formulation and updating of benchmark land price are an intensive and efficient use of land, and the important basis for tax collection and calculation of land value. Therefore, it is very important and necessary to scientifically and reasonably define land price.

4. Results and Analysis

4.1. Connotation, basic condition and expression of benchmark land price

The dynamic land quality grade and land price change with time and affect land factor and land price[9]. Therefore, the results of grading and evaluation should be determined on a unified date so as to ensure the comparability of sample land prices and results[10]. Considering the base data requirements for formulating annual land price index, the urban land grading and valuation reference date of Xichong County is set at January 1st, 2017.

IOP Conf. Series: Earth and Environmental Science **178** (2018) 012036 doi:10.1088/1755-1315/178/1/012036

Taking into account the convenience and availability of the benchmark land price results, the benchmark land price evaluation unifies urban land development degree as "Six-accessible to and one-leveling " beyond the parcel according to the actual condition of Xichong County.
The use year of various types of land in Xichong County is set as the statutory maximum age according to Land Management Law of the People's Republic of China, namely, land use year: 40 years for Commercial land, 70 years for residential land, and 50 years for industrial land. In accordance with parcel control index of Overall Planning of Xichong County, combined with the investigation and measurement of commercial, residential and industrial land samples in Xichong County, the average volume rate of the updated benchmark land price is 1.8, the average volume rate of residential land is 1.8, and the average volume rate of industrial land is 1.

4.2. Analysis of the benchmark land price update results

4.2.1 Analysis of benchmark land price level of different land types With the change of land location, land prices of various land types at different levels show obvious changes, and the land price decreases from the center to the outside. These changes are mainly caused by not only differences in the prosperity and infrastructure of the region, but also market factors. Generally, the grade boundaries of the benchmark land prices in different land types in a town are different because of the different degrees of the influence factors of each land type. As shown in Table 1, the levels of land in Xichong County: commercial land>residential land>industrial land.

Table 1. The results of benchmark land price of commercial and residential, industrial land

Use	Class	Value（RMB/m²）	Area（m²）
Commercial land	I	3150	1110161.5
	II	2070	3640780.2
	III	1260	3932223.5
	IV	825	9113642.7
Residential land	I	1455	3291192.2
	II	1080	4300880.3
	III	825	3482593.7
	IV	540	6721141.7
Industrial land	I	450	1077517.3
	II	330	3593962.7
	III	225	6066661.1
	IV	142	7058661.8

Generally speaking, the land price of commercial land is most sensitive to the location condition[11]. Urban commercial land price in Xichong County changes in the range of 825～3150 CNY/m². Compared with residential land and industrial land, the changing amplitude of commercial land is the largest at each level. The urban area is the commercial center of Xichong County, and the location factors (the prosperity of agglomeration factors and business services) are better. The revenue of commercial land is higher than that of residential land and industrial land. Most residential buildings are mainly built or renovated by self-raised funds, and the affordability of units and individuals is lower. The commercial housing market is not active enough. Most of the industrial enterprises in the town suffer loss, and the re-development ability is poor. Therefore, the price of residential land and industrial land is relatively low.
The variation range of benchmark land price of residential land is 540~1455 CNY /m². Land prices at various classes are smaller than commercial land and larger than industrial land. The first class residential land belongs to the commodity building in the central area. Therefore, the prices are significantly higher than the second, third and fourth classes of residential land. In the first and second classes of land, the benchmark land prices of commercial land and residential land vary greatly. As the

class falls, the gap becomes smaller. This is related to the sensitivity of different land types to the location of lands.

The variation range of benchmark land price of industrial land is 142~450 CNY/m². The sensitivity of industrial land to land location is the lowest, and the industry is concentrated in industrial parks of Xichong County, and is remote from the center area. Therefore, the benchmark land price of industrial land is the lowest, and the variation range is the smallest. According to Notice on Promulgation and Implementation of the Lowest Price Standards of Industrial Land Transfer in China, the lowest standard of industrial land transfer in Xichong County is 84 CNY/m². In the benchmark land price update, the benchmark land price of the last class of industrial land in Xichong County is 142 CNY/m², which is higher than the minimum standard. Therefore, the benchmark land price of industrial land meets the requirements of the document.

4.2.2 Comparison with the previous benchmark land price level In the last round of benchmark land price assessment, the land volume rates of the three types of lands are as follows: the volume ratio of commercial land is 2.1, the volume ratio of residential land is 2.1, and the volume ratio of industrial land is 2.1. The base date of assessment is January 1st, 2011. Because of different volume rates of the two rounds of benchmark land price in Xichong County, in order to make the achievements of the two rounds more comparable, the connotation of the benchmark land price is unified. The benchmark land prices in the two rounds are converted to land price per floor area, and the results can objectively reflect the changes in the level of land use prices.

According to the comparative analysis of the two rounds of benchmark land price and floor price results, as shown in Figure 1, all kinds of land price per floor area in Xichong County have significantly increased. The growth rate of commercial land price decreases first and then increases. In particular, the growth rate of the fourth class of commercial land prices increases the most obviously, with an increase of 21.94%. The growth rate of residential land price gradually decreases with the decline of the class. The growth rate of the first class of residential land prices increases the most obviously, with an increase of 37.58%. The growth rate of industrial land price increases first and then decreases with the decrease of class. The growth rate of the second class of industrial land prices increases the most obviously, with an increase of 39.50%.

After the benchmark price formulation for five years in 2011, the social economy in Xichong County has steadily developed, which promotes the development of real estate market and drives the rise of the price level in Xichong County City. This is mainly reflected in the increase of all levels of land price per floor area of commercial land, residential land, and industrial land. Because of the adjustment of urban planning, the extension of the urban area, the enhanced intensity of land use, and the higher cost of land expropriation and house demolition in recent years, the cost of land acquisition is increased. Therefore, the price of the fourth class of commercial land changes the most obviously. Residential land price growth is most evident in the first class of residential land. Due to the rapid development of real estate market in Xichong County in recent years, which is the most obvious in the residential location (traffic and public infrastructure), the growth rate of the first class of residential land is the largest, and the growth rate of the second class of industrial land is the most obvious. In recent years, since the establishment of Industrial Park in Xichong County, the public facilities are further improved, which enhances the land curing degree and optimizes land location condition. Moreover, the industrial park is mainly located in the second class area with convenient transportation and favorable topographical conditions, and the growth rate of land price in the second class of industrial land is the most obvious.

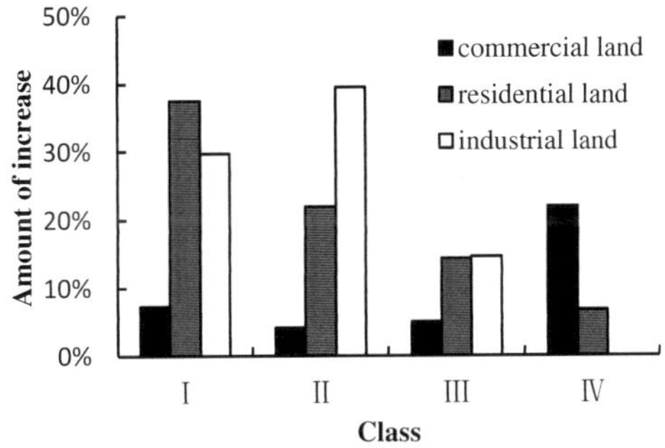

Figure 1. The comparing the results between the benchmark price in 2011 and 2017

4.2.3 Comparative analysis of benchmark land price level in different regions The benchmark land price updates in different counties in Nanchong are basically synchronous with the horizontal comparison. By comparing the overall level of benchmark land prices in some counties in Nanchong, the difference in the overall regional analysis of benchmark land price in Xichong County is analyzed to judge the rationality of the overall level of benchmark land price. As a result of different volume rates in various counties, this paper calculates the land price per floor area to intuitively judge the price difference between Xichong County and other counties. Meantime, the overall land price level of four cities and towns is calculated according to the weighting of areas at each class.

As shown in Table 2, the commercial land Langzhong>Nanbu>Xichong, Langzhong is the county-level city of Nanchong County, and tourism has risen in recent years. Therefore, the price of commercial land is higher than that in Nanbu and Xichong. The price of commercial land in Xichong County is the lowest, and the business prosperity is slightly lower than that in the other two counties. The urban area is smaller, and the population is the least. As for residential land, land Langzhong>Xichong>Nanbu, the development of the ancient city in Langzhong promotes the rapid development of the real estate industry. The beautiful environment attracts most people settled here, and the real estate market in Xichong has become more active in recent years. For industrial land, Langzhong>Xichong>Nanbu, the establishment of Xichong Industrial Park has introduced a large number of industrial enterprises. The development of vegetable and fruit industry increases the processing enterprises of fruits and vegetables. Therefore, the benchmark land price of industrial land in Xichong is higher than that in Nanbu.

Table 2. The comparing the results of the part of county in Nanchong

county	commercial land	residential land	industrial land
Langzhong	1286	1034	463
Nanbu	983	428	279
Xichong	734	497	330

5. Conclusion

According to the statistical results, we can find the difference of benchmark in time and space in Nanchong County, and the necessity to update the benchmark price in Xichong. In the meantime, the research results show that the updating of the benchmark land price in Nanchong County is in line with the actual development of the local economy. The lowest price of industrial land conforms to the relevant regulations of the state. The update result of benchmark land price in Xichong County is scientific and rational, and the update results can provide a reference for latter research. The timely updating of benchmark land price has an inevitable connection with the regional economic development, and the rationality of the price also plays an important role in promoting the process of urbanization. However, the data collection and calculation in the benchmark land price update consume a great deal of manpower and financial resources. Therefore, the method of determining the benchmark land price needs to be further explored so as to provide more accurate and convenient ways to obtain scientific and reasonable results.

6. References

[1] WANG Z, YANG J, QIANG Y. Study on Benchmark Land Price Evaluation in Small-town Cluster [J]. China Land Sciences, 2014, 28(8): 72-79.

[2] WANG L, YANG J, LI Y, et al. Discussion and Assessment on Benchmark Premium of Collective Construction Land Based on the Correction Method [J]. Chinese Agricultural Science Bulletin, 2013, 29(14): 75-80.

[3] CHUAI X, HUANG X, XU Y. The Preliminary Study on Benchmark- Price of the Collective Land in Village—Take LiangYu Village in Anhui Province as an Example [J]. ECONOMIC GEOGRAPHY, 2012, 32(2): 121-126.

[4] Kaier Liao, Jianglong Tang. Study on Benchmark Land Price Update and Some Problems in Small Towns [J]. Journal of Hubei University for Nationalities(Natural Science Edition), 2017, 35(1): 111-113.

[5] Chiqiang Hu, Shiyuan Hu, Xu Tang, Hong Geng. Approach for Evaluating Town Land's Benchmark Prices with Sparse Land Transaction Samples. Journal of Hubei University for Nationalities(Natural Science Edition), 2016, 34(1): 93-98.

[6] Jing Shen, Qun Wu. Approaches to appraising the standard land price in China. Resource and industries, 2007, 9(1): 74-76.

[7] Hongyi Pan, Ling Han, Guiguo Jiang, Wei He. Research on the Key Technology of Updating Cultivated Land Quality Evaluation Combining the Data of Farmland Gradation & the Second National Land Survey: A Case Study of Xichong County. China Land Sciences, 2014, 28(5): 65-72.

[8] LIU Y, CHEN Y, BAI Z, et al. The Discussion of Related Problems of Appraisal of Standard Land Price in Town's-Taking Gangu County of Gansu Province as An Example [J]. Chinese Agricultural Science Bulletin, 2014, 30(17): 62-68.

[9] Yufeng Zhang , Jing Li. Comparative analysis about reference land price and its influencing factor of the towns in Hohhot. Geograghical research, 2007, 26(2): 373-382.

[10] CHEN D, LAN Z, LIU Y, et al. Discussion on the balance of urban datum land prices based on city gravitation model [J]. Science of Surveying and Mapping, 2013, 38(1): 63-67.

[11] WU W, CHEN Y, ZHANG R, et al. Urban Benchmark Land Price Balance from the Perspective of Spatial Diffusion Equation [J]. China Land Sciences, 2014, 28(5): 43-49.

IOP Conf. Series: Earth and Environmental Science **178** (2018) 012037 doi:10.1088/1755-1315/178/1/012037

Analysis on Industrial Land Efficiency in Sichuan Province from the Perspective of Sustainable Development

Xu Mingqiang[1,2] and Wei Wei[3]

1 Chengdu University , Chengdu, China
2 Southwestern University Of Finance And Economics, Chengdu, China
3 Chengdu Teachers College, Chengdu, China
E-mail: thinker127@swufe.edu.cn

Abstract. To promote the efficiency of industrial land in Sichuan province is an inevitable requirement of realizing sustainable development. Through the improvement of the Cobb-Douglas production function to construct the Two-way Fixed Effects panel model of output efficiency of urban industrial land, the data regression analysis of 93 prefecture-level cities since 1998 has found that the growth rate of industrial land efficiency indexes in Sichuan province is higher than the national level and has fallen back in recent years, but the absolute level is always lower than the national average, due to the fact that in the chase of national level, Sichuan industry did not fundamentally eliminate the root cause of low efficiency of industrial land. As a result,it is necessary to realize the improvement of optimal land utilization and land use efficiency through scientific planning, progressive construction steps and industrial transformation and upgrading, etc.

1. Introduction

As is known to all, "Our Common Future" advocates sustainable development which can not only meet needs of contemporary people, but also cause no harm for the future generations to meet their needs. (WCED, 1987). Land resource utilization is an important domain in sustainable development theory. But for a long time, people generally emphasize the prevention of soil degradation in cultivated land use, and pay little attention to sustainable use of industrial land. However,"Empty Towns" and "Commuter Towns" are not rare due to large industrial land area in Sichuan province, the lack of industrial investment and insufficiency of industrial population agglomeration. Both Xi Jinping and Li Keqiang sharply criticized the problem of overuse of urban industrial land and inefficient use of urban land [1]-[2]. Therefore, how to improve efficiency of industrial land under the guidance of sustainable development theory, save land, and do not damage land security needed by future generations because of extensive industrial land, is a question worth studying.

The efficiency of industrial land can be measured by investment intensity, employment density and output efficiency,etc. From the perspective of micro-enterprises to explore the factors affecting the output efficiency of industrial land, Lin Zhang and Yahui Wang found that the output efficiency, elastic coefficient of average capital of industrial land is 0.346 and elastic coefficient of average labor of industrial land is 0.142[3].Antonio Ciccone and Robert E. Hall(1996) found that employment density in counties of the United States was doubled, and average labor productivity could be increased by 6%[4]. Some scholars emphasize that Total Factor Productivity (TFP), especially technological progress, has an important impact on the output rate of industrial land[5]-[6]. Scholars also found that the Plot Ratio and investment intensity cannot reach construction control standards[7]-[8], the proportion of land supply and demand in urban industrial land is too high[9], land and labor input is too much[10], which makes industrial land low in efficiency. After retrieving, it is found that the study on efficiency of industrial land in Sichuan province is very rare and needed to be strengthened.

2. Data and Descriptive Analysis

In order to carry out the evaluation and comparison of present situation, from "Yearbook of China Development Zone"、"China Statistical Yearbook"、"China City Statistical Yearbook" 、"China Urban Construction Statistical Yearbook", "Sichuan Statistical Yearbook" and other channels, the research group collected data. For the sake of comparative convenience, urban industrial land in central cities of provincial capitals were not included, while urban industrial land within urban districts of the prefecture level cities were mainly investigated. From this, 1998-2014 years of balance panel data collected from 23 prefectural cities in Sichuan province have been formed. As a reference, by stratified sampling method from 26 provincial administrative regions (excluding 4 municipalities), they selected 93 prefectural cities sample units, including 33 sample units within the scope of western prefecture-level cities. We obtained the balance panel data of industrial land,industrial employment,industrial enterprises fixed assets and gross industrial output value in 1998 -2014. Since the establishment date of each industry park is different, and many data is missing, the panel data obtained by the research group is unbalanced panel, but it has little effect on the analysis.

The work of data processing includes: (1) according to "China Statistical Yearbook" to calculate the GDP deflator,then gross industrial production over the years and the value of fixed assets are converted in accordance with the price in 1990, in order to achieve comparability; (2)in series of sample individual time, the individual missing data and obviously too high or too low data is inferred as statistical errors, replaced by mean value of sample data before and after the year or inferred by the average growth rate several years ago; (3) if a variable of a sample point for 2 consecutive years shows evident outliers or missing values, then observation value should be deleted of the year ; (4)ordinary Box-plot-- taking "75% quantile+1.5 times quartile spacing" within the side of fence on Box-plot, taking "25% quantile－1.5 times quartile spacing " as within the side of fence below the Box-plot -- find out outliers and delete them.

Observing sample mean over the years (see Table 1), it can be found that the national, the west of China and Sichuan urban industrial land output efficiency and investment intensity in general is in a gradual upward trend and employment density are generally in a downward trend year by year, possibly due to improvement in total factor productivity such as technological progress. It is worth noting that the output efficiency, investment intensity and employment density of industrial land in Sichuan cities are lower than the national average and also lower than the western average (see Figure 1-Figure 3). According to the efficiency index of industrial land, City-Industry Space Integration level in Sichuan province needs to be improved.

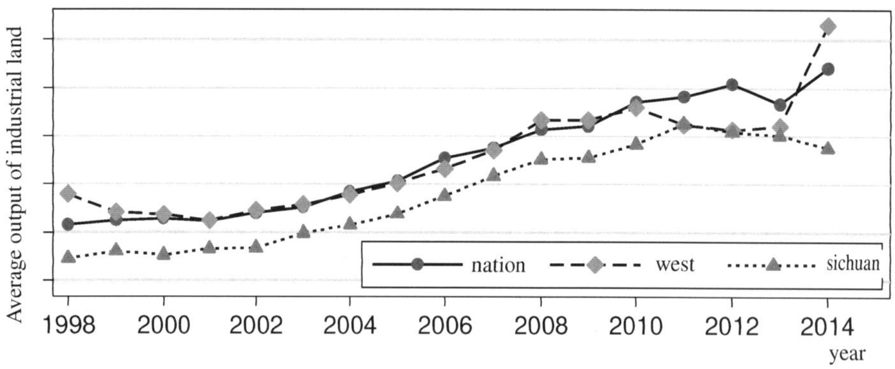

Figure 1. comparison of output rate of urban industrial land between Sichuan province and the nation and the west**Table 1.** comparison of mean value over the years of efficiency index of industrial land in cities[a]

Year	Average output of industrial land (100 million yuan / sq km)			Average investment intensity of industrial land (100 million yuan / sq km)			Average employment intensity of industrial land (10 thousand people/ sq km)		
	whole nation	the west	sichuan	whole nation	the west	sichuan	whole nation	the west	sichuan
1998	5.814	8.939	2.328	3.799	4.056	1.826	0.8	0.756	0.657
1999	6.222	7.075	3.053	4.201	3.815	2.047	0.748	0.683	0.659
2000	6.437	6.852	2.651	4.353	5.837	2.195	0.656	0.567	0.558
2001	6.196	6.207	3.306	3.955	3.656	2.184	0.577	0.486	0.507
2002	7.07	7.287	3.387	4.225	3.943	2.221	0.551	0.453	0.435
2003	7.586	7.889	4.946	4.151	3.733	2.849	0.532	0.449	0.482
2004	9.243	8.877	5.789	3.911	3.695	3.038	0.515	0.435	0.469
2005	10.34	10.11	6.893	3.95	3.919	3.184	0.464	0.424	0.44
2006	12.7	11.61	8.774	4.675	4.588	3.421	0.484	0.451	0.499
2007	13.74	13.5	10.86	4.839	5.001	4.557	0.462	0.452	0.401
2008	15.66	16.68	12.58	5.03	5.493	4.287	0.427	0.443	0.369
2009	16.05	16.68	12.75	6.07	6.191	4.883	0.394	0.388	0.319
2010	18.57	17.97	14.16	6.901	8.13	4.944	0.391	0.365	0.272
2011	19.11	16.2	16.3	6.512	6.531	5.393	0.405	0.377	0.298
2012	20.39	15.67	15.44	7.07	7.587	5.75	0.443	0.436	0.287
2013	18.37	16.05	15.1	6.587	6.498	4.972	0.428	0.365	0.302
2014	22.11	26.5	13.77	6.517	7.235	5.114	0.462	0.424	0.282

[a] note: (1) For longitudinal comparison, output value and asset value have been converted in accordance with 1990 prices, the same below.(2) The data comes from "Yearbook of China Development Zone"、"China City Statistical Yearbook" and "China Urban Construction Statistical Yearbook" or "Annual Report on Urban Development of China".

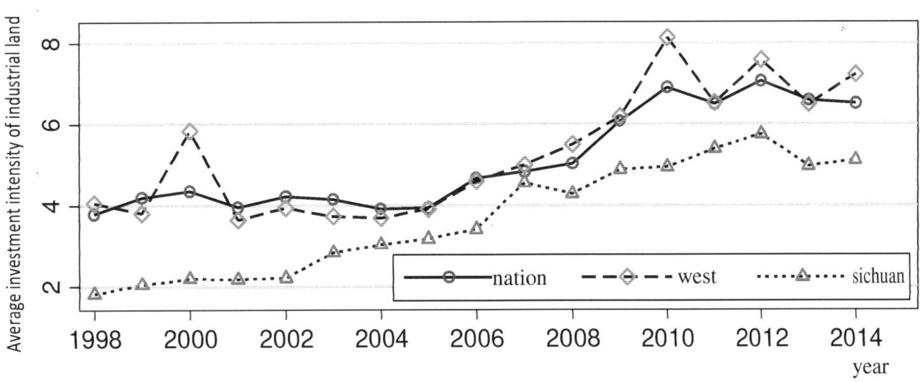

Figure 2. comparison of investment intensity of urban industrial land between Sichuan province and the nation and the west

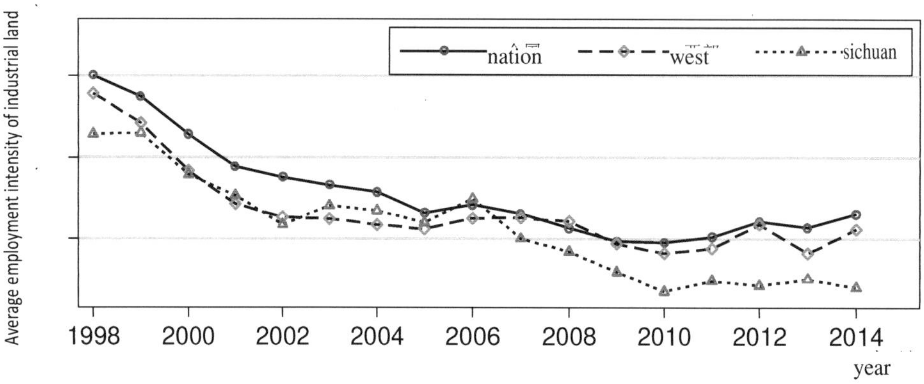

Figure 3. comparison of employment density of urban industrial land between Sichuan province and the nation and the west

3. Variation Model and Regression Analysis of Utilization Efficiency of Industrial Land Use in Sichuan Province

With the aid of expanded Cobb-Douglas (c-d) production function, a further comparison of the total factor productivity between Sichuan and national urban industrial land can be made. In the field of industrial production, land plays a main part of bearing capital and labor, with no substitutional relation with capital or labor. It is proper to make it the denominator of output, capital and labor in the industrial production function, and hereby constructs the efficiency panel model (1) of industrial land output.

$$\left(\frac{y}{m}\right)_{it} = A_{it}\left(\frac{k}{m}\right)_{it}^{\alpha}\left(\frac{l}{m}\right)_{it}^{\beta}\exp(\varepsilon_{it}), \quad i=1,\cdots,n, \quad t=1,\cdots, T \quad (1)$$

Y_{it} represents the total industrial output value of urban industrial land, m_{it} signifies the area of urban industrial land, k_{it} represents the industrial fixed assets of urban industrial land, l_{it} embodies the industrial employment of urban industrial land, α and β indicates output elasticity of the investment intensity and employment density on the urban industrial land. A_{it} is the neutral technical progress, namely the "the abbreviation of variation of production function in any form ", "the economic acceleration or deceleration, the improvement of labor education quality, and the various factors of mobile production function can be classified as' technological change '" (Solow R, 1957). The technical progress changes over time, and may vary between individuals as well.

The discussion of capital, labor and output of the unit area of industrial land in the paper implies that the input of land elements is kept unchanged, which is applicable to the law of diminishing marginal returns, and thence $\alpha + \beta < 1$. In other words, with other conditions being unchanged including the technology, if the investment intensity and density of employment of an industrial land with normal bearing function is increased by ϕ ($\varphi>0$) times, the increase amplitude of output efficiency is in the open interval of ($0, \varphi$).

Considering the individual characteristics, unobservable missing variables and data acquisition problems of sample point, the following Two-way Fixed Effects model of output efficiency of industrial land is designed:

$$\left(\frac{y}{m}\right)_{it} = \left(\frac{k}{m}\right)_{it}^{\alpha}\left(\frac{l}{m}\right)_{it}^{\beta}\exp(z_i'\delta + \lambda_t + u_i + \varepsilon_{it}), \quad i=1,\ldots,n, \quad t=1,\ldots, T \quad (2)$$

Here, λ_t represents the period fixed effect changing over time instead of with the individual, which consists of technological progress, macro-economic changes and industrial policy influences and other variables. z_i is the variable vector of individual characteristic that does not change over time. u_i is the unobservable individual heterogeneity, symbolizing the missing variable changing between individuals but not over time. It is different from z_i in that it is "unobservable". $z_i'\delta + u_i \equiv \eta_i$ is all the individual fixed effects that are unchanged over time, such as location, natural environment, historic and humanities environment and other variables. $u_i+\varepsilon_{it}$ constitutes a composite error term.

Take the logarithm from both sides of the model (2) :

$$\ln\left(\frac{y}{m}\right)_{it} = \alpha\ln\left(\frac{k}{m}\right)_{it} + \beta\ln\left(\frac{l}{m}\right)_{it} + z_i'\delta + \gamma S_t + u_i + \varepsilon_{it}\,,\quad i=1,\dots,n,\quad t=1,\dots,T \quad (3)$$

After the F test, LSDV test, Hausman test and excessive identification test, it is found the above-mentioned Two - way FE is correct in form, clustering robust standard error is adopted in the regression process and the regression parameters can be verified in statistical inspection and economic significance. The impact of the above individual fixed effect, time fixed effect and residual error on the output efficiency of industrial land is merged and recorded as logarithm of total factor productivity (lnA), and the above regression parameter is substituted into the model (3), which can be obtained by appropriate deformation:

$$\frac{y}{m} = A\left(\frac{k}{m}\right)^{\alpha}\left(\frac{l}{m}\right)^{\beta} \quad (4)$$

Based on the Solow's Residual Method, the total factor productivity over the years was calculated by substituting the estimated income parameter, and thus the average growth rate is calculated. Below is derived from further model transformation:

$$\frac{\dot{A}}{A} = \frac{\dot{y/m}}{y/m} - \alpha\frac{\dot{k/m}}{k/m} - \beta\frac{\dot{l/m}}{l/m} \quad (5)$$

From this, the growth rate $(\frac{\dot{A}}{A})$ of total factor productivity (A) is calculated. In combination with the above industrial land output rate, investment intensity and employment density data, the change situation can be found (see Table 2). To note, the growth rate of industrial land output rate, investment intensity and total factor productivity of Sichuan urban industrial land are all higher than the national average. However, it is worth noting that in recent years, the output rate of industrial land in Sichuan cities has experienced negative growth, and the gap with national level has been growing. In addition, as a populous province, Sichuan is endowed with abundant human resources. Employment density has fallen faster than the national in the absence of a fundamental change in the industrial structure, which is worth sharp vigilance.

Table 2. investment intensity,employment density and TFP of industrial land

year	Growth rate of Output efficiency		Growth rate of investment intensity		Growth rate of employment intensity		Growth rate of TFP	
	Whole nation	sichuan	Whole nation	sichuan	Whole nation	sichuan	Whole nation	sichuan
1999	0.0702	0.311	0.106	0.121	-0.0648	0.00261	0.0246	0.252
2000	0.0346	-0.132	0.0363	0.0724	-0.123	-0.154	0.0372	-0.146
2001	-0.0375	0.247	-0.0915	-0.00513	-0.12	-0.0916	0.0338	0.262
2002	0.141	0.0246	0.0684	0.0169	-0.0454	-0.141	0.112	0.0355
2003	0.0729	0.46	-0.0176	0.283	-0.0337	0.107	0.0885	0.308
2004	0.218	0.17	-0.0578	0.0665	-0.0322	-0.0275	0.256	0.142
2005	0.119	0.191	0.01	0.0482	-0.0998	-0.0608	0.132	0.175
2006	0.228	0.273	0.183	0.0744	0.0436	0.134	0.121	0.218
2007	0.0817	0.238	0.0351	0.332	-0.0444	-0.196	0.0708	0.103
2008	0.14	0.159	0.0396	-0.0592	-0.0766	-0.0796	0.132	0.198
2009	0.0254	0.0139	0.207	0.139	-0.0768	-0.137	-0.0725	-0.0352
2010	0.157	0.11	0.137	0.0124	-0.00858	-0.146	0.0841	0.124
2011	0.0291	0.151	-0.0565	0.0909	0.0353	0.0934	0.0532	0.0941
2012	0.067	-0.0528	0.0857	0.0663	0.0938	-0.0347	0.00367	-0.0803
2013	-0.0992	-0.0223	-0.0683	-0.135	-0.0338	0.051	-0.0561	0.0367
2014	0.204	-0.0877	-0.0106	0.0285	0.0808	-0.0679	0.195	-0.0923
Geometric	0.087069	0.117	0.034305	0.066	-0.03373	-0.051	0.074616	0.0932

mean

4. Study on the Causes and Solutions for Growth Speed Change of Sichuan Industrial Land Efficiency and Low Land Use Efficiency

In brief, the growth speed of several indexes in Sichuan industrial land efficiency has been surpassing the national standard since 1998, and in recent years it has dropped down. Nevertheless, the absolute level of industrial land use efficiency is still lower than the national average. It may be explained as follows:

On the one hand, because of poor industrial foundation and low starting, Sichuan has presented a higher speed than the national average in its course of struggling. It can be seen from Figure 1 that the gap of industrial land output capacity between Sichuan and the whole China is narrowed to a certain extent.

On the other hand, the root cause for low land use efficiency of Sichuan fails to be totally eradicated, which is reflected on as follows. Firstly, it doesn' t fully consider the rational expected output value and industrial choice differences of economic, social and industrial developing plan; it plans land scale and structure of development zone on the basis of subjective will or land availability, obviously "launching a large project" and making city planning "an enclosure tool for government". Secondly, the land of development zone is equated with industrial project land; the planned industrial land occupies a too large proportion, while the land for service is insufficient. For one thing, productive service support for industrial enterprises is inadequate, thereby reducing the output efficiency of industrial land; for another, life service support for regional people is also short that impedes regionally agricultural transfer and population citizenization. Thirdly, having ignored the pollution influence, and differences in industry, the mother city service and city level, it considers an industrial park as an independent city and therefore plans land structure in accordance with the Land Use Classification and Construction Land Planning Standard for Cities and Towns (GB 50137, 2011). As a result, residential land and land for commercial and service facilities and public services are left unused or inefficiently utilized.

Accordingly, solutions and suggestions are proposed in this paper. First of all, it' s suggested to make a scientific planning based upon rational evidence and avoid the phenomenon of "launching a large project" on industrial land due to excessive optimism; in planning, achieve a balanced and coordinated development and reasonable layout for the functional zones of industrial production, productive service and life service, in a bid to facilitate enterprises to obtain high-quality productive service support, employees of the urban industrial land to get rich life service support, and promote high efficient use of the land. Second, it should carry out construction step by step but never occupy the land abruptly, adjust industrial land scale according to economic tendency, and lower impact caused by unscientific planning and uncertainty. Third is to actively drive industrial restructuring and upgrading; accelerate technology progress and cleaner production; eliminate backward production capacity; set higher standard in technology, environmental protection and other aspects; take initiative to introduce industrial enterprises characterized by high technology content, low resource consumption, less environmental pollution and good economic benefits; vigorously develop strategic emerging industries, intelligent and high-end manufacturing; build ecological park and arrange productive and life service function zone nearby, thus furthering the industrial land output efficiency and population aggregation.

Fund project: Humanities and Social Science Planning Fund project of Ministry of Education" Research on Key Factors and Influencing Mechanism of City-Industry Integration" (15YJA790074);
China Postdoctoral Science Foundation funded project "Empirical Study on the Implementation Mechanism of City-Industry Integration Based on Empirical Data of China and Foreign industrial parks" (2015M572487).

Biography: Mingqiang Xu (1975-), doctor, post-doctoral of Institute Of Business Administration, Southwestern University Of Finance And Economics, director and associate professor of Regional Economic Research Center of Chengdu University, research direction: urbanization and industrialization. Wei Wei: (1966-) an economic engineer from Chengdu normal college.

5. Reference

[1] CCCPC Party Literature Research Office 2014 *Major literature Selection Since The 18th National Congress Of China* (Beijing: Central Party Literature Press) p 589-607.

[2] CCCPC Party Literature Research Office 2014 *Major literature Selection Since The 18th National Congress Of China* (Beijing: Central Party Literature Press) p 608-622.

[3] Lin Zhang, Yahui Wang 2014 Study on the factors influencing industrial land output efficiency from the perspective of micro-enterprise *East China Economic Management* 28 43-48.

[4] Antonio Ciccone, Robert E. Hall 1996 Productivity and the density of economic activity *The American Economic Review* 86 54-70.

[5] Kaisheng Long , Ligen Chen, etc 2008 Comparative analysis on input and output efficiency of different land types *China Population, Resource and Environment* 18 174-178.

[6] Wang Xirui, Xu Shi and Wu Qun, etc 2015 Temporal and spatial differences of construction land use efficiency and total factor productivity in Jiangsu province *China Land Science* 29 77-83.

[7] Wang Chengxin, Liu Hongyan and Shi Jialu, etc 2014 Evaluation of intensive land use in development zones above provincial level in Shandong province *China Population Resources and Environment* 24 128-135.

[8] Chen Wei, Peng Jianchao and Wu Qun 2015 Loss and efficiency measurement of urban industrial land Use *China Population Resources and Environment* 25 15-22.

[9] Feng Changchun, Liu Sijun and Li Rongwei 2014. Evaluation of urban industrial land efficiency at prefecture level and above in China *Modern Urban Research* 2014(4) 45-49.

[10] Xie Hualin, Wang Wei and Yao Guanrong, etc 2015 Temporal and spatial differences and convergence analysis of urban industrial land efficiency in China's major economic regions *Acta Geographica Sinica* 70 1327-1338.

Income and Conservation Dynamics of the Biological Bee-Fence: A bio-Economic Simulation

A Taruvinga[1] and N Gumpo[2]

1 Department of Agricultural Economics and Extension, University of Fort Hare, Private Bag X1314, Alice, 5700, South Africa
2 Department of Geography, University of Fort Hare, Private Bag X1314, Alice, 5700, South Africa
E-mail: ataruvinga@ufh.ac.za; misvuyo@gmail.com

Abstract. Human-elephant conflicts in most African countries coincide in areas where poverty and natural resources are most profound. Although popularly believed to be an asset capable of generating consumptive and non-consumptive ecotourism revenue, African elephants are in some parts of Africa viewed as pests and predators worth eradicating as a result of high human-elephant conflicts (crop raiding, property damage, human and livestock predation). This has shifted the conservation debate to issues of how much biodiversity (elephants) can be saved in the face of suffering local communities. With that background, we tested the income and conservation premise of the biological bee-fence concept as a complementary problem animal control (PAC) measure from a rural setting where elephants interact with local poor communities using bio-economic simulations. We conclude that the biological bee-fence concept has a significant potential to deter elephants from invading surrounding communities` fields as well as generating the much needed household income. These findings reinforce the conservation and income premise of the biological bee-fence under a typical African rural setting worth up-scaling.

1. Introduction

Problem animal control (PAC) in rural areas remains a big challenge in Africa especially in areas where a significant population of Africa elephants interact directly with local communities [1], [2]. A study conducted in Zimbabwe by Jones [3] revealed that African elephants cause the following damages: (i) crop damage – 78%, (ii) threat to humans – 9%, (iii) property damage – 3%, and (iv) livestock predation – 10%. With respect to problem animal species, Jones [3] also noted that African elephants present the highest share with 87% followed by lions (8%) and, lastly, by other species with a 5% share. Elephants affect the following types of crops: maize – 30%, millet – 30%, cotton - 3%, vegetables – 7%, and sorghum – 30% [3]. Banes [4] based on a study in Okavango Delta area in Botswana estimated that small scale rain-fed crop production profits were reduced by 75% and in some incidences completely eliminated by elephants. Thus far, as African elephants are allowed to exist, they generate social costs as well as potential benefits in the form of ecotourism revenue [5]-[7]. With that background, the paper argues that rural households have a much greater potential to conserve African elephants [6] or assist in their extinction [8], depending on the perceived balance between social costs and potential benefits they derive from elephants. Thus far, available PAC measures normally characterised by "guards, guns and fences" have proved to be expensive and sometimes ineffective towards separating the two species (humans and elephants). Mixed reporting dominates the actual direct benefits to rural households, who share boundaries with elephants [9].

With ineffective PAC measures and limited direct benefits from elephant, local communities find it difficult to support elephant conservation in most rural African countries. This creates a 'negative specie preservation value' for elephants in the eyes of local communities, a scenario that may negatively affect elephant conservation. To complement the existing PAC measures new deterrent cost effective and user friendly methods have been suggested [10]. The *Apis mellifera adansoni* may be a solution worth exploring, based on results from experimental pilot trials conducted in Zimbabwe [11], Kenya [12], [13] and South Africa [2]. The paper therefore appraised the potential of using honey bees to deter elephants and generate household income using bio-economic simulations.

2. Problem statement

High human – elephant conflicts are reported in most rural African communities [6], [9], [14]. This is largely true in cases where elephants cause property damage, invade crop lands, and cause human and livestock predation while offering minimum direct benefits to local communities. Given the sizes of most game parks, physical effective boundaries (game fences and guards) have always been a challenge. Elephants have always managed to take advantage of ineffective boundaries invading crop lands of local communities. This has created a long standing friction between rural households and elephants, the former viewing the latter as pests and life threatening predators [9], [14], [15]. We therefore ask the following question: Can honeybees be trusted by rural households to deter elephants and possibly generate their much needed household income through sale of honey?

2.1. Objectives

- To upraise the potential of honey bees to deter African elephants.
- To estimate the income potential of honey production under the biological bee-fence concept.

3. Related literature

This section presents background literature on the interaction between bees and elephants and the potential of honey bees to deter elephants.

3.1. The interaction between bees and elephants

In Africa, folk tales relating to conflicts (fights) between elephants and bees have been shared from generations to generations [12]. The scientific logic behind these folk tales lies on the fact that although elephants are thick-skinned, Vollrath and Douglas-Hamilton [16] noted that they also have their weak points that can be targeted by bees. Earlier on Sikes [17] reported that ecto-parasites like ticks easily penetrated the thin skinned belly of elephants. Also, behind ears and under the trunk of elephants blood vessels flow close to the surface [18], a crucial weakness factor that is targeted by predators like elephant-lice [19]. Jacobson *et al.,* [20] also noted that the inner trunk membranes and eyes are extremely sensitive capable of causing serious effects from a few bee stings. Since elephants have co-existed with forest honey bees it's logical to assume that some conflicts (fights) occasionally erupts in the wilderness as the two species cross each other`s line of boundary.

The ability of bees to fly and availability of weak points on elephants as revealed above logically makes the African honey bees winners of this possible fight. McComb *et al.,* [21] acknowledges that smell is a crucial sense in the social and foraging decisions for elephants. Building on this idea, Vollrath and Douglas-Hamilton [16] assumed that the bee wax smell may deter elephants from harming a hive and from feeding in its immediate surroundings, otherwise this would start a fight to which an elephant would be the looser. Payne [22] also noted that elephants have a strong sense of hearing, a crucial factor Vollrath and Douglas-Hamilton [16] also assumed to be useful towards explaining why elephants could avoid colonised hives that emit a hissing sound. Since elephants are known to be intelligent with a very good memory, it's logical to assume and believe that elephants will avoid areas with bees, a conclusion that makes the commonly shared African folk tale of bees stinging elephants a true anecdote. This has triggered research towards testing the effectiveness of the concept using practical and scientific approaches.

3.2. Potential of honeybees to deter elephants

Vollrath and Douglas-Hamilton [16] tested this concept in Kenya and discovered that African bees deter elephants from damaging vegetation and trees mounted with hives. Vollrath and Douglas-Hamilton [16] further argued that bees could be employed profitably to protect not only selected trees, but also selected areas, from elephant damage. Contrary to these findings, Karidozo and Osborn [11] tested the same concept by mounting bee hives strategically around crop fields in the Zambezi Valley of Zimbabwe. Their findings revealed that the deterrent effect was not statistically significant. With that background, Karidozo and Osborn [11] argued that bees alone may not stop elephants from crop raiding. Similar conclusions were also forwarded by Ndlovu *et al.*, [2], arguing that most elephant crop raids occur at night and prominently in winter when temperatures are low. The authors therefore conclude that during these times African honey bees will be dormant and unlikely to deter elephants unless there are some trigger mechanisms.

Still pursuing the same idea, King *et al.,* [12] conducted a pilot study to evaluate the concept of using bees to mitigate elephant crop depredation in Laikipia, Kenya. The authors noted that the experimental farm protected by bee hives had fewer raids and consequently had higher productivity. King *et al.,* [12] further noted that socio-economic indicators suggested that not only was the concept popular and desired by the community but also that it could pay for its construction costs through the sale of honey and bee products. Based on their conclusion, King *et al.,* [12] called for a wider experimental test of the concept across different agricultural systems to jointly evaluate its effectiveness.

With such a conflicting background society may therefore query the practical relevance of the concept. In this section the paper looks at the earlier loopholes in the design of the concept against improvement suggested in recent studies. Karidozo and Osborn [11] used three test plots as follows; (a) an artificially planted field strategically located in the bush, (b) paths along elephant refuge areas, and (c) paths around natural water points regularly visited and used by elephants. In their first test plot we applaud their brevity in strategically creating artificial fields in the bush where we expect elephants to be attracted by the green plants. Their hives were also mounted on poles 3m long at a height of 2m and 20m equidistant from each other [11]. Using this design paired with control plots for 2 seasons (2004 & 2005) the study concluded that the claimed deterrent effect was not statistically significant [11]. In test plots two and three, elephants retreated and opened new entrance route a clear indication that they avoided live hives [11]. Similar observations were also noted by Ndlovu *et al.*, [2] for their level 4 responses where 15 out of 21 elephants exposed to the bee noise with honey scent treatment ran away from the source of treatment.

For the success of this concept several factors are crucial to include; (a) bee stinging, (b) bee hissing sound, (c) a triggering mechanism to agitate bees and (d) the bee wax smell. For purposes of introducing the bees to elephants, some repeated stinging should happen to send a threat cue. Once the threat cue has been internalised, then the bee hissing sound and wax smell factors would give a realistic threat cue as elephants approach beehive mounted areas. Comparable conclusions were also noted by Ndlovu *et al.*, [4: pg. 1] highlighting that, "buzzing bees or honey scent as isolated treatments (as may be the case with dormant beehives) were not effective elephant deterrents, but rather an active beehive emitting a combination of auditory and olfactory cues was a viable deterrent".

Recent designs have tried to reduce the space between hives and also to link the hives with a connecting taut wire [12], [13]. Large spacing without a connecting wire may promote free passage of elephants without agitating bees. Also mounting beehives under a bee-hut as suggested in Figure 1, may force approaching elephants to target the space between hives [12], [13]. Once this route is selected, the taut wire will swing attached hives and agitate bees from several nearby hives [12], [13]. In the process bees are most likely to attack the elephants thereby sending a warning threat message.

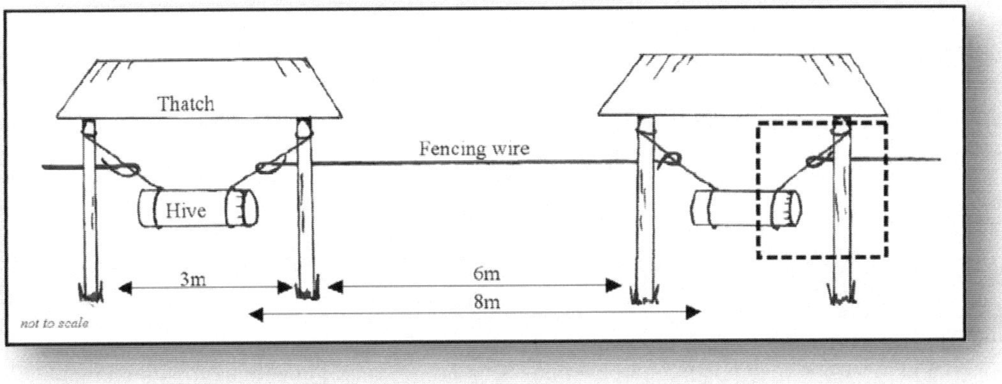

Figure 1. Biological bee fence design [13]

We therefore argue that the biological bee fence may have a positive deterrent impact under the following conditions:

- Colonized (active) bee hives emitting bee hissing sounds and wax smell. This is more critical for elephants that may have previously been exposed to bees since they are better equipped with realistic threat cues [2].
- A bee fence designed with reduced spacing between hives and a connecting taut wire [12], [13] which was largely missing in earlier designs by Karidozo and Osborn, [11].

Literature further questioned the practicality of the above conditions, given that elephants' crop raids are more pronounced during the night and winter seasons when bees are inactive [2]. Future research should therefore focus on understanding the concept's long-term effect, given the possibility that elephants may get habituated to the threat [2]. Also critical to understand across different agro-systems is the potential of the connecting taut wire to trigger bee agitation during the night and winter periods when temperatures are very low. The cost benefit analysis of using night "bee-hissing" drones with sharp spikes needs to be explored to complement natural bees during the night, winter and rain periods when they are inactive. To enhance uptake by potential users (rural households sharing boundaries with game parks) several studies suggest linking the system with direct economic benefits of honey and wax as well as other indirect biodiversity benefits. In the next section the paper upraises the economic potential of the system using bio-economic simulations.

4. Income potential of honey production under the biological bee-fence concept

Several studies point to the merit of linking honey production with the biological bee-fence [11], [12], [13]. Suggested reasons range from; income generation, employment creation, financing the construction of the biological bee fence and strategic livelihoods for rural households [11], [12], [13]. In this section we provide a bio-economic simulation of honey production under the biological bee fence concept. Using standard average yields, running costs and selling prices the paper generated two gross margin budgets (Table 1 & Table 2) for purposes of estimating the economic premise behind honey production under a biological bee fence (1ha). Average yield from a standard Kenyan Top Bar (KTB) hive range between 10kg and 15kg with 2 clearly defined production cycles per year [13]. To fully protect 1ha using a biological bee fence 40 beehives are necessary [13]. Lastly raw honey can be sold at an average price of US$1.5/kg in most rural African countries. Using the above operational assumptions two gross margin budgets were used to estimate the income potential of hectare piece of land protected by a biological bee fence with 40 KTB hives as detailed in Table 1 and Table 2.
The gross margin analysis in Table 1 reveals a gross income value of US$600.00 during year one 1st cycle against an operational cost of US$1,715.00. In Table 2 we generated year one 2nd cycle gross margin estimates for purposes of getting a true picture after removing inputs that can be reused in

subsequent years (hives, harvesting kit and protective clothing). Results reveal that operational costs are likely to be reduced from US$1,715.00 to a mere US$50.00 per cycle against a gross income value of US$600.00.

Table 1. Gross margin budget 1st cycle year 1

Item	**Production Period (Months)**	**Price (US$) Yield (kg)**	**Gross Income (US$)**
Selling price		US$1.5/kg	
Production cycles/year (pc)[a]	2		
80% colonization = 32 hives (c)[b]			
Average yield per hive		12.5kg	
Total yield in 6 months	32 x 12.5	400kg	
Total yield per year	32 x 12.5 x 2	800kg	
Gross Income / hive in 6 months	(12.5kg x $1.5 x 1hive)		18.75
Gross Income / 32 hives in 6 months	(12.5kg x $1.5 x 32hives)		600
Operational Costs	Quantity	/40 hives (US$)	/40 hives (US$)
Hives *	40	1600	
Protective clothing *	1	35	
Harvesting kit (Smoker) *	1	20	
Labour		20	
Transport (to collection point)		30	
Containers (200l) *		10	
Total Variable Costs (Year 1, 1st circle)		1715	**1715**

*: - these costs are incurred once and the equipment can be used for the next 5 years before replacement
(pc)[a] : - harvesting is done 3 times a year during good years and sometimes twice a year during bad years (drought)
(c)[b]:- on average per production cycle only 80% of the hives will be colonized while 20% is normally empty

Table 2. Gross margin budget 2nd cycle year1

Item	**Production Period (Months)**	**Price (US$) Yield (kg)**	**Gross Income (US$)**
Selling price		US$1.5/kg	
Production cycles/year (pc)[a]	2		
80% colonization = 32 hives (c)[b]			
Average yield per hive		12.5kg	
Total yield in 6 months	32 x 12.5	400kg	
Total yield per year	32 x 12.5 x 2	800kg	
Gross Income / hive in 6 months	(12.5kg x $1.5 x 1hive)		18,75
Gross Income / 32 hives in 6 months	(12.5kg x $1.5 x 32hives)		600
Operational Costs	**/ hive (US$)**	**/40 hives (US$)**	**/40 hives (US$)**
Hives *	-	-	
Protective clothing *		-	
Harvesting kit (Smoker) *		-	
Labour		20	

Transport (to collection point)		30	
Containers (200l) *		-	
Total Variable Costs (Year 1, 2nd circle)		50	50

*: - these costs are incurred once and the items can be used for the next 5 years before replacement
(pc)[a] : - harvesting is done 3 times a year during good years and sometimes twice a year during bad years (drought)
(c)[b]:- on average per production cycle only 80% of the hives will be colonized while 20% is normally empty

Table 3 presents the projected linear gross margins from year 1 to year 5. Simulation results reveal an estimate average annual gross margin of US$1,200 against an operational cost of US$433. This translates to an annual average gross margin of US$767, monthly average gross margin of US$64 and a daily average gross margin of US$2. Thus far, per every $ invested towards the biological bee-fence a return (GM/$TVC) of 1.8 is expected as summarised in Table 3.

Table 3. Projected gross margins

Period (Years)	GI / CYCLE (US$)	Total VC / CYCLE (US$)
Year 1	Cycle 1: 600	Cycle 1: 1715
	Cycle 2: 600	Cycle 2: 50
Year 2	Cycle 1: 600	Cycle 1: 50
	Cycle 2: 600	Cycle 2: 50
Year 3	Cycle 1: 600	Cycle 1: 50
	Cycle 2: 600	Cycle 2: 50
Year 4	Cycle 1: 600	Cycle 1: 50
	Cycle 2: 600	Cycle 2: 50
Year 5	Cycle 1: 600	Cycle 1: 50
	Cycle 2: 600	Cycle 2: 50
Average annual Gross Income / Total VC	1200	433
Average annual Gross Margin	US$767	
Average monthly Gross Margin	US$63.92	
Average daily Gross Margin	US$2.13	
GM/$VC	1.8	

Using simple linear gross margin analysis, bio-economic simulations results suggest that on average a biological bee fence for one household surrounding a field of 1ha, has an income generating potential of US$2 per day per household and a return on investment of 1.8. The analysis excludes other high value honey by-products like bee wax and processed liquid honey that require formal markets. Thus far the suggested returns are only based on raw honey easily marketable in most rural areas of Africa.

5. Conclusions

The paper concludes that the biological bee fence (with colonised hives) has a significant potential to deter elephants from invading local communities' crop lands. To trigger agitation of bees more importantly during the night and winter periods when temperatures are low a connecting taut wire is necessary. The paper further concludes that 40 Kenyan Top Bar beehives normally ideal for 1ha (2 production cycles per year) can generate a non-farm household income of US$2 per day per household with a return on investment of 1.8. These findings reinforce the income and conservation premise of the biological bee-fence under a typical African rural setting worth up-scaling.

6. Recommendations

The paper makes the following recommendations: The biological bee fence has a potential to complement available problem animal control measures in areas where human-elephant conflicts exists with economic and conservation benefits. Rural development agencies (NGOs, government and private sector) may target beekeeping promotion to enhance household income and deter elephants from invading crop lands.

7. References

[1] Karidozo M and Osborn FV 2015 Community Based Conflict Mitigation Trials: Results of Field Tests of Chilli as an Elephant Deterrent. *J. Biodivers Endanger Species* 3 (1).
[2] Ndlovu M, Devereux E, Chieffe M, Asklof K and Russo A 2016 Response of African elephants towards a bee threat: Its application in mitigating human-elephant conflict. *S Afr J Sci* 112 (1/2), 1-5.
[3] Jones MA 1994 Safari Operations in Communal Areas in Matabeleland, Harare: Department of National Parks and Wildlife Management.
[4] Barnes JI 2006 Economic analysis of human-elephant conflict in the Okavango delta Ramsar site, Botswana. Unpublished report, Natural Resources and People, Gaborone, Botswana.
[5] Sutton WR 2001 *The Economics of Elephant Management in Namibia*. Ph.D Thesis, Department of Agricultural and Resource Economics. University of California Davis.
[6] Muchapondwa E 2003 *The Economics of Community-Based Wildlife Conservation in Zimbabwe.* PhD Thesis, Department of Economics. Göteborg University Sweden and University of Zimbabwe Harare.
[7] Novelli M, Barnes JI and Humavindu M 2006 The Other Side of the Ecotourism Coin: Consumptive Tourism in Southern Africa. *Journal of Ecotourism,* 5, (1/2), 2006.
[8] Child B, Ward S and Tavengwa T 1997 Zimbabwe's CAMPFIRE programme: Natural Resource Management by the people, IUCN-ROSA Environmental Issues Series No. 2, Harare: IUCN-ROSA.
[9] Taruvinga A 2012 *Adaptation choices, community perceptions, livelihood linkages and income dynamics for district producer communities surrounding Nyatana Game Park in Zimbabwe*, PhD thesis, University of Fort Hare, South Africa.
[10] King LE 2010 *The interaction between the African elephant (Laxodonta africana africana) and the African honey bee (Apis mellifera scutellata) and its potential application as an elephant deterrent.* Ph.D thesis, University of Oxford, UK.
[11] Karidozo M and Osborn FV 2005 Can bees deter elephants from raiding crops? An experiment in the communal lands of Zimbabwe. *Pachyderm* 39: 26 – 32.
[12] King LE, Lawrence A, Douglas-Hamilton I and Vollrath F 2009 Beehive fence deters crop-raiding elephants. *Afr. J. Ecol.,* 47, 131 – 137.
[13] King LE 2011 Beehive fence construction manual: A step by step guide to building a protective beehive fence to deter crop-raiding elephants from farm land. Save the elephants, Nairobi Kenya.
[14] Taruvinga A and Mushunje A 2014 Society`s perceptions of African elephants and their relative influence towards the conservation of elephants, *APCBEE Procedia* 10 (2014) 299 – 304.
[15] Edwards SB 2001 *Legal Trade in African Elephant Ivory: Buying Ivory to Save the Elephant?.* Ph.D Thesis, Graduate School of Law, Nagova University Japan.
[16] Vollrath F and Douglas-Hamilton I 2002 African bees to control African elephants. *Naturwiss 89*, 508–511.
[17] Sikes SK 1969 Habitat and cardiovascular diseases: observations made on elephants (*Loxodonta africana*) and other free-living animals in East Africa. *Trans Zool Soc Lond* 32:1 – 104.
[18] Benedict FG, Fox EL and Baker ML 1921 The surface temperature of the elephant, *rhinoceros* and *hippopotamus*. *Am J Physiol* 56:464 – 474.

[19] Braack LEO 1984 A note on the presence of the louse *Haemotomyzus elephantis* Piaget (*Mallophaga: Rhynchophthirina*) in the Kruger National Park. *Koedoe* 27:139 – 140.
[20] Jacobson E, Sundberg JP, Gaskin JM, Kollias GV and O'Bannion MK 1986 Cutaneous papillomas associated with a herpesvirus-like infection in a herd of captive African elephants. *J Am Vet Med Assoc* 189:1075 – 1078.
[21] McComb K, Moss CJ, Baker L and Sayialel S 2000 Unusually extensive networks of vocal recognition in African elephants. *Anim Behav* 59:1103 – 1109.
[22] Payne K 1998 Silent thunder: in the presence of elephants. Weidenfeld and Nicholson, London.

Acknowledgments
We acknowledge Government Mbeki Research Development Centre (GMRDC) and Southern Africa Climate Change and Biodiversity Research Platform (SACC&BRP) for supporting this study.

IOP Conf. Series: Earth and Environmental Science **178** (2018) 012039 doi:10.1088/1755-1315/178/1/012039

Mount Sinabung Eruption: Impact on Local Economy and Smallholder Farming in KaroRegency, North Sumatra

JongkersTampubolon[1], HLNainggolan[2], AGinting[2] andJAritonang[2]

1 Associate Professor, Department of Agribusiness, HKBP NommensenUniversity, Indonesia
2Lecturer, Department of Agribusiness, HKBP Nommensen University, Indonesia
Email: jtampubolon@yahoo.com

Abstract. Indonesia is a disaster prone country. One of the geo physical disaster is the prolonged of Mount Sinabung Eruption in Karo Regency, North Sumatra. This area is famous for horticultural (vegetable and fruit) producer in the region. The eruption has displaced people, destroyed infrastructure and livelihood. This research aims: (i) to investigate the magnitude of the economic losses, (ii) to find out impact on agricultural production and (iii) to investigate the adaptation pattern undertaken by farmers. The results of research showed that (i) the extent of losses on the regional level is bigger than the annual local government budget, whereas the feasibility of farm business deteriorated significantly, (iii) the disaster has jeopardized village self help organization, worsened access to farm credit as well as caused labour shortage in agriculture. Based on empirical finding, recommendation for the rehabilitation and mitigation is proposed.

1. Introduction

After 400 years of quiescence, Mount Sinabung roars back to life in 2010 with its activity up to alert level *warning* (level IV). The eruption has been ongoing until today (December 2017) which has lead to the displacement of 12,000 people. Karo Regency at the foot of Mount Sinabung is famous as a fruit and vegetable production area. Fifty nine percent (59%) of the damage and losses caused by the disaster is borne by the agricultural sector, especially the horticultural subsector. This research aims to (i) investigate the extent of losses due to the Sinabung volcano disaster, (ii) to analyze the losses on the level of farm business, (iii) to investigate the condition of supporting factors after eruption and (iv) to investigate the adaptation pattern undertaken by farmers in response to the risk of farming disruption due to volcanic ash rain. The empirical findings of this research will be used to propose measures that can be taken by the local government and the Regional Disaster Management Agency (BPBD) to assist farmers through appropriate recovery programs (rehabilitation) and to plan the development of agricultural sector as part of volcanic disaster mitigation program in the long-term (reconstruction phase).

2. The Magnitude of Damages and Losses

The eruption of Mount Sinabung has displaced 12,000 people (3,000 households) and 2,670 HHs have to be permanently relocated to new settlements. As of September 2017, only 370 HHs were successfully relocated to new settlements that have just been constructed. In the second phase, 1,683 households were relocated, but since the land for village development was not available, only 1,350 households were processed under an independent relocation scheme in which the victim's family have to find a new settlement location on their own with funds provided by the Regional

Disaster Management Agency. In addition, there are 648 families to be processed in the third phase of the relocation plan.

As stated by numerous studies [1]-[4], natural disasters in developing countries will directly affect agriculture and food security, as the greater the contributions of agricultural sector to Gross Domestic Product (GDP) and so long as it provides employment,the higher the level of losses that this sector experiences. Similar thing is found in Karo Regency, where the contribution of agriculture sector to Regional GDP and to employment was 61 % and 67% as the disaster struck [5].The agricultural sector bears almost 2/3 of the total losses, followed by the housing sector (28%). Considering that the household and farming are run under a farm household system, the losses in agriculture and housing sector are basically borne by one single family.

In line with experienced in New Zealand [6], that horticultural crops are most vulnerable against volcanic ash, the assessment conducted by the Regional Disaster Management Agency in 2015 [7] found out that in agricultural sector, the biggest loss was experienced by horticultural subsector (88.57% of total monetary value), especially vegetable crops where the heavy damage attacked 3,550.55 ha of plant with the main commodities affected include Chilli (47.91 %), Tomato (18.33%) and Potato (11.63%). The damage of fruit plants on 1,512.87 ha includes oranges (77.82%). Staple crops suffered damage of 1,221.91 ha with heavily damaged crops of dry land rice (62.80%) and maize (20.58%). Plantations are also affected by the exposure of volcanic ash but most of the plants have only mild/moderate damage that allows them to recover. The crop area damaged include 2,474.67 ha with main commodities of coffee (67.97%) and cocoa (23.57%).

The value of losses due to Sinabung volcano disaster will be more noticeable, if compared to the financial condition of Regional Government of Karo Regency. Total value of losses estimated by the Regional Disaster Management Agency[7]amounts to IDR 1.35 trillion and the value of agricultural losses amounts to IDR 1.06 trillion. Whereas the cost required for the rehabilitation and reconstruction amounts to IDR 3.6 trillion (especially agricultural sector spends Rp 1.21 trillion). In 2016, the Regional Budget Revenues and Expenditures of Karo Regency amounts to IDR 1.53 trillion [5].

As such, the cost needed for the recovery of agricultural sector due to eruption of Mount Sinabung nearly absorbed all regional expenditure within a year (80%). This underlines the view of UNDP[8], that disaster is an impediment to development in which limited budget that should be allocated for development purposes is otherwise spent on emergency responses and reconstructing damaged or lost facilities and infrastructure.

At the regional level, in addition to the destruction of thousands of hectares of planted areas, the eruption of Mount Sinabung that spewed volcanic ash has lead to the decrease of total cropping area especially staple crops area (dry land rice, corn and sweet potato) and horticultural crops where chilli, carrots, tomatoes and cabbage as most popular commodities cultivated by farmers. Figure 1 shows the development of cultivation of staple crops and horticultural cropping area which shows continuous improvement since 2002 with peak total area in 2010.

Due to eruption, the area of cultivation decreases and fluctuates according to volcanic condition. The chronology of eruptions was documented as follows[9]: (i) 29 August to 7 September 2010, the eruption produced volcanic ash reaching as high as 1000 - 5000 meters, (ii) 2011 to the end of 2013, nearly three years with relatively quiet conditions (non-eruptive), (iii) 18 December 2013 - January 10, 2014, the eruption and cold lava floods occurred, and the biggest threat is lava flood with limited impact on mountain foot of which it has actually been relocated and (iv) eruption with volcanic ash roared back to life in August 2015. By associating volcanic activity with the total staple crops and horticultural harvested area, it is indicated that the eruption will cause schock in the form of decrease in total cultivation area with one year lag.

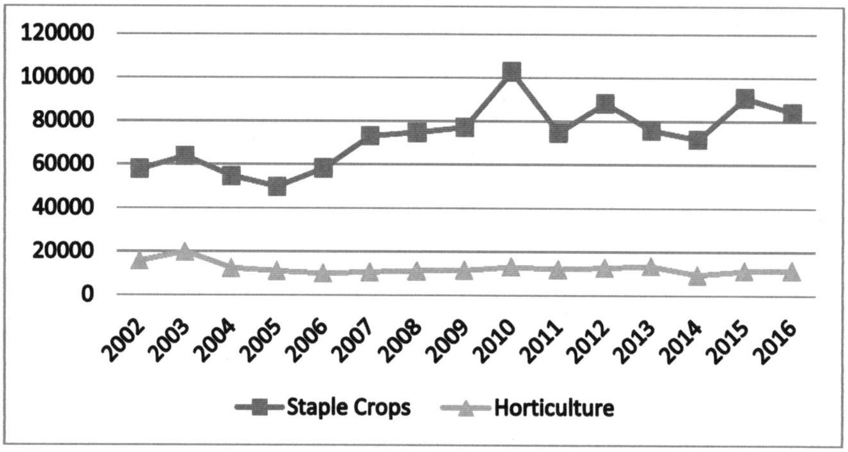

Figure 1. Development of Total Staple Crops and Horticultural Harvested Area, Karo Regency, 2002 - 2016 (in Ha)

3. Losses at the Level of Farm Family Business

In line with agricultural conditions at the regional level, there is a decrease in the total cultivation area at the farm business unit level. From interviews with 34 farmers in Simpang Empat District (one of the districts in the hazard zone II), it was revealed that there is a significant change in horticultural business. By comparing before and after situation of the eruption of Mount Sinabung (in 2009 vs. 2017), it was found that the total cropping area has decreased, productivity also decreased while on the other hand the production cost increased. The logical consequence of the changes of these three variables is the decrease in farm income.

The reduction in the cropping area and decrease in the production led to the decrease in the market supply, which eventually raised prices. However, the rise of the selling prices and thus the revenue, were far below the level of cost increase, thus the farm income experienced a substantial decrease. Two cost components that did not occur in a normal situation are technical actions to deal with volcanic ash and avoidance of the outbreak of certain pests after eruptions. To get rid of the ash covering plant leaves, farmers flushed them, which required pumping machine as well as its operational cost. Meanwhile, to prevent pest outbreak (especially against tomato crops), farmers set up nets surrounding the whole cropping area.

The financial damage caused by Mount Sinabung eruptions will be more conspicuous in the results of further farming analysis. While the business feasibility of carrot, cabbage, tomato and chilli farming was good before the eruptions (R/C ratio between 1.43 and 4.79), after the eruptions, only cabbage farming was economically feasible to be conducted (R/C ratio = 2.46), while tomato, carrot and chilli farming incurred loss with an R/C ratio < 1 (see Figure 2).

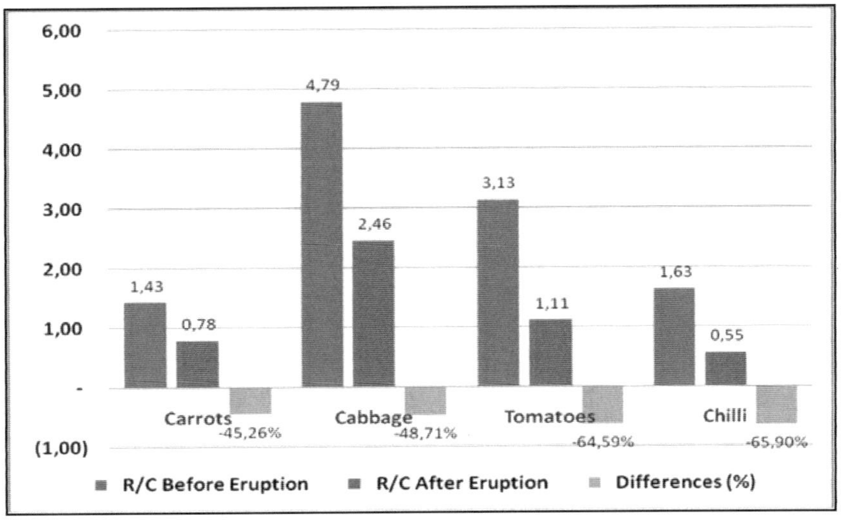

Figure 2. Feasibility of Horticulture Production Before and After Mt Sinabung Eruption (R/C ratio Comparison).

4. The Impact of Eruption on Farm Management

In running their farm businesses, farmers are relying on at least five factors that help support the success of their businesses[10]-[14], including (i) availability of production factors such as seeds, fertilizers and pesticides, (ii) availability of agricultural wage workers, (iii) availability of credit as a source of funding to finance production facilities and pay the workers, (iv) marketing guarantee for agricultural products and (v) the condition of social infrastructures (local institution) in agricultural sector such as self help organization and cooperative.

The results of interviews with farmers, with a scale of 0 to 100 discovered that the condition they face today is categorized as bad with a total score of 52.35. Likewise there are two factors that are still categorized in a good condition including the availability of production facilities and marketing of products, each with a value of 68.82 and 89.41. Distributors of production facilities and products wholesalers are located outside the village (in the Capital Regency), which is why they still exist despite the eruption. As for the other three factors, the condition is very bad after the disaster. Those poorest scores are the availability of credit for farm business (25.88), local social support system (34.71) and availability of wage workers (42.94).

5. Farmers Adaptation to the Volcano Disaster

In response to potential volcanic ash disrupting the crops, farmers react by maintaining existing farms and providing water pumps to water the plants on a daily basis in order to get rid of the ash that covers the leaves. However, in the long run, farmers reduce the total farm area in general.

The fluctuation depends mainly on the intensity of eruption of Mount Sinabung with a one-year lag. Similarly, in the short term (last four years), there has been changes in patterns of the cultivated commodity. Long duration production crops such as chilli, tomato and corn are consistently reduced and replaced by plants with shorter production cycle, such as carrots, beans and cabbage, which can be harvested less than three months (Figure 3).

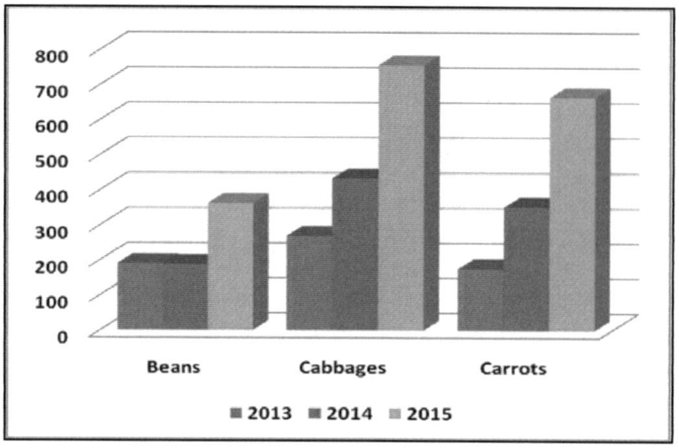

Figure 3. Development of Harvest Area of Horticultural Commodity with Shorter Production Cycle at Simpang Empat District, Karo Regency, 2013 – 2015 (in Ha).

6. Conclusions and recommendations

This study indicates that the eruption of Mount Sinabung has resulted in huge losses not only on the area directly exposed to volcanic ash (hazard zone I and II) but also on the whole area of Karo Regency as it deals with the damages and losses whose value is greater than the total regional expenditures in a year. Funds allocated from the Regional Budget Revenues and Expenditures for rehabilitation and construction should be used for other programs to improve community welfare.

At the farm level, the eruption of Mount Sinabung has weakened the feasibility of horticultural farming with a decrease of R/C value of 45 - 66% as compared to the period before the eruption. Some commodities such as chili and carrot are cultivated even in a loss condition with an R/C value <1.

In response to farming disturbance, farmers reduce the total cultivated area so that the trend of staple crops and horticultural cropping area significantly decrease, despite the fluctuation in which the area increases after the period of quiet volcanoes and the total area decreases after the eruption gets back alive within a year time lag.

To maintain the existing farms, farmers equip themselves with water pump machines to flush plants on a daily basis so that the leaves of the plants are free from volcanic ash. These efforts incur additional costs. Similarly, in the short term to reduce the risk of crop failure, there seems to be a clear pattern of adaptation in which farmers expand the harvest area of short-period-cultivation vegetable crops (harvestable before three months) such as beans, cabbage and carrot replacing chilli and tomato whose production period can reach 16 months.

Of the five factors supporting the farming, only two factors are still well functioning, namely the availability of production facilities and marketing of agricultural products, because business players in these two areas are located outside disaster-stricken area (the capital of regency). While other factors such as access to farming credit, availability of agricultural paid worker and local institution such as self-help organization are all in bad condition because the actors involved in them constitute the victims of volcano disaster.

Based on the findings of this research, it is recommended that in the rehabilitation phase, the Regional Government and Regional Disaster Management Agency (BPBD) to include pump machines and farm credit in the distribution package, but the provision of this assistance begins by forming community organization at the village level so that the provision of equipment is adjusted to the capacity of farm. In addition, village community organizations may help monitor the distribution and utilization of farm credits. For the long term, it is expected that government produce high technology and high value types of crops that can still produce well despite being exposed to volcanic ash, beside changing in agricultural technique. This type of specific commodity development effort can only be realized through experiments and collaborations with research institutes and universities.

7. References

[1] Israel D C and Briones R M 2012: Impacts of Natural Disasters on Agriculture, Food Security, and Natural Resources and Environment in the Philippines, in Sawada, Y. and S. Oum (eds.), *Economic and Welfare Impacts of Disasters in East Asia and Policy Responses*. ERIA Research Project Report 2011-8 (Jakarta: ERIA) pp.553-599.
[2] Lebon S L G 2009: *Volcanic Activity and Environment: Impacts on Agriculture and use of geological data to improve recovery processes*. Faculty of Earth Science, University of Iceland (Master's Thesis) chapter 4pp 29 - 52.
[3] FAO 2014: *Mainstreaming Disaster Risk Reduction in Agriculture: An Assessment of Progress Made Against the Hyogo Framework of Action*. (Rome: UN-FAO) pp 8 – 10.
[4] Chapagain T and Raizada MN 2017: Impact of Natural Disasters on Smallholder Farmers : Gaps and Recommendation. *Agriculture and Food Security* 6 (39): 1 – 16.
[5] BPS (Central Statistic Agency), various years: KaroRecency in Figure.
[6] Neild J, O'Flaherty P, Hedley P and Underwood R 1998: Impact of a Volcanic Eruption on Agriculture and Forestry in New Zealand. *Ministry of Agriculture and Forestry Technical Paper 99/2*. Ministry of Agriculture and Forestry New Zealand.
[7] BPBD (Local Agency for Disaster Management) 2015: *Action Plan for Rehabilitation and Reconstruction After Eruption of Mount Sinabung*, Karo Regency, North Sumatra.
[8] UNDP 2004: *Reducing Disaster Risk – A Challenge for Development* (New York: UNDP) p 19-25.
[9] Gunawan H, Surono, Budianto A, Kristianto, Prambada O, McCausland W, Pallister J and Iguchi M 2017: Overview of the eruptions of Sinabung eruption, 2010 and 2013–present and details of the 2013 phreatomagmatic phase, *Journal of Volcanology and Geothermal Research* (forthcoming).
[10] Mosher A T 1966: *Getting Agricultural Moving: Essentials for Development and Modernization* (New York,Praeger).
[11] Upton, M., 1996: The *Economics of Tropical Farming Systems.*Cambridge: Cambridge University Press) pp 34 - 44.
[12] Barry P J and Robinson L J 2001: *Agricultural Finance: Credit, Credit Constraint, and Consequences. Handbook of Agricultural Economics*Vol I, ed Gardner B L and Rausser G C (Amsterdam: Elsevier) chapter 10pp 513 – 571.
[13] De Silva L R and Kodithuwakku K A SS 2005: Pluriactivity and Socio-economic Success of Rural Households. *SrilankaJurnal of Agricultural Economics* Vol 7 (5): 85 – 108.
[14] Whitman Z 2014: *Rural Organization Impacts, Responses, and Recoveries to Natural Disasters: Case Studies from the Canterbury Earthquake Sequence and the 2010 Southland Snowstorm.*Department of Geological Science, University of Canterbury (Doctoral Thesis) chapter 5 pp 135 - 166.

Acknowledgement. The authors would like to thank The Ministry of Research, Technology and Higher Education of Republic Indonesia for providing grant to fund the research.

IOP Conf. Series: Earth and Environmental Science **178** (2018) 012040 doi:10.1088/1755-1315/178/1/012040

The Statistics of Recyclable Resources in Beijing: Satus, Problems and Countermeasures

Tingting Liu[1], Yufeng Wu[1,*], Xi Tian[1], Yu Gong[1] and Yanmei Li[1]

1 Institute of Circular Economy, Beijing University of Technology, No.100 Pingleyuan Road, Beijing, China, 100124
Email: circular_economy@163.com

Abstract. Utilization of recyclable resources not only contributes to resource conservation, but also to the reduction of environmental pollution. The statistics on recyclable resources provides effective data support for the promotion of waste recycling and utilization. However, the research and practice concerning recyclable resources statistics are still at initial stage. By investigating and analysing the statistical status quo of Beijing's recyclable resources production and recycling, this study pointed out problems such as undefined statistical types, incomplete statistical scope, nonstandard statistical methods, absence of supervision, and low integration of statistical resources among different departments. Based on the problems mentioned above, the implications for recyclable resources management were discussed. Establishing and issuing recycling catalogue, standardizing the statistical methods and process, building a sharing mechanism for the data monitoring, innovating the data collection method, and establishing statistical supervision system and appraisal mechanism are suggested to improve the recyclable resources management.

1. Introduction

According to the Guidance on Comprehensive Utilization of Resources during the 11th Five-Year Plan, launch by National Development and Reform Commission (NCDR) on December 24, 2006, and Regulations for Recyclable resources issued by Ministry of Commerce People's Republic of China (MOFCOM) on March 27, 2007, recyclable resources refer to all kinds of wastes produced in production and living consumption, which have lost original use values but could obtain new use values through recycling, classification, and processing. With the exacerbating resource shortage and environmental pollution, the government is gradually focusing on the recycling and utilization of recyclable resources[1-2]. The rational utilization of recyclable resources and the recycling of waste material not only contribute to resource conservation, but also to the reduction of environmental pollution[3-4]. The precondition for the recycling and utilization of recyclable resources is a clear understanding of current situation of recyclable resources production and recycling, which includes information such as production & recycling types, quantity, main body and circulation channel. However, the current statistical works of recyclable resources still remain at initial stage, which means that it is in urgent need of carrying out related statistical works in a systematic way and establishing statistical system. In light of the statistical status quo of recyclable resources, Beijing Municipal Commission of Development and Reform, Beijing Municipal Bureau of Statistics and Beijing Municipal Bureau of Finance together issued the Notes on Related Issues of the Further Construction of Beijing's Energy Conservation and Emission Reduction Statistical System in 2013, which has raised requirements for the construction of recyclable resources statistical system. Based on the Notes, after having a clear understanding of the statistical status quo of Beijing's recyclable resources production and recycling, existing problems in Beijing's recyclable resources statistics are pointed out and some countermeasures and suggestions are put forward. This study will provide support for the establishing statistical system of recyclable resources in Beijing.

2. Review of Literature

The statistics of recyclable resources can be divided into direct and indirect statistics. The typical direct statistical ways include general census, statistical statement, sampling survey, key-point investigation and typical survey. Among them, statistical statement is usually used by government to gather the recycling data and sampling survey is often used by scholars to estimate the recycling amount of recyclable resources[5-6]. Liu used sampling survey to estimate the total quantity of recyclable household solid waste in Beijing and mentioned the statistical system for waste management needed to be improved[6]. The indirect statistical way is mainly through the estimation by various models to get the generation data, such as market supply A method, Stanford model, Carnegie Mellon model, time series or material flow analysis (MFA)[7-10]. Stanford model and MFA are widely used for the estimation of waste electrical and electronic equipment (WEEE). Steubing et al. (2010) used the MFA to estimate the E-waste generation in Chile from 1996 to 2020 [11]. Meanwhile, some studies combined the direct with indirect way to estimate the waste generation and recycling data. Population balance model were used[12-13]. Kim et al. conducted a survey with questionnaires and population balance model estimate the e-waste for 8 products in South Korea from 2000 to 2020[13]. Although the statistics of recyclable resources should be studied as a system[6], all the studies mentioned above are focused on the estimation or forecasting one type of several types of recyclable without showing the whole picture. The study on recyclable resources status, existing problems and suggestions for building the statistics system is needed.

The statistics data are generally issued by the government and industry association. For example, the Environmental Protection Department of Hong Kong government issued the waste recycling and disposal data on Hong Kong Waste Reduction Website. The Statistics Finland released the generation and treatment data of municipal waste every year on the Statistics Finland official website. However, the statistical method and channel were not indicated. Due to the variety and complication of recyclable resources generation and collection, the statistical system should be developed. Policy-makers need sound waste statistics to assess and develop waste management policies. The premise is to identify the current statistical status and problems.

3. Statistical Status Quo of Beijing's Recyclable Resources

3.1. Statistical Types

As a megacity driven by consumption, Beijing has various and complicated production and recycling types of recyclable resources. According to the Medium and Long Term Planning for the Construction of Recyclable resources Recycling System (2014-2020) compiled by Ministry of Commerce People's Republic of China in 2014, recyclable resources consist of the following nine major types: waste metal, scraped car, WEEE, waste plastics, waster paper, waste tire, lead-acid battery, discarded energy saving lamp and waste glass. Each of the above-mentioned type could also be subdivided into different varieties. For instance, the waste paper includes waste magazines, newspapers, books, etc. It is difficult to conduct statistics due to the diversity and complexity of recyclable resources types.

Currently, the statistical types of Beijing's recyclable resources production mainly include household refuse, WEEE and scraped car. According to the data provided by Beijing Municipal Commission of Development and Reform, in 2010, the total amount of Beijing's household refuse is 6.35 million tons, the amount of WEEE is 4.6 million, and the amount of scraped car is 140,000.

The statistical types of Beijing's recyclable resources recycling mainly consist of waste paper, waste metal, waste plastics, waste glass and others. According to the estimated data provided by Beijing Municipal Commission of Commerce, the recycling amount of Beijing's recyclable resources is 4.67 million tons in 2010, 4.8 million tons in 2011, 4.9 million tons in 2012 and 5 million tons in 2013. The proportion of waste recycling types is shown in figure 1.

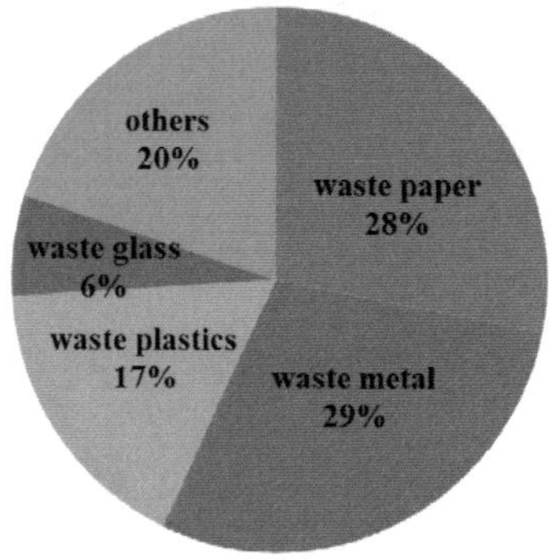

Figure 1. The proportion of Beijing's recyclable resources types in 2013

3.2. *Statistical Methods*

It is difficult to conduct statistics due to complicated recyclable resources type, various main production body and extensive circulation channel. In particular, the most important problem faced by statistical works of recyclable resources is how to ensure scientific statistical methods, to enable statistical results to reflect the actual situation to a greater degree.

At present, the main statistical methods applied recyclable resources production amount in Beijing are investigation and registration. For example, the investigation of the recyclable resources amount of household refuse is calculated based on the statistics of material recycling system. According to the recycling system, the average production amount of recyclable resources of every household per week in Beijing is 7.5 kilograms. In 2010, the household registration amount of Beijing is 4.88 million, and the average amount of per household is 2.55. Moreover, there are about 8.74 million temporary residents. If the average recyclable resources production amount of temporary population is calculated according to the 50% of registered population, the total amount of recyclable resources produced by Beijing's households (namely the source of life) is around 2.6 million tons. The WEEE comes from offices, public institutions, enterprises, social organizations and individuals. Among which, the amount of WEEE that come from offices and public institutions is obtained based on the data related to fixed asset management registered in Bureau of Finance. Systematic statistical methods concerning the amount of WEEE that come from enterprises, public institutions and individuals could not be found. But government sectors and research institutions calculate the theoretical amount of waste production partly on the basis of product sales volume and theoretically-discarded life. The amount of scraped car mainly comes from the data registered in Beijing Municipal Vehicle Administration.

The main statistical methods adopted in the statistical woks of recyclable resources recycling amount in Beijing are investigation and statistical statement. Beijing Municipal Commission of Commerce uses the following method to calculate the recycling amount of recyclable resources: First, it asks the commissions of commerce from 16 districts and counties to report the recycling amount of recyclable resources that come from major recycling enterprises and several social unrecorded enterprises. According to the practical situation, the recycling data reported by district and county commission of commerce roughly account for 50% of the practical total recycling amount, and the recycling data of social unrecorded enterprises roughly account for 50%, based on which Beijing Municipal Commission of Commerce calculates statistics. At last, it calculates the total recycling amount of recyclable resources in Beijing. In addition, industry institutions, colleges and universities also use investigation and mathematical model to calculate the recycling amount of recyclable resources. For instance, by investigating Beijing Tiantianjie Recyclable resources Recycling & Utilization Limited Company, Beijing Recyclable resources and Secondhand Goods Industry Association have found out that in 2013 the monthly average recyclable resources recycling amount of per household in Beijing is 7 kilograms. Based on the 50,310,000 households in Beijing, the recycling amount of recyclable resources produced by residents is calculated to be 422,600 tons.

3.3. *Main Body and Statistical Channel*

IOP Conf. Series: Earth and Environmental Science **178** (2018) 012040 doi:10.1088/1755-1315/178/1/012040

Statistics of recyclable resources involves various main bodies and channels (as shown in figure 2). The main body of statistics includes governmental department, industry association, advanced research institutes, etc., among which many government departments are engaged in statistics of recyclable resources. The statistical channel of recyclable resources production amount could be divided into resident and organization according to the main production body of recyclable resources, and the organization includes enterprise, public institution, social organization, office, army, etc. The statistical channel of recyclable resources recycling amount could be divided into formal recycling enterprise and informal recycling enterprise, among which the former includes recycling companies and individual businesses for recyclable resources registered in industry and commerce department, while the latter mainly refers to private itinerant vendors that do not have business licenses.

Currently, the main statistical body concerning recyclable resources production amount in Beijing includes government function department such as Beijing Municipal Commission of Development and Reform, Statistical Bureau, Finance Bureau and Vehicle Administration. In addition, associations, colleges and universities also investigate and calculate to provide statistical data. Beijing Municipal Commission of Development and Reform counts the production amount of recyclable resources that come from household refuse, the amount of WEEE and of scraped car. The production amount of household refuse is counted and reported by Beijing Municipal Commission of City Management; the amount of WEEE is partly counted and reported by Finance Bureau, and partly obtained from investigation. The amount of scraped car is counted and reported by Vehicle Administration.

The main statistical body concerning recyclable resources recycling amount in Beijing includes governmental department such as Beijing Municipal Commission of Development and Reform, Statistical Bureau, Commerce Commission and Environmental Protection Bureau. In addition, Beijing Recyclable resources and Secondhand Goods Industry Association, colleges and universities also investigate and calculate to provide statistical data. Beijing Municipal Commission of Development and Reform is also in charge of National Urban Mining Base, so the Commission controls the statistical data of "urban mining" that come from Beijing Green Focus Industrial Base of recyclable resources. According to the statistics, the total recycling amount of recyclable resources in the base is around 630,000 tons, among which, the amount of waste paper is 250,000 tons, of waste steel is 200,000 tons, of waste plastics is nearly 80,000 tons, of WEEE is 1.15 million (about 26,000 tons), of scraped car is 24,000 (about 24,000 tons) and of other recyclable resources is 50,000 tons. Statistical Bureau has started to count municipal recyclable resources enterprise above a certain scale since 2005. However, because the Bureau only counts recyclable resources enterprises above a certain scale instead of recycling enterprises, the annual statistical amount of enterprise is around 10. In 2012, only nine enterprises in Beijing were counted. Beijing Municipal Commission of Commerce has developed the management information system for recyclable resources recycling enterprise since 2010, thus recycling enterprises could report statistics conveniently through Internet. The main recycling enterprises are required to report statistics, while other medium and small sized recycling enterprises registered in Industrial and Commercial Bureau can report voluntarily. The statistical types of recyclable resources mainly include the following five types: waste paper, plastics, glass, metal, household appliances, etc. In addition, related enterprises will systematically report their monthly changes of recycling websites, the recycling amount of all kinds of resources and business volume, thus providing relatively authoritative data. Furthermore, Beijing Municipal Commission of Commerce also takes charge of the "old for new service" related to products such as household appliances. When institutions or individuals purchase appliances in this way, the discarded products will be sent to recycling enterprises, and the data will be reported to Commission of Commerce.

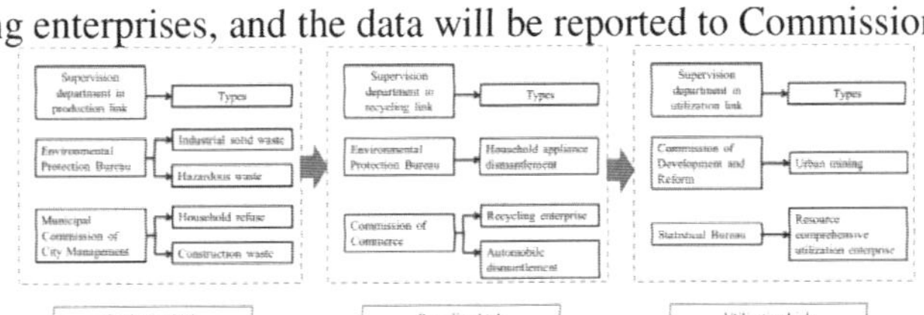

Figure 2. The main statistical department and channel of Beijing's recyclable resources

4. Existing Major Problems in Beijing's Recyclable Resources Statistics

4.1. Undefined Statistical Types and Complicated Statistical Contents of Recyclable Resources

There are various types of recyclable resources. However, the government of Beijing has not established or issued unified illustration and catalogue regarding the statistical types of recyclable resources, which is a major problem in statistical works. In other words, the government of Beijing has not stipulated what kinds of recyclable resources are required to be counted. There are disagreements over the existing recyclable resources type too. Taking the waste metal as an example, waste metal could be divided into waste aluminium, waste steel, waste copper, etc. The waste aluminium could be subdivided into aluminium come from pop-top cans and from scraped doors and windows. The waste steel could also be subdivided into construction waste steel and waste steel come from automobile dismantlement. In different recycling links, different enterprises and departments divide recyclable resources according to their own standards. In addition, the statistical types of the production amount could not correspond to those of the recycling amount. The undefined statistical types led to the disordered statistical contents. The statistical statement could be created with a definite purpose and the statistical works could be conducted in systematic way only when the government has a clear understanding of the recyclable resources types that are required to be counted and built the recycling catalogue.

4.2. Incomprehensive Statistical Scopes and Poor Traceability of Data

Although the related government's functional departments and industry associations in Beijing have carried out part of statistical woks of recyclable resources, the overall statistical works have presented a problem that the statistical scope is relatively incomprehensive. For instance, in 2013, there were altogether 1479 recycling enterprises that registered in industrial and commercial departments and 3475 individual businesses in Beijing. However, when it came to the recycling amount of recyclable resources, standard recycling enterprises reported statistics autonomously while some registered enterprises and individual businesses did not report statistics at all. Meanwhile, the existing statistics lack of systematic statistics and investigation of itinerant nonstandard retailers. It is difficult to trace the source of statistics even if the related statistics could be referred to. As a result, the incomprehensive statistical scope could have a negative influence on the accuracy of statistical results.

4.3. Relatively Nonstandard Statistical Methods and Existing Cross Statistics

The sources of recyclable resources are very extensive and huge in amount. Many complicated factors such as the types of waste products, space-time condition, consumption, recycling, and the features of utilization group have combined influences on the production and recycling amount, which has inevitably brought great challenges to the construction of statistical model of recyclable resources. When conducting statistical works of recyclable resources, the government mainly depends on the data reported by enterprises, and calculates statistics based on related information. The standard statistical process and scientific calculating methods have not been built yet. The statistics reported by enterprises conclude the recycling amount of some itinerant retailers and non-principal enterprises, which are not removed when calculating the total recycling amount of the society, thus leading to the problem of cross statistics. How to ensure scientific statistical methods and make statistical results reflect actual situation to a greater degree are the major problems faced by statistical works.

4.4. The Absence of Supervision and Difficulties in Verifying the Accuracy of Data

The statistics of recyclable resources recycling mainly come from the report of enterprises, and there are no specialized functional organizations or personnel to verify the reported statistics. The statistical information is closely related to the interests of some departments and large-scale enterprises. For example, to enjoy preferential policies such as government subsidies, some departments or enterprises make a false report of the recycling amount. In addition, since the majority of recycling enterprises object to the existing tax revenue system and half of the enterprises have been transformed into privately-owned ones, it is even more difficult to conduct statistical work now than previous days.

Moreover, some enterprises do not record the recyclable resources into the account in light of their own economic benefits, thus influencing the accuracy of statistics.

4.5. The Imperfect Statistical System and Low Integration of Statistical Resources among Different Departments

The statistical system and methods of recyclable resources are not unified. There are no responsible organs and professional personnel to take charge of the statistical woks on a national scale. The related statistical works are conducted only by scattered administrative units that count and calculate statistics according to their benefit demands. Thus it leads to major problems in aspects such as the standard of statistical types, calculating methods and accounting principles. In addition, the related responsible organs do not share information with each other and there is no leading department to coordinate. As a result, it is very difficult to obtain complete and effective statistics.

5. Suggestions on the Construction of Beijing's Statistical System in Recyclable Resources

Currently, the statistical works of recyclable resources in Beijing are still at initial stage. When it comes to aspects such as statistical types, methods and mechanism, there is room for improvement and perfection. After studying the statistical methods and analyzing those methods in practical situations, the following suggestions have been put forward for the implementation and perfection of statistical works in recyclable resources.

5.1. To Establish and Issue Recycling Catalogue

The establishment and issue of recycling catalogue contribute to the classification of recyclable resources, thus laying a foundation for the creation of the statistical statement. At the same time, the recycling catalogue also plays an important guiding role in employees and related interest bodies in recyclable resources industry. It is suggested that Beijing Municipal Commission of Commerce should join hands with departments such as Municipal Commission of Development and Reform, Municipal Commission of City Management, Environmental Protection Bureau, Department of Transportation and Construction and Department of Information-based Economy to establish and issue the recycling catalogue of recyclable resources, in order to determine matters such as the recycling type, recycling standard, and utilization guidance of recyclable resources. Meanwhile, based on the recycling catalogue, Beijing Municipal Commission of Commerce should work together with departments that participate in the statistical works of recyclable resources, such as Municipal Commission of Development and Reform and Statistical Bureau, to establish the statistical statement, in order to determine the types and contents of recyclable resources that are required to be counted.

5.2. To Standardize the Statistical Methods and Process

Scientific statistical methods can ensure the effective obtainment of statistics and increase the reliability of statistics. The statistical works of recyclable resources could be conducted according to different major bodies in production and recycling link. It is suggested to count production amount by using the method that combines sampling statistics, statistical statement and model-based estimation together. When it comes to the recyclable resources such as waste paper, plastics, clothes and household appliances that come from households and retail businesses, it could be estimated and counted on the basis of Statistical Bureau's sampling survey and according to the per capita production amount and population size. When it comes to recyclable resources such as WEEE and discarded office furniture that come from offices, public institutions and commercial office buildings, it could be counted on the basis of statistical statement. The recycling amount could be counted by using the method that combines sampling statistics and statistical statement together. For example, the recycling data of formal enterprises (including principal enterprises) could be counted by using statistical statement, and the recycling data of informal retailers could be counted by using sampling survey. Furthermore, the whole statistical process should also be standardized, which requires the leading departments of statistical works to take the lead, and cooperate with other related departments and industry associations to complete statistical and investigation works.

5.3. To Build a Sharing Mechanism for the Data Monitoring

Due to the various statistical types of recyclable resources and complicated statistical channels, the statistical works need the cooperation and coordination among different departments, in order to broaden statistical scope to a greater degree, distribute resources effectively, share existing statistics, reduce repetitive statistics, and increase efficiency. According to the function analysis of related functional departments and their participation in existing statistical works of recyclable resources, it suggested that different departments should work together to determine the statistical mechanism. In addition, Municipal Commission of Commerce and Statistical Bureau should take the lead, and join hands with other related departments such as Municipal Commission of City Management and Environmental Protection Bureau to build a sharing and supervision platform for the statistics.

5.4. To Innovate the Data Collection Method
Considering the ever-changing life style of residents, the recycling model should be innovated by using the technology of mobile internet. It is suggested to design a specialized mobile application to track the recycling trading information and obtain trading statistics. For instance, after downloading and installing such a mobile application, residents could search nearby recycling enterprises that have recycling qualification on their phones. Recycling enterprises could offer their prices online. After recycling enterprises providing door-to-door collecting or residents sending waste goods to recycling enterprises, both the recycling enterprises and residents could directly trade online. As a result, the trading statistics could be updated online, and information such as production types, recycling amount and distribution scope of recyclable resources could also be obtained online. At the same time, by strengthening the qualification management of recycling enterprises that participate in the online trading, the disruption of illegal enterprises to the market could also be lessened.

5.5. To Establish Statistical Supervision System and Appraisal Mechanism
It is suggested to enforce the supervision over enterprises that report statistics and departments that participate in statistical works, and establish statistical supervision system. The commission of commerce of all districts and counties should send specially-assigned personnel to regularly verify the statistics reported by enterprises. At the same time, Municipal Commission of Commerce should regularly go to all districts and counties, to inspect and supervise the reported recyclable resources statistics. An appraisal mechanism should also be established, to reward those enterprises that regularly report accurate and effective statistics, and punish those enterprises that do not report or make a false report. In addition, it is suggested to make statistical work as a performance index when assessing the performance of leaders that directly take charge of statistical works of recyclable resources, in order to advance and improve statistical works.

6. References
[1] Liu Y.P, Li Y.J and Xu Y.F 2008 Beijing municipal recyclable resources recycling system construction plan. *J. Environment and Sustainable Development* 5 1-3
[2] Zhu T, Gao S and Zhang M 2013 Recyclable resource industry: how to get out of the prosperity façade. *J. Environmental Protection* 6 39-41
[3] Wang R and Yin H.P 2009 An empirical analysis of development of recycling resources industry in Beijing. *J. Journal of China Agricultural University* 1 77-83
[4] Zhou H.C 2008 Development and problems of renewable resources industry in China. *J. Renewable Resources and Recycling Economy* 5 5-8
[5] Gu B, Wang H, Chen Z, Jiang S, Zhu W, Liu M, Chen Y, Wu Y, He S, Cheng R, Yang J and Bi J 2015 Characterization, quantification and management of household solid waste: A case study in China, Resources. *J. Conservation and Recycling* 98 67-75
[6] Liu T, Wu Y, Tian Z and Gong Y 2015 Urban household solid waste generation and collection in Beijing, China, Resources. *J. Conservation and Recycling* 4 31-37
[7] Ikhlayel M 2016 Differences of methods to estimate generation of waste electrical and electronic equipment for developing countries: Jordan as a case study. *J. Resources Conservation and Recyclable* 8 134-139
[8] Guo X and Yan K 2017 Estimation of obsolete cellular phones generation: A case study of China. *J. Science of the Total Environment* 5 321-329

[9] Petridis N.E, Stiakakis E, Petridis K and Dey P 2016 Estimation of computer waste quantities using forecasting techniques. *J. Journal of Cleaner Production* 8 3072-3085

[10] Li B, Yang J, Lü B and Song X 2015 Generation estimation of waste electrical and electronic equipment: methods review and selection strategy. *J. Acta Ecologica Sinica* 24 7965-7973

[11] Steubing N, Böni H, Schluep M, Silva U and Ludwig C 2010 Assessing computer waste generation in Chile using material flow analysis. *J. Waste Management* 3 473-482

[12] Nguyen D.Q, Yamasue E, Okumura H and Ishihara K 2009 Use and disposal of large home electronic appliances in Vietnam. *J. Journal of Material Cycles and Waste Management* 11 358-366

[13] Kim S, Oguchi M, Yoshida A and Terazono A 2013 Estimating the amount of WEEE generated in South Korea by using the population balance model. *J. Waste Management* 2 474-483

Acknowledgements

This study was funded by Beijing Natural Science Foundation (9164022), and also supported by the Beijing municipal commission of education social science project (SM201610005011), Beijing outstanding talent young scientists training program (2015000020124G036), Beijing Social Science Fund (16YJC042) and Beijing Natural Science Foundation (9172001).

IOP Conf. Series: Earth and Environmental Science **178** (2018) 012041 doi:10.1088/1755-1315/178/1/012041

Research on Feasibility and Development Path of Ecological Poverty Alleviation of Yi Nationality in Luquan, Yunnan

Ying Zhao[1]

1 Southwest Minzu University,Chengdu, China

Abstract. Based on the field investigation of Yi and Miao of Luquan Autonomous County in Yunnan Province, this paper chose the ecological poverty alleviation as the angle of view, and analyzed the generation process of the poverty problem of the Yi Nationality. At present, many poverty alleviation problems faced by China are caused by the imbalance of economic development and the single mode of development in the process of social transformation. In particular, poverty alleviation in ethnic areas needs to incorporate local knowledge such as traditional ecological ethics, which is closely related to the management system of ethnic work affairs. The article regarded the precise poverty and ecological poverty as the value guidance, starting from the particularity of the ecological environment of Yunnan, strengthening exchanges and communication between the government and the Yi nationality compatriots, enhancing the cultural penetration and cultural influence, making it effectively used in multi governance work, in order to promote the national governance system and improve the poverty alleviation work.

1. Feasibility Analysis of Ecological Poverty Alleviation of Yi Nationality

1.1. The Challenge of Yi Nationality's Ecological Poverty Alleviation

Firstly, the natural environment is fragile, and it is obvious to have poverty situation because of disaster. Luquan is located in Dian Zhong Plateau with complex terrain and undulating surface[1]. In addition, there are rolling hills and rivers and streams extending in all directions in Luquan. For the terrain in Luquan County, the south is more complete, while the northern and central areas are undulating[2]. The mountainous area of Luquan County accounts for 98.4% of the total area, so the mountain falls greatly and the slope is steep. At the same time, it is on the west side of the Xiaojiang Fault Zone with intense crustal movement, facing long-term landslides, floods, landslides and other natural disasters. This leads to poverty situation. On 22th to 25th , August in 2015, continuous heavy rain caused different degrees of disasters in Ping Town, Sayingpan Town, Jiao Ping Town, Cui Hua Town, Ze Hei Countryside and Tang Lang Countryside. The direct economic loss is 2 million and 980 thousand yuan[3]. Among them, crops affected area are 7065 acres, and 1526 tons of grain yield are reduced, resulting in economic crops loss of 447 thousand. Natural disasters lead to huge losses in temporary social and economic construction, and bring new challenges to poverty alleviation. At the same time, the agricultural foundation in poor areas is weak with traffic block, and information is difficult to gained. Moreover, the cost of poverty alleviation is relatively high, and poverty alleviation is difficult.

1.2. Lower Labor Force

Poverty alleviation work is not limited to economic income support, but also lies in the level of education and the assistance of talent knowledge. "According to statistics in 2009, only 40% of junior high school graduates in poor areas can continue to receive high school education." The backwardness of the education level also brings new pressure to the ecological environment. Due to the lack of proficiency in a particular line, people only depend on the ideology of "if you live on a mountain, you live off the mountain; if you live by the water, you live off the water" to maintain the basic production and living, which will undoubtedly cause the new pressure to the ecological environment and natural environment. The ability of self poverty alleviation is weak, which brings new difficulties to poverty alleviation. Shortage of available resources for poverty alleviation in poor area is difficult for surplus labor, especially the sick and elderly and other vulnerable groups to arrange suitable jobs. The poor villages also lack the sustainable development mechanism of economic development and the sustainable development mechanism of ecological civilization construction, which undoubtedly creates new problems for the ecological poverty alleviation.

1.3. Poverty Alleviation Work is Seriously Divorced from Protection of Ecological Environment

In recent years, precise poverty alleviation and ecological civilization construction have become the two major tasks of social and economic development in Yunnan province. The two are highly integrated in space and time, both of which are unified[4]. But in the specific implementation of the project, both of individuals are separate, and sometimes there are frictions and collisions. In the work of poverty alleviation, there is lack of awareness of ecological environment protection, and the elements of poverty alleviation in ecological civilization construction are also less considered. Taking Luquan County as an example, some areas are located in mountainous areas. The economic base is weak, and the environmental carrying capacity is weak and the ability of economic self-development is weak. Moreover, the ecological environment self repair ability is poor, and the contradiction between economic and social development and ecological environment is heavy. In view of this, it is particularly important to adhere to the poverty alleviation and poverty alleviation and improve the ecological environment, so it is urgent to explore a new path for ecological poverty alleviation suitable for the local.

1.4. Scattered Poverty Alleviation Projects and Funds

Poverty alleviation projects and ecological civilization construction project funds belong to different government departments[5]. For example, working in relief funds belonging to the NDRC system management, The capital of urban and rural infrastructure construction belonging to the management of the traffic department and the projects and funds involved in the construction of ecological civilization are mainly managed by the government's science and technology departments. As for the government departments in the implementation of poverty alleviation projects, the use of funds are independent of each other with poor coordination, thus, the phenomenon of decentralization and overlapping of project implementation and fund use is common, which indirectly restricts the effect of poverty alleviation.

1.5. Unperfect Supervision System of Ecological Poverty Alleviation

Ecological poverty alleviation project is a major sustainable project which is " beneficial to the present and to the future". During the specific development period of ecological poverty alleviation, there must be many unknown factors, and it is extremely important to carry out effective supervision and management. For example, in the process of poverty alleviation, the poverty alleviation subject is prone to use the poor funds, and it is easy to destroy the ecological environment in the process of poverty alleviation and development, and the ecological benefits and economic benefits can not develop simultaneously; when there is a problem in the poverty alleviation work, if the social supervision and the internal supervision system of the government have not been established yet or to be waited for improving, this will affect the effectiveness of poverty alleviation projects.

2. The Way of Ecological Poverty Alleviation of Yi Nationality

2.1. Promoting Ecological Agriculture and Tourism

Ecological agriculture refers to the protection and improvement of the ecological environment in the rural areas. The agricultural economic system and the agricultural ecosystem should be taken into account in order to maximize the ecological and economic benefits[6]. This is to consider the agricultural production, processing, sales and other aspects, in order to adapt to the modern economy in the agricultural requirements. Ecological agriculture is conducive to promoting accurate ecological poverty alleviation in an orderly way. For example, pesticide residue is damage to the ecological balance of the local, so the use of advanced science and technology can find alternatives to pesticides, which can avoid such phenomena. A large number of straw burning causes serious air pollution. We can promote straw returning technology, which can increase the yield and avoid environmental pollution. The trend of eco agriculture industrialization is the key to the development of eco agriculture. According to the local regional advantages and geographical advantages, it is tailor-made for its ecological agriculture pillar industry, so that to achieve the goal of agricultural sustainable development, and promote the optimization and arrangement of its production materials. It uses regional overall layout and enterprise operation management to achieve the development of ecological agriculture. In addition, we should combine ecological agriculture with ecological tourism, and give full play to the advantages of local ecological culture. For example, Luquan county should combine the cultural background of the two long march of the Red Army to explore a red tourism route, including Cui Hua Chian Bridge - Red Army Martyrs Monument - Shuang Hua Stone Bridge Army Battlefield - Jiao Xi Xiao Mi, Ma Jinhua Martyrs - General Tree - Red Army Cross the River Memorial Hall in Jiao Ping. We should develop the local ecotourism resources deeply, and promote the sightseeing tourism agriculture of "close to nature, return to nature". In the same way, we will set up experiential farming and experiential fruit and vegetable planting courses to inject new vitality into ecological sightseeing agriculture and alleviate the employment pressure in rural areas.

2.2. Improving the Ecological Compensation Mechanism

Ecological compensation is the local or economic subject's support given by the government through the way of economic compensation to a certain extent. For example, the implementation of the policy of returning farmland to forests. It is necessary for farmers to abandon the instinctive demand for planting grain in this area, and farmers should plant economic and ecological forest. For farmers, it is undoubtedly the sacrifice of material production to seek long-term development of the ecological environment. However, the direct economic losses caused by returning farmland to forests should be compensated by policy subsidy or by ecological purchase.

2.3. Improving Social Security System and Public Service System

In thoughts of precise poverty alleviation, there is a mention about "reveal all the details of a number of social security" from "the five batch", which refers to establish and improve the social security system in the poor areas, and bring the poor people who have lost some or all of their labor capacity into the scope of protection. We can accurately identify the poverty status of special groups through local conditions, and provide different types of social insurance or state subsidies, so that to improve the living conditions of local poor groups. The implementation of the new rural cooperative medical system has greatly alleviated the problem of "difficult medical treatment and expensive medical service" for the poor. For patients with chronic diseases or serious illness, it is recommended to popularize the serious illness medical assistance system to solve its worries. The social security system and universal coverage can effectively reduce poverty depends on the environment, so that to avoid the expense of the ecological environment for the basic life need, which is be of great advantage to poverty reduction. In order to alleviate the social conflicts, the basic living allowance should be provided to the families who suffer from illness or those who have lost the ability to work by accident.

2.4. Encouraging Full Employment of Farmers

Training agricultural technical talents according to local conditions and combining them into local agricultural production can be combined with remote teaching and on-the-spot guidance. We should mobilize the production enthusiasm of the poverty alleviation object, and guide the development of poverty alleviation from "blood transfusion" to "poverty alleviation". We should strengthen the support of vocational education in poor areas, and change the concept of destruction of ecological environment due to life. For the ecological environment or ecological fragile areas, the government should do a good job of ecological migration in time, and adjust the work strategy of poverty alleviation and development. In order to protect the ecological environment as the premise, the government should be combined with the electricity supplier or micro business development mode, and promote the superior industries in poverty-stricken areas.

2.5. Constructing Concept of Ecological Civilization

The precise ecological poverty is based on the ecological causes of poverty in poor areas to develop a series of measures and to improve the ecological environment, so that to achieve effects of poverty alleviation. Therefore, the construction of ecological civilization is related to the well-being of the people and the future of the nation, and is related to the "two one hundred years" struggle goal and the realization of the great rejuvenation of the Chinese dream of the Chinese nation. Because of the restriction of education level and cultural level, the concept of ecological civilization construction is not clear, and it is easy to fall into the dilemma of ecology and poverty. Therefore, we should strengthen the education of poverty alleviation in the region, and not only in the nine years of compulsory education; it should retrain young people with no proficiency, so that to establish the ecological environment protection and construction of ecological civilization consciousness, and to seek a sustainable development road of the ecological environment. We should improve the farmers' participation in the local environmental governance, protect their environmental supervision power and punish the illegal activities that destroy the environment. Moreover, we should combine the construction of ecological civilization with the work of poverty reduction, and make sure the unity of ecological civilization construction and poverty alleviation work.

3. Development Path Choice of Ecological Poverty Alleviation of Yi Nationality

3.1. Model Innovation of Ecological Poverty Alleviation of Yi Nationality

First, it is the design of ecological poverty alleviation model. It mainly includes the mode of ecological poverty alleviation in situ and the ecological poverty alleviation model from the land. The model of ecological poverty alleviation in situ is aimed at poor groups, which are difficult to move out of their own development because of education and culture, or in poor areas with good agricultural and livestock production environment. The ecological poverty alleviation model from the land is aimed at the ecological fragile areas, promoting ecological migration and ecological subsidies measures to achieve poverty alleviation and poverty reduction purposes. We should guide the farmers and herdsmen to transform into the ecological population, focusing on financial and material resources to carry out ecological civilization construction and poverty alleviation project, so that to find jobs for ecological construction and promote employment and provide employment opportunities for local farmers and herdsmen. We should encourage local farmers and herdsmen to directly participate in the ecological governance and poverty alleviation work. In addition, they should contribute to the works of improving income, promoting employment and improving the mechanism and ecological poverty alleviation. However, at the same time, we should combine the protection of local culture and the development of new ways of living, taking into account the local ecological environment carrying capacity and the economic life of farmers; moreover, encouraging farmers and herdsmen appropriate to give up is not conducive to the sustainable development of environment production and life style. It is necessary to give appropriate subsidies and patient guidance for the sacrifices they make. We should guide the local ecological agriculture to form an industrial chain. We should actively utilize the advantages of the natural environment in the poor areas, and actively cultivate Chinese herbal medicines, vegetables, fruits and other pollution-free agricultural products, organic food and green food. We should help to establish local leading enterprises in rural areas, and cooperate with small farmers to build cooperatives in poor areas to develop local industries, so as to eliminate poverty as soon as possible. According to the actual situation of the whole village, we should make the rural natural landscape and humanistic landscape organically integrate. Moreover, we should scientifically grasp the relationship between poverty alleviation and development and environmental protection projects, make unified layout and planning, focusing on capital investment in infrastructure construction and implementing water, electricity, access, housing construction and other projects. For the concentrated areas of poor villages, we can focus on the whole piece of planning and reconstruction, and enhance the scale effect of the local poverty alleviation. But for the characteristic buildings or houses, we should preserve and protect them, and restore the local elements through repair. In the central document of the new ten years of The Outline of China's Rural Poverty Alleviation and Development (2011 - 2020) issued by the state council, one of the main contents of the national special poverty alleviation is the whole village, and the key is to improve the comprehensive quality of the labor force of the poor. It has a dual role for the realization of ecological protection and the promotion of employment. In view of this, we should improve the poverty stricken areas especially the rural poverty education system, strengthen the quality of youth education process and promote youth occupation education, so that to accelerate the pace of education for the elderly elderly speed, and better balance the relationship between ecological civilization construction and poverty alleviation.

3.2. Establishing a "Low Carbon Reduce Poverty" Security System
In 1997, after a series of international climate treaty negotiations, the Kyoto Protocol was formally signed, which specifies the importance of afforestation for carbon dioxide emissions in the clean development mechanism. In principle, it agreed that developed countries or developing countries could offset the environmental pollution caused by greenhouse gases through the implementation of forestry carbon sequestration projects. We should integrate the overall goal of low-carbon poverty reduction into the overall plan of social development, decompose all its objectives into the work plan of the government departments, and coordinate the relationship among the government, enterprises and society. According to the local actual situation, we should implement green low-carbon financing mechanism and supporting measures to ensure the steady operation of the low-carbon poverty alleviation funds, strengthen the administrative supervision and regulation and enforcement mechanism, encourage professional companies to participate in the operation of low-carbon poverty alleviation projects and promote the industrialization development. We should also support the preferential development of carbon sequestration projects in poor areas with limited conditions. Moreover, to gain the domestic and international financial support, we should also focus on the development of low-carbon technology projects and the implementation of public welfare low-carbon plan, and provide subsidies and support for special projects.

3.3. Policy Proposal

3.3.1. Paying Attention to Comprehensive Governance in Poor Areas. The key to ecological poverty alleviation lies in comprehensive management. We need to mobilize the enthusiasm of the object of poverty alleviation and make them participate in the "ecological agriculture industrialization" and the work of poverty alleviation, which will turn the extensive economic construction into paying attention to the development of poverty alleviation subject. We can take the relevant township pilot ecological poverty alleviation project, and the government and project authorities need to work together to make ecological poverty alleviation programs and objectives. At the same time, the government and the project management department and the poverty alleviation subject should make positive communication. For the problems, they should coordinate and make the integration of resources, so that to avoid unhappy things.

3.3.2. Strengthening Protection of National Ecological Ethics Culture. In addition to the Han nationality, there are 25 ethnic minorities in Yunnan, among which Yi nationality accounts for 5 million and 28 thousand, accounting for 10.94% of the total population. The broad and profound ecological ethics culture of Yi nationality, mainly includes text art, religion, taboo pacts, cultural and other aspects. For example, the parts of ecological ethics culture in collected and published ancient books and annuals of Yi Characters in Luquan county are integrated and protected. This provides a theoretical basis for the realization of ecological civilization construction, and provides reference for the implementation of ecological poverty alleviation planning and policy, and provides a new perspective for promoting the sustainable development of economy, culture and environment.

3.3.3. Adhering to Education of Ecological Awareness. The main body of ecological poverty alleviation is poor farmers, and the actual effect of ecological poverty alleviation depends on the ecological awareness of the theme of poverty alleviation and even ecological concept. This not only relies on the government or law enforcement force to implement the goal of ecological poverty alleviation, but also depends on the moral constraints, public opinion pressure and so on, so that the concept of ecological poverty alleviation can deeply go into the minds. Initially, we can encourage farmers to participate in the protection of ecological environment through encouraging measures. The village committee may organize the training course of ecological poverty alleviation education regularly, and invite relevant experts to go to the countryside to guide the local farmers, and the school can also add the content of ecological civilization construction in the course of science or science. This can gradually form the consciousness of respecting nature, fearing nature and protecting nature.

3.3.4. Adhering to Leadership of Government. In response to the relevant requirements of the national poverty alleviation, since the end of 1985, Luquan began to tackle poverty. In 1986, it was listed as one of the 592 key poverty counties in China. In 2001, it was listed as one of the key counties of poverty alleviation and development in 73 counties of Yunnan province. The county committee formulated the "8321" poverty alleviation overall plan, that is, each person must have about 8 points of high and stable farmland, and each household has more than 300 yuan of livestock and poultry income per year. Each household has 2 acres of economic forest fruit, and each household transfer a labor force to the non food industry up; moreover, the production policy of "long catching on forest, short catching on grazing, not long and not short catching on fruit trees and then grasp well on the agricultural and industrial deputy" was implemented. After years of continuous construction, the rural poor population dropped from 386 thousand at the end of 1985 to 119 thousand at the end of 2013 (the total population of the county is 397111, of which the agricultural population is 385982, and the per capita net income of farmers is 142.33 yuan, which is much lower than the poverty line of 200 yuan issued by the State Council's Poverty Alleviation Office in 1985. And the poverty line is that the per capita net income is less than 2300 yuan). Thus, the proportion dropped from 97.19% in 1985 to 28.3% at the end of 2013. The per capita net income of farmers increased from 142 yuan in 1985 to 5438 yuan in 2013, which is an increase of 38.3 times, and is an average annual growth of 13.9%. This effectively realized the parallel strategy of human resources development and ecological environment protection, and comprehensively developed the economic living standards in poor areas. The ecological idea of sustainable development should be based on the protection of the ecological environment, and we should put emphasis on poverty alleviation of poverty reduction and the construction of ecological civilization development, and promote the poverty stricken areas of returning farmland to forest, and afforestation projects, so that to avoid the deterioration of the ecological environment. And the ecological compensation mechanism should be established and perfected.

4. Conclusions

Generally speaking, ecological poverty alleviation takes the ecological environment and poverty alleviation as the basic framework, and takes the ecological environment sustainable development as the core goal. At the same time, it considers the particularity of ecological ethics and local society, which has the practical significance of diversity and dynamics. Combined with the actual situation of Luquan County, its poverty problem also has the particularity of ecological protection, religious culture, cultural construction and historical origin. This is due to they explored a way suitable for the survival and development of the local. This is not the reproduction in the modern sense or be independent in the local cultural background. It is determined by this basic point that precise ecological poverty alleviation must combine the different ecological environment and poverty situation in each region. In addition to pursuing the economic growth point of poverty alleviation, we should also consider its evaluation in the traditional social ethics and social norms. We should give full play to the initiative of the theme of poverty alleviation, and properly supervise and manage it. In view of this, when examining the poverty problem of ethnic minorities in contemporary society, we must consider the construction of local local culture and social network mechanism.

5. References

IOP Conf. Series: Earth and Environmental Science **178** (2018) 012041 doi:10.1088/1755-1315/178/1/012041

[1] Anonymous 2007 General Situation of Luquan in Yi and Miao Autonomous County *J. Ethnic Publishing House.* 5 91-92
[2] Wanlong, Lin, Ling, Zhong, Hanwen and Lu 2008. Cooperative Anti Poverty Theory and Practice in Yilong *J. Agricultural Economic Problems.*11 59-65
[3] Xueping, Li, Meng adn Wang 2014 Livelihood Vulnerability under the Framework of Multidimensional Poverty "Action Structure" -- Based on Empirical Investigation and Theoretical Analysis in Wuling Mountainous Area *J. Journal of Huazhong Normal University (Humanities and Social Sciences).* 5 2
[4] Xiaobing, Wang, Chengfang, Liu, Linxiu and Zhang 2011 College education and the poor in China: documenting the hurdles to educational attainment and college matriculation *J. Asia Pacific Education Review.* 4 533
[5] Shunbao, Jie Wu and Luo 2012 Textual Research on Humanism and History *J. Youth and Society (Special Embroidery Art).* 9 32
[6] Lixia Tang, Jiangyue Luo and Xiaoyun Li 2015 Policy and Practice Dilemma of Implementing Precise Poverty Alleviation Mechanism *J. Guizhou Social Sciences.* 5 151-156

Acknowledgements
The initial results of western project of National Social Science Fund: "Yi nationality annual: Investigation and research based on four provinces of Yunnan, Sichuan, Guizhou and Guangxi" (13XMZ067)
The initial results of the historical value and military culture of sub item of "The Classic Translation and Annotation on Bi Mo of Yi Nationality" from the 2014 national project of the National Social Science Foundation (second batch) of "Yun Guichuan's 100 works of The Classic Translation and Annotation on Bi Mo of Yi Nationality" (14ZDB119)
The periodic results of "The Study on the Ancestral Memory of Yi Nationality in Southern Yunnan under the Background of Cultural Integration" from the 2016 youth fund project of the Ministry of Education Humanities and Social Science Research Program (16YJC850009)

The periodic results of "The Research on National Religion and National Governance" from the 2015 National Major Social Science Fund (15ZDB123)

IOP Conf. Series: Earth and Environmental Science **178** (2018) 012042 doi:10.1088/1755-1315/178/1/012042

Research on the Method of Emission Trading in Harbin

A Long Yang[1] , Chong Du[2]andGuang Bo Ma[3]

1 First author,College of Water Resources and Electric Engineering of Heilongjiang University, Harbin, China
2 Corresponding author,College of Water Resources and Electric Engineering, Heilongjiang University, Harbin, China
3 Third author,China Petroleum Jidong Oilfield Branch Office,Tangshan city,China
E-mail:1119776309@qq.com

Abstract: In order to alleviate the environmental problems in Harbin, we need to establish a more perfect system of emissions trading. Although China's introduction of emissions trading system in 2014, after several years of development has made great progress, but because of geographical, environmental, economic differences can not form a unified model, in the initial distribution method of emissions trading, the initial distribution price , trading years and secondary market trading prices are different. Therefore, in the establishment of emissions trading system should be based on Harbin's environmental indicators,economic development and etc situation,analysis of the study the initial distribution of emissions permits,the initial price customization,and trading years ,to draw up a fair and reasonable, sustainable development methods to solve environmental pollution problems.

1. Introduction

With the continuous development of emission trading in China's pilot cities, it is also urgent to establish a relatively perfect emission trading system in the governance of environmental problems in Harbin.Although China has introduced emissions trading in 1997, it was not until 2014 that the country introduced the first administrative regulations on the use and trading of emissions permits:<Guiding Opinions of the General Office of the State Council on Further Promoting Paid Use and Pilot Work on Trading of Emission permits>(State Council issued [2014] No. 38).Although has obtained certain achievements,Although some achievements have been made, the initial allocation, but the initial price, the trading period, the secondary market transaction mode and the price method are different in the emission trading, which can not form a unified model.These factors will have a great impact on the emission trading market, develop a reasonable method to the maximum extent stimulated the activity of the market[1].

2. Determination of Total Emission Permits

2.1. Allow the Total Amount of Emission

In the establishment of emission trading of a primary market of In the Songhua River Harbin municipal area control unit, allow the pollutants emission amount according to < Harbin city "12th Five-Year" energy-saving emission reduction requirements in planning >,by 2015, COD emissions controls within 67930t/a, ammonia nitrogen emission control is within 12897t/a,to determine the total allowable emission amount of the primary market of emission trading in Harbin. Before the next term of use of the emission permits, the total amount of emission permits should be re-approved in accordance with the Harbin Environmental Standards and Economic Development Goals.

2.2. *Emission Permits That Can be Initially Allocated*

According to the determination of the total amount of emissions allowed, at the same time, in order to ensure the normal economic development of the next term of emission permits, according to the economic development data. Calculate the government should reserve the emission permits, at the same time, after deducting the emission permit of the social public utilities, the initial emission permits shall be allocated.The government reserve can be calculated according to the growth rate of GDP in Harbin, the calculation formula is as follows:

$$W_{reserved} = W \bullet (\frac{D_{target\,year}}{D_{planning year}}) \tag{1}$$

In the formula:$W_{reserved}$:the amount of emissions reserved by the government;W:the total area of emission permits;$D_{target\ year}$:the area GDP of the target year;$D_{planning\ year}$:the area GDP of the planning year.

At the same time, based on the investigation of pollution source data, calculation of public utilities emissions and reserved. Among the 26 sewage draining exit in the Harbin municipal area control unit, the public utilities which are not for profit are mainly urban sewage treatment plants, and the total pollutant emission is the emission amount of public utilities required.The calculation formula is as follows:

$$W_{\text{dispensability}} = W - W_{reserved} - W_{\text{utilities}} \tag{2}$$

In the formula:W:the total area of emission permits;$W_{reserved}$:the amount of emissions reserved by the government;$W_{utilities}$:emission permits of public utilities in the region;$W_{dispensability}$:the amount of emission permits that can be initially allocated.

3. Initial Allocation Method of Emission Permits and Computing Method of Allocation Quantity

3.1. *Initial Allocation Method of Emission Permits*

The emission permit is a property in the definition of economics, emission permit allocation means the distribution of property interests. So, how to carry on the initial allocation of emission permits is an important issue of emissions trading.The initial allocation scheme of emission permits has three categories: free allocation, public auction allocation and fixed price selling[2].

Free allocation has certain rationality, in a longer period of time will exist,but contrary to the basic principles of polluter-paid environmental law,making the enterprise without the power and pressure to reduce emissions, not only make the environment resources not optimal allocation,and but also hinder the emission reduction technology research and development innovation,from the perspective of efficiency is not desirable. In the meantime,enterprises from the government departments in charge of free distribution of emissions, often can't satisfy the production scale of enterprises and want to get more free emission permits, so as to give the"enterprises rent seeking" and "government rent-creating" creation opportunities, resulting in emissions of corruption , And ultimately can not guarantee the allocation of fairness, and even affect the normal production of enterprises[3].

Auction allocation is compared with free allocation, the advantage of auction is very obvious: Firstly, the auction can obtain the emission reduction marginal cost information of the auction enterprises, so that the enterprises with higher economic benefit have priority to obtain environmental resources, so as to achieve the optimal allocation of emission permits; Second, the auction price of the auction can also be used as a reference standard for the secondary market trading of the emission permits, that is the price signal; Third, the auction process automatically realized by the market, effectively avoiding the "rent-seeking and rent-creating " behaviors inherent in the free allocation. However, due to the auction will increase the cost burden of polluters, and therefore face political resistance; Meanwhile, if the government does not limit the management, it will make a small number of strong financial strength enterprises to auction and hoard, and then monopolize the emission permits,so as to achieve the monopoly of the product market. Therefore, despite the great advantage of the theory of auction, but in practice the development is relatively slow.

The fixed price refers to: The emission enterprises shall purchase the emission permit of by the price standard for the use of the paid use emission permits set by the government, The establishment of the price standard is very important for the paid use of emission permits and the pilot work of emission trading. Considering condition that China's current emission collection cost is difficult to encourage enterprises to treasure the environment constraints and incentive mechanisms, the environmental capacity of resource value based paid use of the standard price to effectively establish paid use price standard machine-made of emission permits including reflects the relationship between market supply and demand, the degree of scarcity of environmental resources and pollution governance cost . Adhere to the scarcity of environmental resources and the ability to withstand combine the affordability of the principle. According to my national conditions and the various regions different industries,adaptation to local condition to establish the use of emission pricing standards.

Each one has its own merits of three ways of distribution,but according to<Guiding Opinions of the General Office of the State Council on Further Promoting Paid Use and Pilot Work on Trading of Emission permits>:"The introduction of emission permits paid to obtain " provisions of the initial allocation of emission permits. The emission permits free of charge to those social utilities, small sewage and not more than a certain emission standards of polluters, but they can not be traded on the market; For the business unit,the emission large amount of polluters,government should implement the distribution in the form of a fixed price to sell at the present stage, after the theory and the system are gradually improved then use the fixed price and the auction of the mixed way distribution. In the recent years, the central government has put forward the policy of "revitalizing the old industrial base", Harbin, as the old industrial base in northeast China,so in some projects to support enterprises in the allocation of emission permits should be given certain concessions, such as preferential price: only pay 1/2 of the price to use the same emission permits, but can not be traded.

3.2. Computing Method of Allocation Quantity

It is should to calculate allocation the emission permits of each enterprise after the allocation method is determined,There are two initial distribution accounting methods for emission permits:The first is the "grandfather system" model， it shows the recognition of the work and respect for the history. It is easy to operate and the enterprise has high recognition; The second is the "output value" model, reflecting the current rules of occupation, relatively fair, but also can promote the enterprise pollution control and emission reduction work. Although it is necessary to rely on a large number of pollution sources survey data, but the allocation of accounting methods are relatively objective, close to the reality.

Harbin has a lot of old industrial enterprises that have good credit, also have a comprehensive range of pollution data,so for these enterprises to adopt the first method, this method is convenient and quick, and easy to accounting calculation,but also affirmation and continuation of past EIA approval work, enterprise recognition is higher; As for the rest of the emission information is not complete, credit degree is not high and some emerging enterprises in accordance with the second methods of accounting.

4. Initial Allocation of Prices and Secondary Market Transaction Prices

4.1. The Initial Price and the Determination of Secondary Market Price Method

Based on the emission permits allocation method already established above, the government considers the endurance capacity of existing emission units and the local environmental quality improvement requirements to price the initial emission permit. There are three methods of initial pricing： the first is the use of (social average or the region of some pollutants) unit pollutant management costs to calculate the initial price of emissions (called the social average cost of governance); The second is through the different industries and enterprises in the emission of shadow price (sewage performance) accounting to determine the price of initial emission permits; The third is the use of auction value discovery function, through the average auction price over the years to develop guidance prices [4].

The above three kinds of initial allocation of emission permits pricing methods have advantages and disadvantages, The first method is through the various industries units of pollutants management costs accounting for the price of emission permits , is the emission of the lowest price, if the emission trading price below this price will not help to promote the enterprise independent reduction, technological innovation and control unit Water pollution control; The third method is through the average auction price over the years to develop the initial allocation price, because the emission permits is an essential resource in the production process, is a scarce resource, and in the auction process has increased the cost, so the auction price must gradually increase high trend, but not conducive to the initial allocation of emissions and trading to promote the work. At the same time, for some poor strength, smaller enterprises, facing enormous pressure, that is not conducive to the growth of small and medium-sized enterprises and the sustainable development of the economy, so will the price as the highest price of emission trading; The second method is through different industries and enterprises of the sewage performance that the shadow price (unit amount of emissions effectiveness) accounting for the initial price, relatively close to the enterprise production and management of the actual situation, but in the actual operation process, such as the same industry to the emission enterprises pricing , will affect the development of certain pollutants larger enterprises; according to different industries on the emission enterprises were priced, it will affect the initial allocation of emissions permits fairness.

According to the above points, in the Songhua River Harbin section of the control unit emission permits in the initial allocation of price development,adopt the hybrid approach:at the early stage of implementation emission permits mechanism ,the initial allocation of emission permits pricing method should be used unit pollutant management cost,When the emission permit useful life period expires,the government sets the initial allocation price according to the average auction value over the years [5].But the price of the secondary market is determined by the market:if the two enterprises have already negotiated,the transaction will be carried out according to the agreement price;There is did not reach an agreement,the government reserves and other parts of the emission permits are free auction, according to supply and demand conditional free bid.

4.2. Calculation Method of Initial Price

Environmental capacity is a kind of natural resources, but there is no price in the market,also can't use other resources pricing method for pricing,and recovery cost method is the existing method used in the pricing of emissions trading [6].This method is to restore the environment has been damaged to the original state, the cost of recovery process to measure the cost of environmental capacity has the value, because this method only restore environment to original state and without causing other costs, so this value is the lowest value of environmental resources.

Owing to environmental pollution is a complex pollution, is a variety of pollutants common pollution,so this method is to take the average cost of different pollutants governance as this pollution factor emission permit initial transaction price. Considering the economic development, industry level and enterprise fairness etc factor of Harbin,the initial price calculation of emission permits is calculated as follows:

$$\text{Initial Price of Emission Rights (ten-thousand-yuan}/t) = \sum_{i=1}^{i=n}\left[\frac{\text{Social Average Pollution Control Cost}}{(1+\text{Discount Rate})^{n-1}}\right] \quad (3)$$

(Note: n: The useful life of emission permits, The discount rate is the central bank's current n-year loan benchmark interest rate)

The initial price of the emission permits is calculated from the formula 3-1, and the social pollution control cost is a certain pollution factor for a certain region.In order to reduce this pollution factor by 1 tonne, the various industries of the region need to invest in unit pollutant management costs, the unit is: （ten-thousand-yuan / t • year） .There are two ways to calculate the cost of unit pollutant treatment:The first one is according to the region to control pollutant the consumption total operating costs and the total handling capacity of the ratio obtained;The second is to calculate the cost of all the enterprises in the region to deal with this pollutant, and then weighted average, which is the average cost of the treatment of this pollutant in this area.Owing to each enterprise sewage discharge in time, quantity is different,so the former calculation method is more difficult, the latter is relatively simple.

The calculation formula of the unit governance cost of a certain pollutant for certain industry enterprise in the region is as follows:

$$C_a = C_{a,i} / T_{a,i} \tag{4}$$

(note: a :is an enterprise; i : the length of time, years or months;C_a:is the 'a' enterprise pollutant management costs;$C_{a,i}$:is 'a' enterprises in the 'i' period of time to run the total cost of governance;$T_{a,i}$:is 'a' enterprises in the 'i' period of time to total handling capacity)

By the formula 3-2 to calculate the region of all industries of a business of this pollutant management costs, and then weighted average, obtain the industry to some of the pollutant management costs; Similarly calculation of other industries, then, the weighted average of some pollutant treatment costs in different industries, which is the average cost of treatment of some pollution factor in the region [7].

5. Determination of the Useful Life of the Emission Permits

Many provinces and cities in China have launched a pilot emissions trading in 2014, but there are great differences in the rules for the use of the emission permits, there's one year, three years, five years, twenty years, and there's a permanent use of emission permits. In the guidance of<Guiding Opinions of the General Office of the State Council on Further Promoting Paid Use and Pilot Work on Trading of Emission permits>(State Council issued [2014] No. 38), after the initial approval of the emission permits of all sewage enterprises at the end of 2015, "once every five years in principle" is approved again [8]. Therefore, in the Songhua River Harbin section control unit emissions trading, the proposed emission permits useful life for five years, and synchronized with the "Five Year Plan", is conducive to the production of national economy and the construction of the combination, and promote the development and construction of Harbin city.

6. References

[1] Xianfeng Huang and Yi Wu and Baohai Xu 2014 *Study on initial pricing strategy of emission trade in Guizhou* [J] Environmental Protection and Technology (01):6-9

[2] Wenyan Zhao 2013 *Study on Pricing Model of Initial Emission Permits* [D] Jilin University

[3] Wu-Pu zhao and Wang qian 2010 *An Economic Analysis of Emission Trading* [J]China population resources and environment S2:55-58

[4] Ting Liu 2013 *Study on Compensated Use Price Calculation of Initial Emission Permits in Jilin Province* [D] Jilin University

[5] Ningliao Chao 2009 *Average Shadow price and equilibrium price : A case study of tradable pollution permit markets* [J] European Journal of Operational Research (196):1207-1213

[6] Jays Coggins and John R Swinton 1996 *The Price of Pollution: A Dual Approach of Valuing SO2Allowances* [J] Journal of Environmental Economics and Management (1):58-72

[7] Yiping Chu 2011 *Study on the Initial Pricing Scheme of Emission Trading* [J] Environmental Science and Technology S2:380-382+400

[8] Qing Chen and Miao Chang 2014 *Comparative Analysis of Emission Trading Periodsetting Scheme* [J] Chinese Journal of Environmental Management (05):6-10

Exploring the Response of Combined Urban Sewer System to the Implementation of Green Roof

Yuqin Gao[1], Dongdong Wang[1], A Schmidt[2] and Yun Tang[2]

1 College of Water Resource and Hydropower Engineering, Hohai University, Nanjing 210098, China;
2 Department of Civil and Environmental Engineering, University of Illinois at Urbana-Champaign, Urbana, Illinois, USA
E-mail:178164576@qq.com

Abstract. In order to find an efficient and reliable method for assessing the impact of green roofs on runoff in urban areas, IUHM-GreenRoof combination model is established via Matlab. Taking CDS-51 catchment located in Dolton, Illinois as a case, and considering different green roof coverage ratios as well as runoff conditions, the influence on the catchment outlet runoff because of the application of green roofs is analysed by running the model in two typical storms. The results show that green roofs have good water storage capacity, which can effectively decrease the catchment runoff in terms of peak discharge and volume. Moreover, the green roof coverage ratio has more influence on runoff than green roof runoff conditions.

1. Introduction

With the high-speed development of urbanization, urban landscapes are constantly changing, with far more impervious surfaces than natural watersheds. The sealing effects of impervious surface result in a series of environmental problems within urban catchments, such as storm water flooding, which has become an issue of major concerns in the US, especially in some big cities with dense population and frequent rainfall events[1].

Due to the side effects brought by urban expansion, various sustainable development strategies has appeared, such as rain gardens, green roofs and permeable pavements, to help manage storm water. Green roof technology has become a popular strategy for retaining rainfall and reducing storm runoff, especially in ultra-urban areas where space is too limited to implement other best management practices (BMPs). Green roof is a relatively independent vegetation system built on building roofs, which is conventionally composed of six components: (1) plants; (2) a growing media; (3) a geotextile filter fabric; (4) a drainage layer; (5) a waterproofing; (6) the roof deck[2] (Figure 1). Green roofs have many environmental and financial benefits for urban catchment, focusing on managing stormwater, green roofs can: postpone the starting time of runoff; mitigate stormwater runoff; distribute the runoff over a longer period. Thus green roofs make urban hydrological cycle tend to approach the natural one, and lower urban flooding risk as well.

Up to now, various models have been developed for the hydrological process of green roofs. Hilten et al.[1] conducted a study on the effectiveness of green roofs to mitigate stormwater runoff by using HYDRUS-1D. Kasmin et al.[3] built a simple conceptual model composed of a substrate moisture storage component and a transient storage component, in order to find a robust method to identify monthly evapotranspiration values. Versini et al.[4] developed a conceptual model and integrated it into SWMM to reproduce the hydrological behavior of green roofs from building scale to basin scale. A commonality among these methods is that they are not physically based in their representation of

IOP Conf. Series: Earth and Environmental Science **178** (2018) 012043 doi:10.1088/1755-1315/178/1/012043

the underlying hydrologic and hydraulic processes of green roofs, meanwhile require burdensome input data, which restrict their use.

The scope of this paper is to develop a mathematical model for green roofs, incorporate it into IUHM appropriately to explore the influence of green roof application on urban runoff.

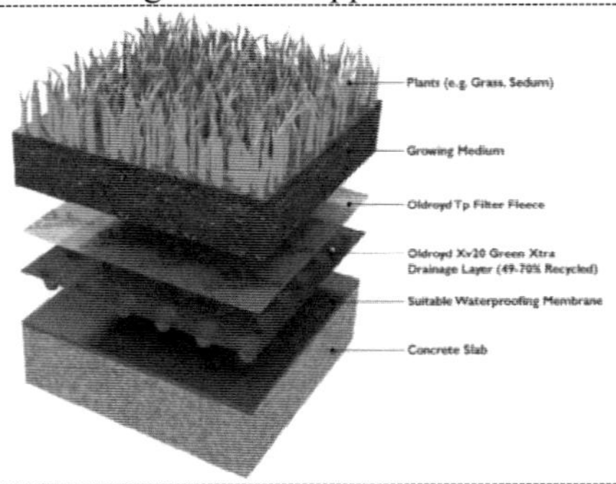

Figure 1. A typical example structure of a green roof

2. Modeling

2.1. Illinois Urban Hydrologic Model (IUHM)

The Illinois Urban Hydrologic Model (IUHM) is a mathematical model built by Cantone and Schmidt (2010) for simulating complex urban sewer system. It is rooted in the initial concept of the Geomorphologic Instantaneous Unit Hydrograph (GIUH) developed by Rodriguez-Iturbe and Valdes (1979)[5]. By using the Strahler ordering procedure, an urban catchment of order, Ω, can be defined into either overland state or conduit state. In each state, the travel times are calculated based on the kinematic wave assumption. When a drop of water falls on the catchment, it will flow from lower order to higher order, until reaches the outlet of the sewer system[6]. With the overland area being sub-divided into pervious and impervious regions, the total number of water possible flow path would be 2^{Ω}. For each path there is a corresponding exponential distribution of travel time in each state. Combining the travel time probability-density-function with the probability of a drop of water taking a certain pathway, the network impulse response function could be derived. The catchment runoff hydrograph is then determined by combining the network impulse response function and the excess runoff for all possible paths.

Unlike traditional deterministic models, IUHM can still give a highly accurate output without detailed parameters of all pipes and overland signatures which are not always available. Also, IUHM has been setup to allow the incorporation of other hydrologic and hydraulic processes, for example, the best management practices (BMPs), which makes the subsequent integration easy to implement.

2.2 Green Roof Model

Green-Ampt method was used to calculate infiltration capacity in the model. This model divided the whole rainfall-runoff process into four states.

2.2.1. Initial state. At the beginning of a precipitation event, soil water content keeps increasing, but does not reaches field capacity. All rainfall is infiltrated and stored in soil. The runoff intensity is 0.

2.2.2. Pre-saturated state. In this phase, soil moisture content exceeds field capacity but soil is not saturated yet. As rainfall continues infiltrating, water flow out of the soil layer and runoff occurs. Green roof was assumed to be a linear reservoir in this state, thus discharge rate would be proportional to the realtime water storage in the soil column. The runoff intensity is calculated by equation 2.1.

$$q(t) = a * S(t) / dt \tag{2.1}$$

$$S(t + \Delta t) = S(t) + f(t) * dt - q(t) * dt \tag{2.2}$$

q(t): runoff intensity at time *t*, inch/hour; *S*: real-time soil water storage, inch; *a*: calibration factor; *dt*: time step, hour; *f*: infiltration rate, inch/hour.

2.2.3. Saturated state. Runoff at the second state is much smaller compared with water infiltrated in, so soil water storage continues increasing. When soil water content reaches saturated moisture content, soil becomes saturated. In this state, the whole system reaches a relative balanced state. Darcy's Law is used to calculate runoff intensity.

$$q(t) = K * \frac{D + h(t)}{D} \tag{2.3}$$

K: saturated hydraulic conductivity, inch/hour; *D*: soil depth, inch; *h*: ponding depth, inch.

2.2.4. Recession state. In this state, rainfall stops and ponding water is also consumed up by infiltration or evapotranspiration, but runoff still continues because of the water stored in the soil column in previous phases. However, the discharge rate will decrease as time goes. In this study, runoff was assumed to follow an exponential decay. The runoff intensity is calculated by equation 2.4.

$$q(t + \Delta t) = q(t) * e^{-\lambda} \tag{2.4}$$

λ: calibration factor.

2.3. Combined IUHM-GreenRoof Model

Green roof model was embedded into IUHM as a function without changing the IUHM code too much. Important changes of the original code are listed as follows compared with the original IUHM:

(1) A variable of green roof coverage ratio is defined based on the impervious area, denoted as "grratio". In other words, if it is given a value of 20%, then 20% of the impervious area will be replaced by green roof. Users can change its value to test the effect of different green roof coverage.

(2) A variable of routing part of green roof runoff to impervious area is introduced into the model, with a value ranging from 0 to 1, denoted as "grimp". If its value is set to 0, that means all green roof runoff will be routed to pervious area; if indicated as 1, then all green roof runoff will be routed to impervious area. Users can alter its value for testing the difference of various routing ways.

(3) Due to the introduction of green roof coverage, the original impervious area will be correspondingly decreased by the same ratio. Meanwhile, the average impervious overland flow length will be decreased due to the reduction of impervious area. The excess runoff on both impervious and pervious overland regions will also be changed due to the inflow of green roof runoff

3. Case Study

3.1. Calumet Drop Shaft 51 catchment (CDS-51)

The CDS-51 catchment is a 5th-order complex urban system with an area of 782 acre, located in the Village of Dolton, Illinois. Figure 2 shows a plan view of pipe system in CDS-51. The catchment

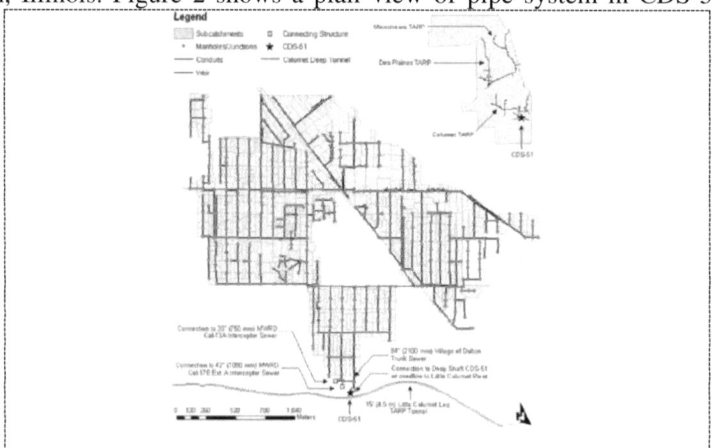

Figure 2. Pipe System Plan view of CDS-51 catchment

captures combined storm and sanitary flows and delivers them to Calumet system of Chicago's Tunnel and Reservoir Plan (TARP)[6]. The combined sewers flow from lower-order pipes to higher- order ones until end up in a 5th order pipe with a diameter of 2150mm. In this study, catchment outlet runoff will be used as the comparing variable to testify the effect of green roof application.

3.2. Rainfall Events

Two typical rainfall storms were chosen to test the behavior of the combined IUHM-GreenRoof Model: (1) January storm in 2008, a low intensity, long duration event, started at 18:00 and lasted 15 hours, with a peak intensity of 0.46 in/h and a total precipitation is 3.12 in.; (2) July storm in 2007, a high intensity, short duration event, started at 02:00 and was 7 hours long, with a peak intensity of 0.74 in/h and a total precipitation of 1.84 in. Figure 3 and 4 shows the rainfall intensity of January 2008 storm and July 2007 storm respectively.

The reason for choosing these two storms is that Cantone (2010) has simulated the same rainfall events using IUHM, and proved that IUHM-derived runoff matching the actual flow data for CDS-51 well, which ensures the credibility of the comparison between the runoff behavior of green-roof-implemented catchment with that of the original catchment.

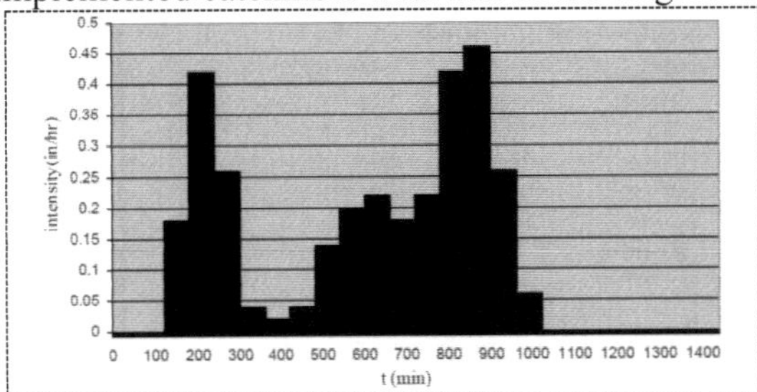

Figure 3. Rainfall intensity of January Storm

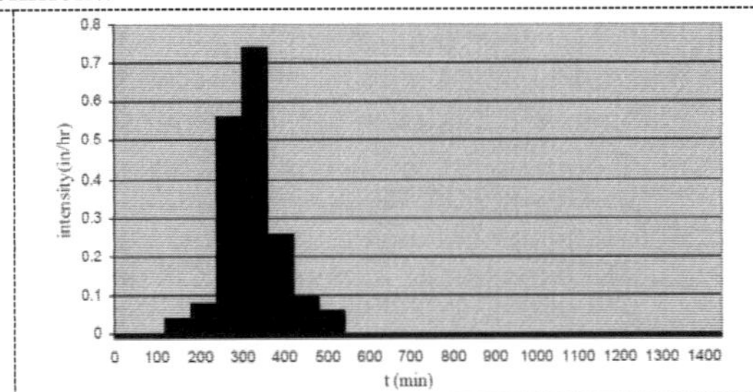

Figure 4. Rainfall intensity of July Storm

3.3. Extensive Green Roof

Table 1. Input parameters for green roof.

parameter	value	unit
growing medium depth	8	in
saturated hydraulic conductivity	0.43	in/h
soil water suction head	4.33	in
initial soil moisture content	0.041	
field capacity	0.18	
saturated soil moisture content	0.453	
overflow collar height	4.72	in
evapotranspiration rate	0.0018	in/h
depression storage	0.15	in
linear reservoir calibration factor "a"	0.0025	
recession state calibration factor "λ"	0.002	

An extensive green roof with 8 in deep sandy loam was simulated. Most hydraulic parameters for the growing medium are listed in Table 1. For effective comparison, depression storage of 0.15 in was assumed for green roof as it was the same in IUHM. Evapotranspiration rate was set as 0.0018 inch/hour according to a study done by Rezaei (2005). This value was used in recession state.

3.4. Simulation

As mentioned before, there are two kinds of routing method for overland flow in IUHM, one is routing overland runoff from pervious regions to impervious regions (referred as the pervious-impervious case), and the other is inverted (referred as impervious- pervious case). For both of these two cases, the original IUHM has first been run for later comparison.

In pervious-impervious case: (1) To research the influence from different green roof coverage, assuming all green roof runoff would be routed to previous area, green roof coverage ratio was set for

20%, 30%, 40%, and 50%, respectively. (2) To research the effect of altering green roof runoff routing method, a simulation of routing all green roof runoff to impervious area, or pervious area, or 50% to each, has all been run respectively, with 20% green roof coverage.

In impervious-pervious case: only different green roof coverage (20%, 30%, 40%, and 50%) would be explored for the catchment to test for different impact to catchment outlet runoff.

Combination case: the pre-proposed methods are extreme conditions, the combination of these two conditions can be realized by simply giving a percentage based on the actual situation.

4. Results and Discussion

4.1. Green Roof Rainfall-Runoff Relationship

The infiltration mechanism of green roofs leads to a unique rainfall-runoff relationship (Figure 5). Compared with rainfall, green roof runoff starts at a later time with a lower peak. For the January storm, green roof peak discharge is about 0.25 in/h, which is 45.7% less than peak rainfall intensity. This retention effect is even better during the July storm, in which green roof peak discharge is about 73% less at only 0.2 in/h. This shows that green roof performs very well in retaining water.

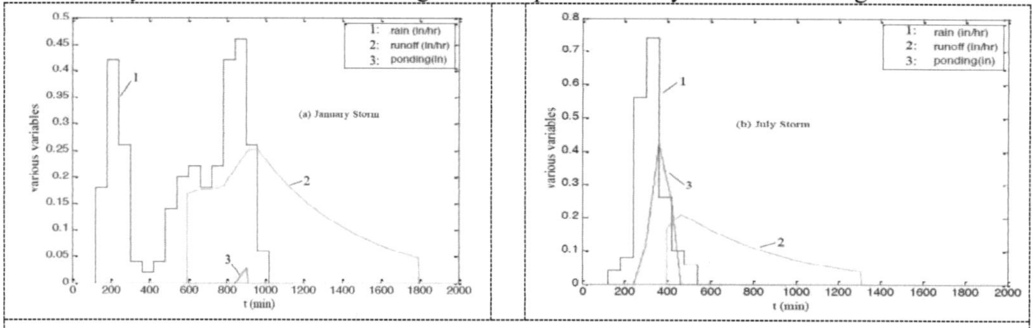

Figure 5. Green roof rainfall-runoff relationship for two storms

4.2. Green Roof Impact in Pervious-Impervious Condition

Pervious-impervious condition is more prevalent in urban areas. In January storm, for every 10% increase in green roof coverage, the first and second peak runoff is decreased by nearly 10%. For July storm, peak decreasing rate is similar to what happened in January storm (Figure 6).

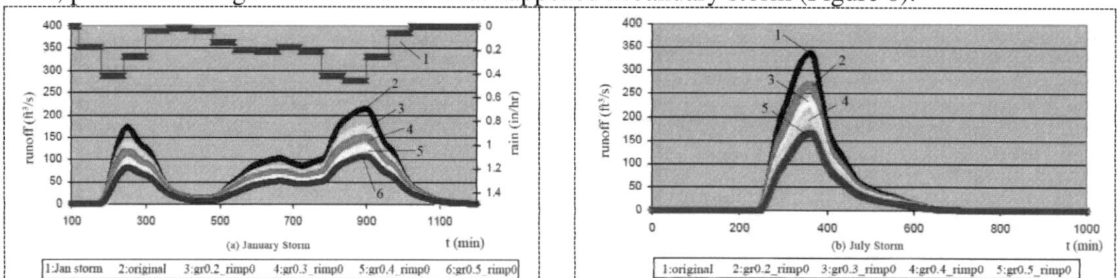

Figure 6. Catchment runoff comparison with different green roof coverage for two storms

This shows that green roofs have a similar retention effect for both two types of storm under pervious-impervious case.

Except for the overland flow routing method, the green roof runoff routing condition would also affect catchment outlet runoff (Figure 7). For the January storm, peak discharge tends to be smaller when more green roof runoff is routed to the permeable area. This is reasonable because permeable area can absorb and allow more water infiltration. However, it should be noted that the first peak discharge almost doesn't change while the second peak apparently decreases. This reflects the retention effect of the green roof. At the beginning of a rainfall event, the green roof stores all water infiltrated and slowly releases it after the soil capacity is exceeded. At the time of the first rainfall peak, the green roof still holds all water falling on it, but when the second rainfall peak comes, it has already begun to release flow.

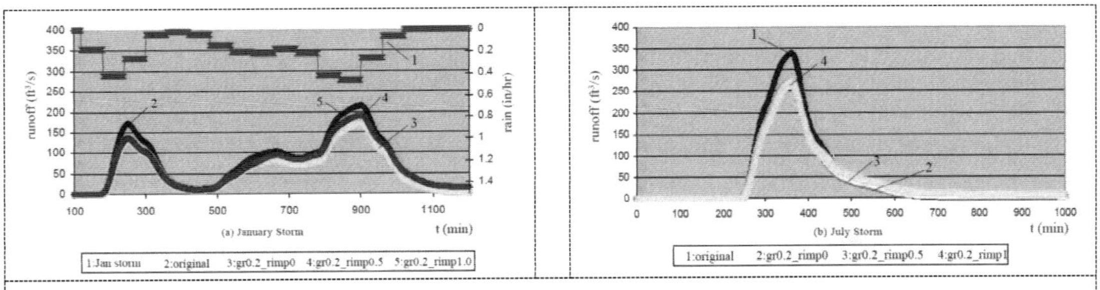

Figure 7. Catchment runoff comparison with different green roof runoff routing methods for two storms (20% green roof application).

For the July storm, the entire catchment runoff graphs seem to overlap with each other. Possible reason is that the storm intensity is too high compared with infiltration capacity, even if green roof runoff is routed to permeable regions, it still cannot be infiltrated and becomes the excess runoff.

For both storms, the routing ratio of green roof runoff has a smaller effect to catchment runoff than green roof coverage does.

4.3. Green Roof Impact in Impervious-Pervious Condition

Under impervious-pervious condition, catchment runoff volume and peak both decrease as green roof coverage increases (Figure 8). However, the peak discharge decreasing rate of the July storm is almost twice than that of the January storm. This shows that green roofs have a better retention effect for the high-intensity/short-duration storm in impervious-pervious condition.

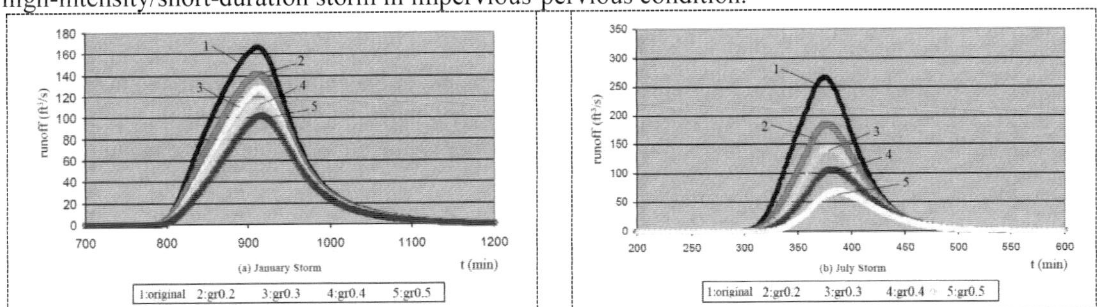

Figure 8. Catchment runoff comparison with different green roof coverage for two storms.

4.4. Green Roof Impact in Combined Condition

Figure 9 shows how 20% green roof works on an urban catchment with 60% overland flow routing from pervious to impervious area and 40% in the inverted way. In January storm, the first and second runoff peak decreased 22% and 18.6%, respectively. In July storm, peak reduction was around 22%.

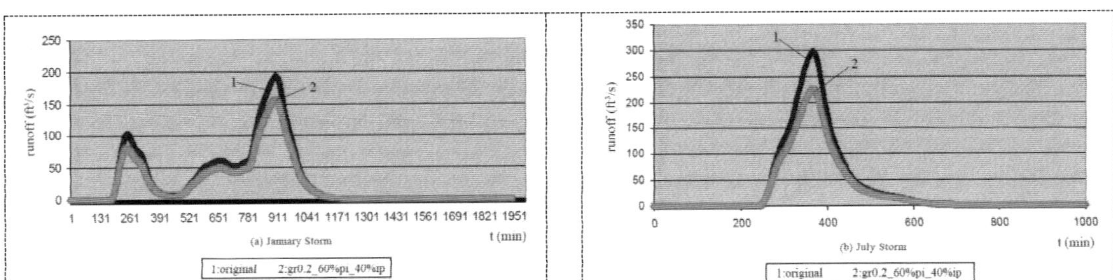

Figure 9. Catchment runoff comparison with combined condition for two storms

5. Conclusion and Future Work

Form the results analysis above, several conclusions can be drawn:

(1) Under pervious-impervious condition, green roofs have similar retention effect for both two types of storm. Catchment runoff tends to have higher peak if more green roof runoff is routed to impervious area. However, this effect diminishes for the high-intensity/short-duration rainfall event. In addition, green roof coverage has bigger effect to catchment runoff than green roof runoff routing ratio.
(2) Under impervious-pervious condition, catchment runoff volume and peak both decrease as green roof coverage increases. Meanwhile, green roofs have a better retention effect for high-intensity/short-duration storm than low-intensity/long-duration rainfall event.
(3) In combined condition, 20% green roof application could decrease catchment runoff peak by about 22% for both storm types. However, this value could change with the percentage of overland flow routing methods, resulting in a more complicated condition.
The incorporation method of IUHM and green roof developed in this study provide a tool for understanding how the implementation of green roofs can affect urban catchment under different conditions. With modification, it could be extended to more BMPs. The framework of adapting IUHM to kinds of BMPs could also be a research topic.

6. Reference

[1] Hilten R N, Lawrence T M and Tollner E W, 2008 *J. Hydrol.* 358, 288-93.
[2] Mentens J, Raes D and Hermy M, 2006 *LandSc. Urban Plann.* 77, 217-226.
[3] Kasmin H, Stovin V R and Hathway E A, 2010 *Water Sci. Technol.* 62(4), 898-905.
[4] Versini P A, Ramier D, Berthier E, Gouvello B de, 2015 *J. Hydrol.* 524, 562-75.
[5] Rodriguez-Iturbe I and Valdes J B, 1979 *Water Res. Res.* 15(5), 1409-20.
[6] Cantone J and Schmidt A, 2010 *Water Resour. Res.* 47,W08538,doi:10.1029/2010WR009330.

Acknowledgments

The study was supported by the National Nature Science Foundation of China (No.51309076), the Fundamental Research Funds for the Central Universities (No.2014B05814). The Project was also funded by the Priority Academic Program Development of Jiangsu Higher Education Institutions.

Study on Division of Landscape Character Areas of River Corridor in Northwest Arid Area Based on LCA Method——Take Zhangye Section in Heihe River Basin as an Example

Yanli Tao[1], Xuesong Xi[1*], Yabei Xin[1], Fangxin Liu[1], Bingbing Zhang[2]

1 China Agricultural University, No. 17 Tsinghua east road, Beijing, China
2 Huachengboyuan Engineering Technology Grope Co.,Ltd, No.10 Xuanwumenwai Street, Beijing, China
*E-mall: 466938953@qq.com

Abstract: The division of landscape character areas is an effective way to plan the corridor of river corridors with rich natural and human resources. Based on the method and advanced experience of landscape character assessment in Europe, this paper summarizes the steps of landscape character areas classification and the evaluation index system of landscape character assessment for river corridors in the arid region of Northwest China. Taking Zhangye section in Heihe River Basin as an example, the area was divided into 11 landscape character areas in watershed scale. At the same time, the mapping of landscape character areas and the description of the landscape character areas were carried out.

1. Introduction

In recent years, the level of urbanization in our country has increased at an average annual rate of 1.2%. The rapid development of urbanization has led to great changes in the regional landscape. The overall performance is fragmentation and simplification of the land mosaic, the diversity of landscape types and the continuity of landscape pattern have been destroyed, as well as the development of modern society led to the regional landscape planning out of the traditional landscape characters and land use patterns [1]-[2]. The Heihe River Basin is located in the intersection of the Qinghai-Tibet Plateau and the Mongol Plateau where the weather is arid and dry. The natural and cultural resources along the river are abundant. The landscape type of the whole basin is extremely complex with alpine, oasis, desert and Gobi deserts [3], but the ecological environment is fragile, the middle reaches of the river basin face the homogeneity of a variety of landscapes and the loss of regional characteristics. Meanwhile, as an important part of the national ecological security barrier and eco-economic function area, Heihe's healthy development of the environment is crucial to the economic development and ecological security of the entire Hexi Corridor [4]. The division of the landscape character area of Heihe River Basin is of great strategic significance for more scientific and targeted protection of the ecological health of the river basin, building the base of Heihe River Basin harnessing and ecological planning as well as providing reference for the planning of river-type corridor with rich natural and human resources.

2. Research Area Overview

The Heihe River flows 185 kilometers from Yingluo Gorge to Zhengyi Gorge and flows through the Hexi corridor irrigation area (Gaotai County, Linze County and Ganzhou District), and is called "the middle reaches of the Heihe River." The scope of the study area includes the entire middle reaches of the Heihe River basin, namely Gaotai County, Linze County and Ganzhou District of Zhangye City, Gansu Province, with the administrative scope as the boundary.

3. Methodology - Landscape Characters Assessment

Landscape character assessment (LCA) method began in the United Kingdom where Carys Swanwick and Christine Tudor clearly defined the landscape character assessment in an official document issued by the agency, in which said Landscape Character Assessment as addressing both the characterisation process, involving identifying, mapping, classifying and describing landscape character, and the process of making judgements based on landscape character to inform a range of different decisions [5]. Similar concepts Landscape character and landscape character areas often come along, where landscape character is a distinct, recognisable and consistent pattern of elements in the landscape that makes one landscape different from another, rather than better or worse While Landscape Character Areas are single unique areas which are the discrete geographical areas of a particular landscape type. Each has its own individual character and identity, even though it shares the same generic character with other types [6]. The LCA methodology is now mainly used in areas such as regional development, urban planning, land use, conservation of nature and landscapes, sectoral resource planning and impact assessment of sustainable development.

3.1. Landscape Character Assessment Steps

It does not involve the quantitative appraisal of the value, but pays more attention to the regional history and culture, the field investigation and the sensory evaluation of the experts and related stakeholders in the landscape character assessment. Based on this principle, and with reference to the assessment steps of universal landscape characters in Europe, the division of landscape character areas in the Zhangye section of the Heihe River Basin mainly includes the following processes (Figure 1).

3.2. Landscape Character Assessment Index System

The European Landscape Convention states that landscape is an area, as perceived by people, whose character is the result of the action and interaction of natural and/or human factors[7], Swanwick C. also believes that landscape includes natural factors (geology, soil, climate, hydrology, vegetation, etc.), human factors (residential form, enclosure, land use and other human intervention) and cognitive aesthetics (smell, memories, associations, preferences, sounds, Touch, symbols, etc.)[5].It can be seen that river landscape is a comprehensive concept. Human factors, natural elements and visual perception are all important evaluation indicators for the assessment of landscape characters of river corridors. Therefore, on the basis of natural landscape characters, cultural landscape characters and perceptual characters, the evaluation index system for landscape characters at watershed scale consists of two levels of indicators, including eight first-level indicators and eight second-level indicators (Table 1).

Table 1. The evaluation index system of landscape characters Assessment.

Regional general indicators	First-class indicators	Second-class indicators
Natural landscape characters	Terrain	Landform type
	soil	Soil erosion
	Hydrology	Average annual groundwater depth distribution
	creature	Biodiversity
Cultural landscape characters	Land use	Evolutionary characters of land use function
	cultural heritage	The degree of cultural heritage gathering

IOP Conf. Series: Earth and Environmental Science **178** (2018) 012044 doi:10.1088/1755-1315/178/1/012044

	Economy	**Rural resident disposable income**
Perception characters	**visual perception**	**Landscape view of the water body boundary**

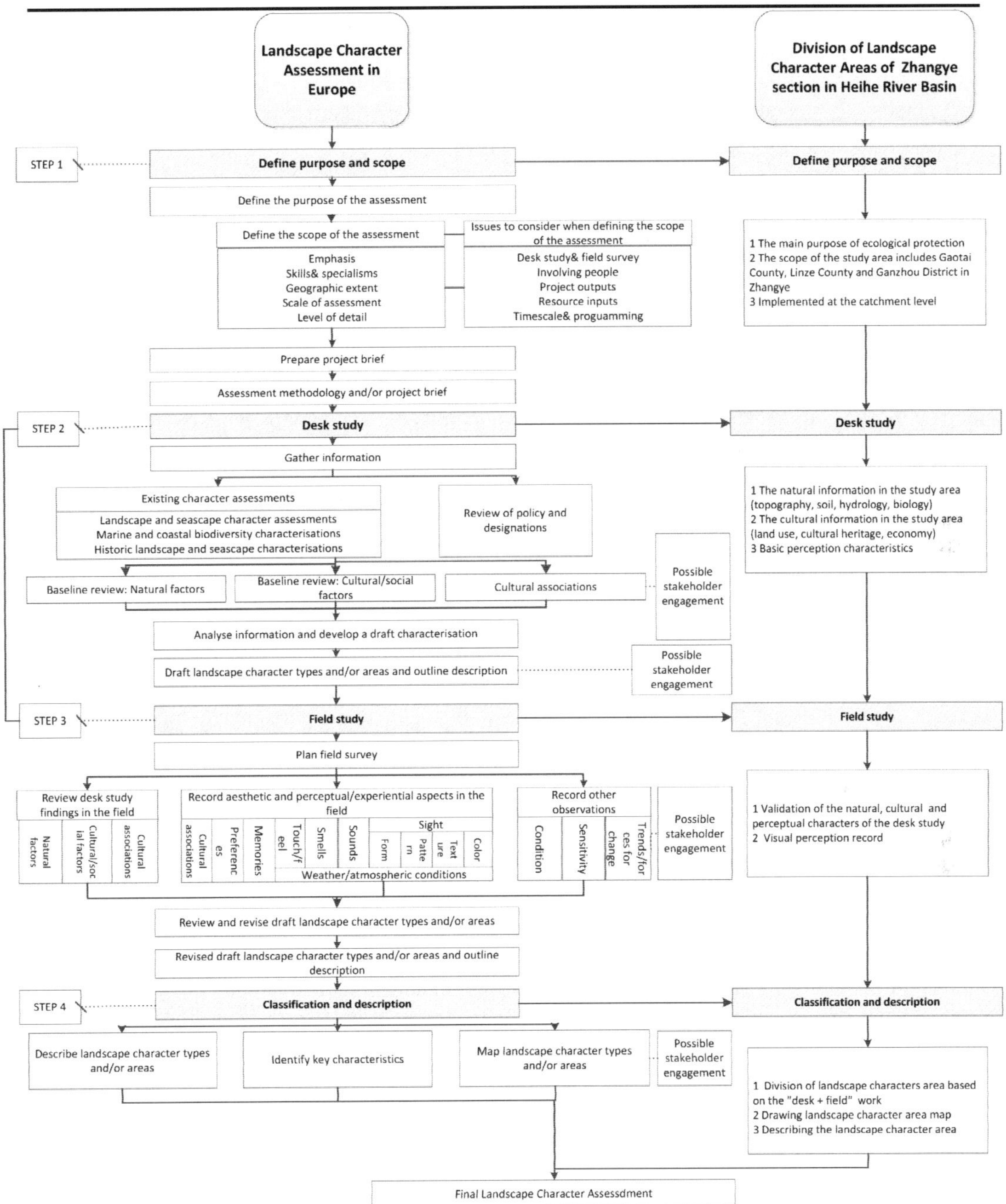

Figure 1. The flow chart of LCA steps.

4. Results - Category and Description

4.1. Layered Description

4.1.1. Landform type. According to the topographic data, the landform type of the study area is plotted (Figure 2). Most of the Heihe River basin is alluvial plain in the study area. Only the northern part of Gaotai and Liyuan River and the southern part of Heihe River in Ganzhou are flood plains. Specific circumstances as shown in Table 2:

Table 2. Landform type in study area [a].

Name	Landform type	Elevation (m)	Slope (%)
Gaotai County	Alluvial plain	1283-1300	0%—15%
Linze County	Alluvial plain	1283-1500	0%—15%
Ganzhou District	Alluvial plain	1300-1600	0%—15%

[a] Source: Zhangye City Land Resources Bureau.

4.1.2. Soil. Drawing a map of the state of soil erosion in the study area (Figure 3), we can see that the Heihe River Basin is strongly eroded in Linze County, and most of them are moderately eroded in Ganzhou District and Gaotai County. The north and south of Gaotai County are extremely eroded and the central part of the plateau is slightly eroded. The Heihe River basin is slightly eroded in the east of Heihe, Linze County, moderately eroded in the west, moderately eroded in the north of Liyuan, and strongly eroded in the middle While the southern part is slightly eroded. The Heihe River basin is moderately eroded in northern Ganzhou, the central part is slightly eroded and the south is moderately eroded.

4.1.3. Hydrology. According to the map of groundwater depth (Figure 4) and river distribution (Figure 5), the depth of groundwater in the study area is shallow, most of which is in the range of 50m and the depth of the area away from the main trunk of Heihe River is more than 300m.There is a narrow transitional section and a great depth gap.

4.1.4. Creatures. There are about 3 animals in the west of the northern Heihe River basin in the northern part of the Gaotai County while a rich distribution of animals on the west side of the Center, with about 37 species of animals; There are about 4 kinds of western part of Linze County and about 5 kinds of animals in the Liyuan River, including Black Stork, Short-eared Owl, Barn Owl, Goshawk and Redtail.At the same time, about seven species of animals are distributed in the northern part of the Ganzhou District of the Heihe River Basin (Figure 6).
The Heihe River basin distributes the desert vegetation such as Sympegma regelii and Reaumuria in the northern part of Gaotai County and Salt meadow such as Glycyrrhiza inflata in the middle part of the Heihe River floodplains, while the southern part distributes more plant species including tall grass meadow like Calamagrostis brachytricha and the desert vegetation like Sympegma regelii and Reaumuria. In addition, there is a distribution of tall grass and Calligomum mongolicum in part of the south of Linze County while Ganzhou District prefers to grow apples, pears and other fruit trees and farmland-based (Figure 7).

4.1.5. Land use. According to the detailed classification data from the Land and Resources Bureau, the land use map (Figure 8) was drawn. It can be seen that the middle reaches of the Heihe River Basin are the main farmland planting area.

4.1.6. Cultural heritage. The study area is rich in cultural protection resources where distributed a variety of types of cultural relics, ancient sites, ancient tombs, modern architecture, cave temples and stone carvings. There are altogether 201 cultural protection units of various types in Ganzhou District with concentrated distribution in or around Ganzhou City, 113 units in Linze County and 168 in Gaotai County. The cultural heritage sub-concentration map (Figure 9) drawn by the distribution of cultural heritage points shows that Ganzhou District concentrates most of the national and provincial cultural protection units.

4.1.7. Economy. With the town as the smallest unit of statistical size, residents living along the Heihe River have higher incomes, where per capita income of rural residents is above 11,000 yuan (Figure 10).

4.1.8. Perceptual characters. According to field investigation records, the Heihe River Basin has hilly and low mountains in the vicinity of Zhengyi Gorge in northern Gaotai County, with a semi-enclosed field of view. The central area is composed of trees and shrubs where the terrain is flat and the field of vision is wide, and the south area is far from the centre of Gaotai County with a wide view; Most of the farmland and villages in the western part of Linze County have a wide field of vision and many hilly areas in the east have semi-closed horizons, and there is a flat terrain and wide field of vision in Liyuan River; On both sides of the northern part of Ganzhou District, the Heihe River Basin is characterized by an open horizon of farmland and rural residential land. On the east and west sides of the centre are urban parks, farmland and industrial parks, with a broader field of vision and there is no tall buildings in the south with a broader horizon.

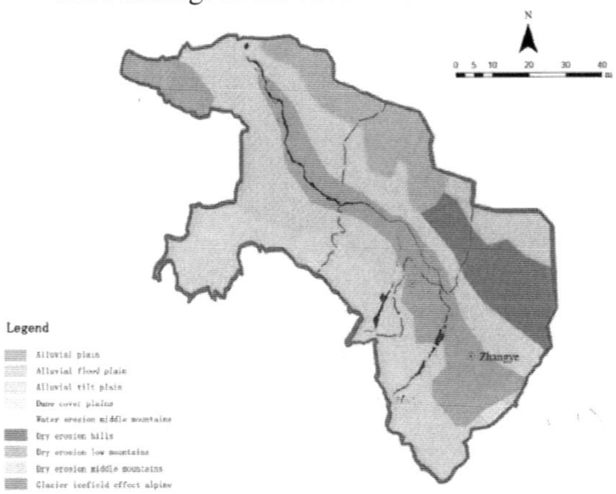

Figure 2. Landform type map. (Data source: Land and Resources Bureau)

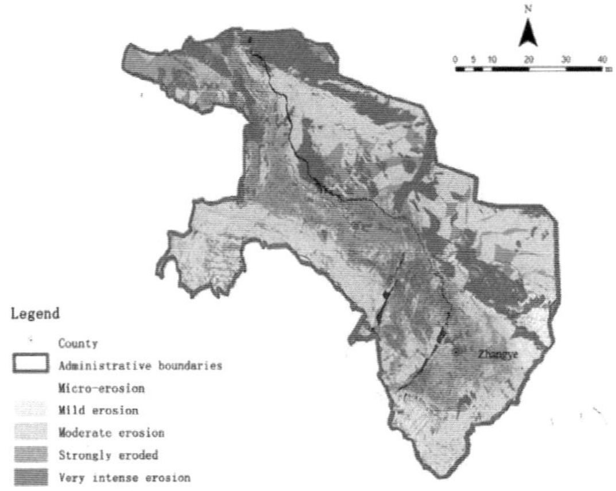

Figure 3. Soil erosion status map. (Data source: Zhangye Eco-environmental Protection Planning [5])

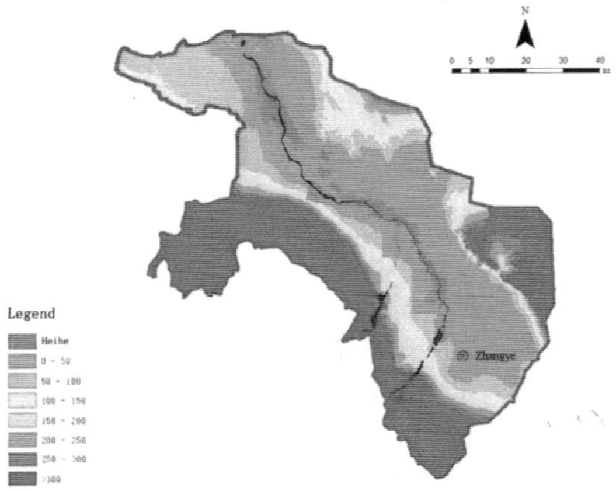

Figure 4. Annual groundwater depth map. (Data source: Multi-process Coupling of Agricultural Water Conversion and Efficient Water Regulation Project in the Middle Heihe River Basin)

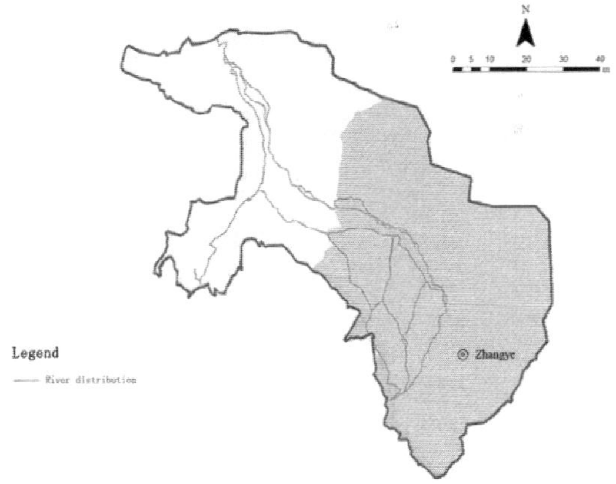

Figure 5. River distribution map. (Data source: Zhangye Water Authority)

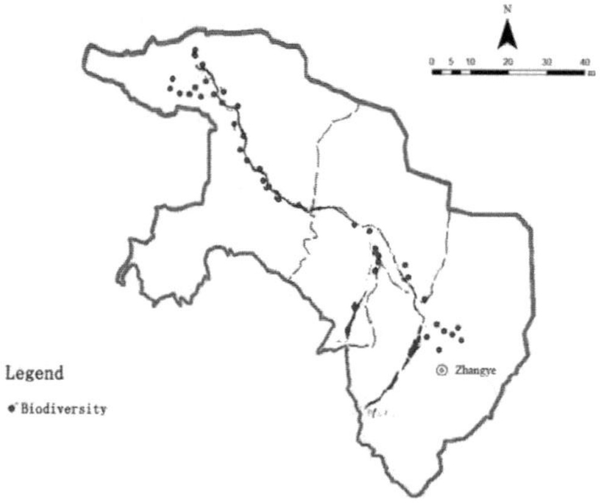

Figure 6. Animal Distribution. (Data source: Zhangye Wetland Nature Reserve Planning)

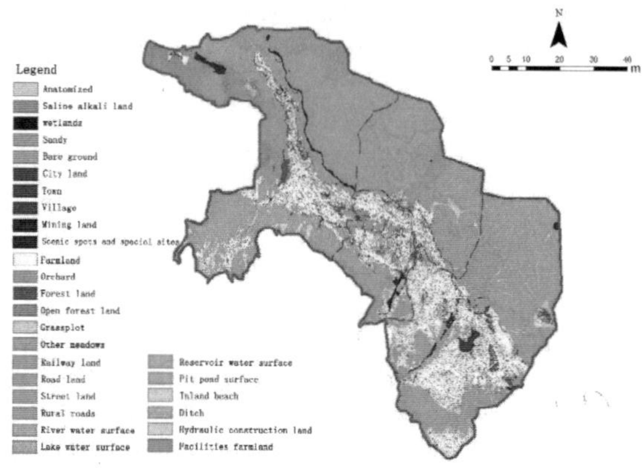

Figure 7. Plant Distribution. (Data source: Zhangye Eco-environmental Protection Planning)

Figure 8. Land use map. (Data source: Land Bureau of Zhangye)

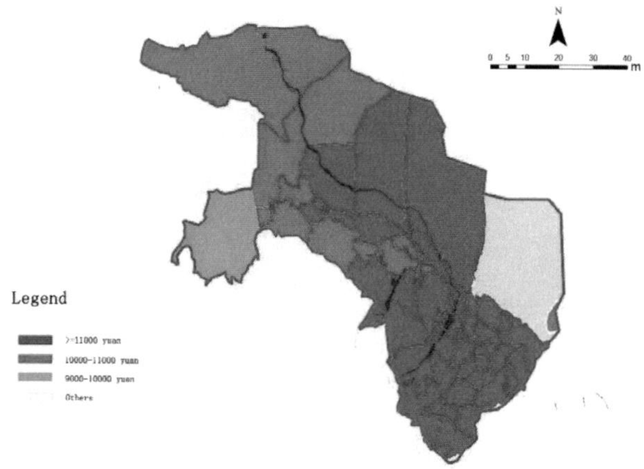

Figure 9. Cultural heritage sub-concentration map. (Data source: Land Bureau of Zhangye)

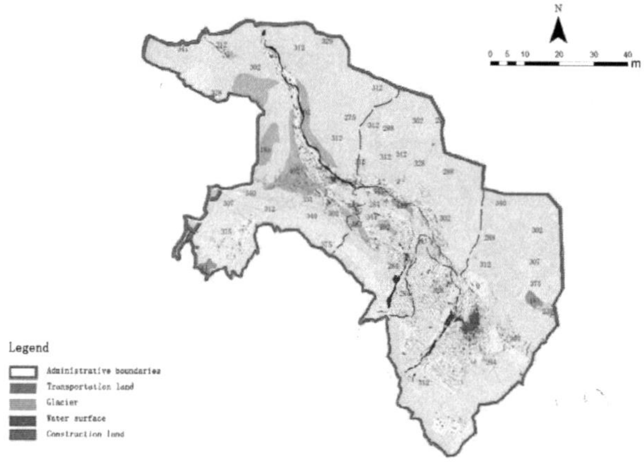

Figure 10. Rural resident per capita disposable income distribution. (Data source: Statistical Yearbook of all counties)

4.2. Watershed-Scale (Superimposed) Landscape Characters Zoning and Description

The middle reaches of the Heihe River basin in the study area were summed up into the above-mentioned evaluation index system of LCA. At last, the final watershed-scale landscape characters

IOP Conf. Series: Earth and Environmental Science **178** (2018) 012044 doi:10.1088/1755-1315/178/1/012044

were formed according to the elements of the landscape characters in each layer. Demarcation is based on distinguishable characters on the ground such as roads, geographical elements, and land-use types [8].

Landform type
Soil erosion
Average annual groundwater depth distribution
Biodiversity
Evolutionary characters of land use function
The degree of cultural heritage gathering
Rural residents disposable income
Landscape view of the water body boundary

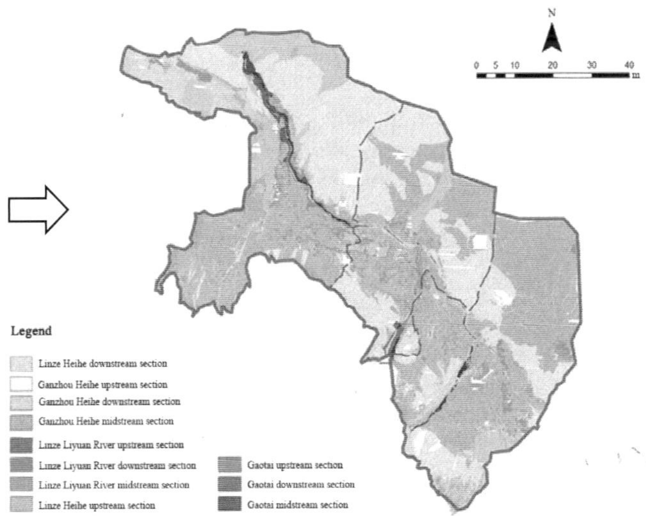

Figure 11. Watershed- scale landscape character area.

Gaotai County is divided into three watershed-scale landscape character areas: Gaotai downstream section, Gaotai midstream section, Gaotai upstream section, of which the demarcation line between Gaotai downstream section and midstream section is the boundary of county road 221 while the midstream and upstream section line is Heihe Bridge in Heiquan Township, Gaotai County and County Road 214;Linze County is divided into 5 landscape character areas: Linze Heihe upstream segment, Linze Liyuan River downstream segment, Linze Liyuan River midstream segment, Linze Liyuan River upstream segment, of which the dividing lines are respectively Heihe Bridge in Pingchuan Town and county road 214, county road 217, forest land boundary, 218 Linli County Road, Linze County; Ganzhou is divided into three watershed-scale landscape character areas that Ganzhou downstream segment, Ganzhou midstream segment, Ganzhou upstream segment, and the dividing lines are respectively Ganxin Heihe River Bridge and Provincial Highway 213.

The specific scope of the landscape character areas covered by each landscape character type was drawn and drawn to form the watershed-scale landscape character areas of the Heihe River (Figure 11):

On the basis of zoning, the landscape characters were corresponding described in each landscape character area (Table 3):

Table 3. Watershed-scale landscape character area description of Zhangye Section in the Heihe River.

Name of Landscape Character area based on Watershed Scale	Description
Gaotai downstream section North Shore: Justice Gap - Tiancheng Village South Shore: Justice Gap - Tiancheng Village	Landform type: Alluvial flood plain; Soil: Very strongly eroded area; Hydrology: Average annual groundwater depth of 0-50 meters; Creatures: 3 kinds of animal distribution, plant distribution of desert vegetation such as Sympegma regelii and Reaumuria; Land use: mainly farmland; Cultural heritage: Distribution of 9 cultural heritage protection units, which are mainly beacon towers ruins and temple sites; Economic: 10000- 11000 yuan（per capita income of rural residents）; Visual perception: More hilly and low mountains, vision closed.
Gaotai midstream section North Shore: Tiancheng Village - Great Bay scenery South Shore: Tiancheng Village - Heihe Bridge in Heiquan Township	Landform type: Alluvial plain; Soil: Lightly eroded area; Hydrology: Average annual groundwater depth below 100 meters; Creatures: 3 kinds of animal distribution, plant distribution of saline meadow such as Glycyrrhiza inflata Batal; Land use: mainly woodland; Cultural heritage: Distribution of 9 cultural heritage protection units, which are mainly beacon towers ruins; Economic: More than 11000 yuan（per capita income of rural residents）; Visual perception: flat terrain, wide field of vision.
Gaotai upstream section North Shore: Great Lakes Bay Scenic Area - East boundary of Gaotai County South Bank: Heihe Bridge in Heiquanv Township - East boundary of Gaotai County	Landform type: Alluvial plain; Soil: Very strongly eroded area; Hydrology: Average annual groundwater depth of 0-50 meters; Creatures: 4 kinds of animal distribution, plant distribution of High grass meadow; Land use: mainly farmland; Cultural heritage: Distribution of 9 cultural heritage protection units, which are mainly the Red Army cultural sites; Economic: More than 11000 yuan（per capita income of rural residents）; Visual perception: Large farmland, flat terrain, wide field of vision.
Linze Heihe downstream section North Shore: West Boundary of Linze County - Heihe Bridge in Banqiao, Linze South Shore:West Boundary of Linze County - Heihe Bridge in Yanuan Township, Linze	Landform type: Alluvial plain; Soil: Strongly eroded area; Hydrology: The average annual groundwater depth is shallow; Creatures: 2 kinds of animal distribution, plant distribution of High grass meadow; Land use: mainly farmland; Cultural heritage: Distribution of 9 cultural heritage protection units, which are mainly beacon towers ruins; Economic: More than 11000 yuan（per capita income of rural residents）; Visual perception: Large farmland, flat terrain, wide field of vision.

Linze Heihe upstream section North Shore: Heihe Bridge in Banqiao, Linze - Eest Boundary of Linze County South Shore:Heihe Bridge in Yanuan Township, Linze- Eest Boundary of Linze County	Landform type: Alluvial plain; Soil: Strongly eroded area; Hydrology: The average annual groundwater depth is shallow; Creatures: 2 kinds of animal distribution, plant distribution of desert vegetation such as Reaumuria; Land use: mainly farmland; Cultural heritage: Distribution of 19 cultural heritage protection units, which are mainly beacon towers ruins; Economic: More than 11000 yuan（per capita income of rural residents）; Visual perception: Large area of farmland with broad field of vision, east of farmland is short hill with semi-closed field of vision.
Linze Liyuan River downstream section West Shore: Heihe Bridge in Yanuan Township, Linze - Heihe Bridge in Xiaotun Township, Linze East Shore: Heihe Bridge in Yanuan Township, Linze - Heihe Bridge in Xiaotun Township, Linze	Landform type: Alluvial plain; Soil: Strongly eroded area; Hydrology: Average annual groundwater depth below 100 meters; Creatures: 5 kinds of animal distribution, plant distribution of desert vegetation such as Kalidium; Land use: mainly woodland; Cultural heritage: Distribution of 5 cultural heritage protection units, which are mainly ancient village ruins; Economic: More than 11000 yuan（per capita income of rural residents）; Visual perception: low trees and desert vegetation, broad vision.
Linze Liyuan River midstream section West Shore: Heihe Bridge in Xiaotun Township, Linze- Lianhuo highway East Shore: Heihe Bridge in Xiaotun Township, Linze- Lianhuo highway	Landform type: Alluvial plain; Soil: Lightly eroded area; Hydrology: Average annual groundwater depth below 150 meters; Creatures: Poor animals distribution, plant distribution of desert vegetation; Land use: mainly urban land; Cultural heritage: Distribution of 5 cultural heritage protection units, which are mainly ancient village ruins; Economic: More than 11000 yuan（per capita income of rural residents）; Visual perception: urban construction lands are located on both sides of the river, with a closed field of view.
Linze Liyuan River upstream section West Shore: Lianhuo highway -South Boundary of Linze County East Shore: Lianhuo highway -South Boundary of Linze County	Landform type: Alluvial plain; Soil: moderately eroded area; Hydrology: Average annual groundwater depth below 200 meters; Creatures: Poor animals distribution, plant distribution of desert vegetation; Land use: mainly urban land; Land use: mainly urban land; Cultural heritage: Distribution of 3 cultural heritage protection units, which are mainly the Red Army battle sites; Economic: More than 11000 yuan（per capita income of rural residents）; Visual perception: urban construction lands are located on both sides of the river, with a closed field of view.
Ganzhou Heihe downstream section West Shore: West Boundary of Ganzhou District -National Road 312 East Shore: West Boundary of	Landform type:Alluvial plain; Soil: moderately eroded area; Hydrology: The average annual groundwater depth is shallow; Creatures: 5 kinds of animal distribution, no plant distribution;

Ganzhou District -National Road 312	Land use: mainly farmland; Cultural heritage: Distribution of 3 cultural heritage protection units, which are mainlyTomb ruins; Economic: More than 11000 yuan（ per capita income of rural residents）; Visual perception: Farmlands are located on both sides of the river, with a closed field of view.
Ganzhou Heihe midstream section West Shore: National Road 312 - County Road 202 East Shore: National Road 312 - County Road 202	Landform type:Alluvial flood plain; Soil: lightly eroded area; Hydrology: Average annual groundwater depth below 50 meters; Creatures: 1 kinds of animal distribution, no plant distribution; Land use: mainly urban land; Cultural heritage: Distribution of 25 cultural heritage protection units, which are mainlyTemple ruins; Economic: More than 11000 yuan（ per capita income of rural residents）; Visual perception: Buildings are located on both sides of the river, with a closed field of view.
Ganzhou Heihe upstream section West Shore: County Road 202 - South Boundary of Ganzhou District East Shore: County Road 202 - South Boundary of Ganzhou District	Landform type:Alluvial flood plain; Soil: Strongly eroded area; Hydrology: Average annual groundwater depth below 200 meters; Creatures: no animal distribution; Land use: mainly urban land; plant distribution of desert vegetation such as Sympegma regelii Land use: mainly farmland; Cultural heritage: Distribution of 8 cultural heritage protection units, which are mainly beacon towers ruins and temple sites; Economic: More than 11000 yuan（ per capita income of rural residents）; Visual perception: large farmland, flat terrain, wide field of vision.

5. Discussion

In the European assessment method, in very general terms a national / regional assessment may be carried out at 1: 250,000 scale, a county assessment at approximately 1: 50,000 scale, and a district assessment at approximately 1; 20,000-1: 10,000 scale, a site level / project based assessment might be carried out at around 1: 2,500 scale [6]. In the overall research plan for the division of landscape character areas in the Heihe River Basin, the scale of division is carried out from two aspects: watershed scale and local scale. Watershed-scale landscape character area refers to the relatively same types of special landscapes in the characteristics with similar combinations of landform type, drainage landform type, vegetation, land use and settlement morphology [9]. At the local scale, the landscape character area is the smaller scale division of the watershed-scale landscape character area, where each division has commonality and also its own unique characteristics and definitions [10]. The results of this phase are reflected in the analysis of the history, landscape formation, changing trends and pressures of a region, forming a landscape character map and the description of features that are relatively non-value related [11]. The research on the division of watershed scale has been relatively complete at present and this paper also discussed the watershed scale only. The division of landscape character areas at the local scale will be the focus in the next research work.

6. Conclusion

Based on the assessment methods and advanced experience of landscape character assessment in Europe, this paper summarized the watershed-scale evaluation index system of landscape character assessment from three aspects of nature, culture and perceptions in the arid region of Northwest China, and divided the landscape character areas in the Zhangye section of Heihe River. These drew to the following conclusion:

a) From the perspective of methodology, it is scientific and effective to use the LCA method based on the three aspects of nature, culture and perception characters, as well as eight factors including terrain, soil, hydrology, creature, land use, cultural heritage, economy, visual perception to divide the landscape character areas according to the multi-dimensional and complicated geographical conditions of the Zhangye section of Heihe River basin.

b) From the perspective of geographical practice, the study area was divided into 11 landscape character areas in the watershed scale after stratification and superimposition. The landscape character areas in the study area have similar landform type, soil, hydrology, flora and fauna, land use, Cultural heritage, the combination of economy. Each landscape character area has a shared landscape form that can be identified on site and on the map and each area can be identified in different areas of evaluation. The division of the landscape character area of Zhangye section in Heihe River basin provides the scientific guidance for the next step of the forest planning, spatial planning, ecological restoration, land consolidation and sustainable utilization in Zhangye based on regional characteristics.

7. References

[1] Kongjian Yu, Dihua Li, Hailong Liu and Jin Cheng 2005 Urban spatial development pattern based on ecological infrastructure - Taizhou case of "anti-planning" *City Planning* 29 76-80

[2] Yi Yuan, Feng Xie and Peijun Shi 2003 Characteristics of landscape patches of urban land and agricultural land during rapid urbanization: A case study of Shenzhen *Journal of Beijing Normal University (Natural Science)* 39 835-842

[3] Zhangye Regional Annals Compilation Committee 2003 *Zhangye Regional Annals* vol 1 (Zhangye: Author) p 1066

[4] Zhangye City Environmental Protection Bureau 2013 *Zhangye City Ecological Environment Protection Planning* (Zhangye: Author) pp 30-40

[5] Carys S 2002 *Landscape Character Assessment Guidance for England and Scotland* (London: The Countryside Agency and Scottish Natural Heritage)

[6] Tudor C 2014 *An Approach to Landscape Character Assessment* (London: Natural England)

[7] Council of Europe 2000 *European Landscape Convention* (Florence: Author) http://www.coe.int/en/web/conventions/full-list/-/conventions/rms/0900001680080621

[8] KongjianYu , Dihua Li and Xili Han 2005 "Anti-planning" *Urban Planning* 29 64-69

[9] UNESCO 2008 *Operational guidelines for the implementation of the world heritage convention* (Paris： UNESCO world Heritage Center) http://www. unesco. com

[10] Wekerle G 2001 Planning for urban agriculture in suburban development in Canada *Urban Agriculture Magazine* 36-7

[11] Martín B, Ortega E, Otero I and Arce RM 2016 Landscape character assessment with GIS using map-based indicators and photographs in the relationship between landscape and roads *Journal of Environmental Management* 180 324-34

Effects of Shallow-deep Drainage Ditch on Soil Chemical Properties and Soil Enzyme Activities in Sanhu Farm

Wenchao Zhang[1], Fahu Li[1*], Jianqiang Zhu[2] and Xinguo Zhou[3]

1 College of Water Resources and Civil Engineering, China Agriculture University, Beijing 100083, China
2 College of Agriculture, Yangtze University, Jingzhou 434025,China
3 Farmland Irrigation Research Institute, Chinese Academy of Agricultural Sciences, Xinxiang 453003, China
E-mail address: lifahu@cau.edu.cn

Abstract. Waterlogging has become a serious agricultural problem in Jianghan Plain. To lower groundwater level and mitigate the damage of waterlogging to crop growth, shallow-deep drainage ditch has been applied in agricultural production. The objectives of this study were to discover the effects of shallow-deep drainage ditch on soil chemical property and enzyme activity, and their correlations with each other. Winter wheat experiment was conducted from October to May in Sanhu Farm. Soil samples (0-20cm depth) were collected from five plots that were 1m, 3m, 5m, 7m, 9m away from deep drainage ditch, respectively. Soil water content (SWC), soil chemical properties and enzyme activities were measured in three growth stages of winter wheat. The results showed that SWC was increased with distance away from deep ditch. Soil chemical properties were lower at 1m and 7m away from deep ditch. Phosphatase and sucrase at 3m away from deep ditch were 11.31% – 21.49% and 42.95% – 64.04% increase, and urease at 7m was 6.24% – 44.29% increase compared to other sampling points, respectively. Soil nutrient contents and enzyme activities were positively correlated with each other except the relationship of total nitrogen and urease. Our studies demonstrated the improvement of soil chemical properties and soil enzyme activities in shallow-deep drainage z

1. Introduction

Soil waterlogging created by seepage from irrigated uplands and canal water with inadequate water management practices[1], is estimated to affect adversely about 10 % of the global land area[2]. With the exception of rice, most crops are sensitive to waterlogging and show significant decline in yield when grown in flooded soils[3]. During waterlogging, gas exchange decreases between soil and air. In this condition, oxygen in soil is depleted rapidly, and the soil may become hypoxic (low oxygen) or anoxic (no oxygen) within a day[4], [5]. Lack of soil oxygen can limit plant yield directly by altering root metabolism or indirectly by changing plant nutrient availability[4], [6]. Also, high level of soil water content influence soil chemical and physical properties to decrease soil quality. Plant root growth and microbe activity are limited by oxygen deficit at high level of soil water content and nutrient cycling can be affected because of limited enzyme activities[5], [7]. So it is important to lower groundwater level for improving crop growth and soil quality[8].

In order to control groundwater level, drainage measures have been adoped in agricultural areas for many years. Drainage ditch, pipe drainage, drainage by pumping station and so on are typical waterlogged farmland drainage engineering measures[9], [10]. Drainage ditch is developed earliest and most widely used in these drainage measures[9], [11]. Drain depth and spacing are important factors in design of drainage ditch system. Hydraulic conductivity and drainable porosity are basic parameters required in design of field drain spacing, and these parameters obtained from laboratory analysis are to establish drainage equations to figure out drain spacing and drain depth[12]. But relationships between geophysical property and hydraulic conductivity are complex, non-unique and scale dependent[13]. A drain depth will be specified and the spacing will be calculated based on recharge schedule and midpoint water table depth. Subsequently, drain depth will be varied to calculate a range of depths and spacings for an economic analysis[14]. In contrast with ordinary drainage ditch, shallow-deep drainage ditch could achieve better drainage effects and be more cost-effective by combining deep and shallow ditches. To reduce soil moisture content, shallow-deep drainage ditch has been utilized to lower underground water level in Sanhu Farm where crop yield and quality can be decreased by waterlogging. Studies have demonstrated that shallow-deep drainage ditch is more efficiency than pipe drainage system, ranging from 6.5% to 16.3% increase in drainability[15]. As an important component of soil fertility, soil enzymes are mediators and catalysts in matter transformation and stabilized by colloids which are primarily from microbial origin[16]. After soil enzyme-mediated degradation processes, soil organic matter complex can be decomposed , then assimilated by microbe and plant[17]. Soil enzymes have been reported to be sensitive to changes and affected by many factors, including tillage treatment[18], crop species and climatic change[7], etc. Decades of research has revealed that mineral compound generally has significant effects on enzyme activity[19]. The relationships between soil enzyme activities and soil environmental factors contain the electrical conductivity of saturated soil extract, pH value, available nutrient, and organic carbon have been turned out to be significant correlative[20]. Therefore, soil chemical property and soil enzyme activity are indexes of soil activity and productivity to estimate soil fertility and health. Ditches play important roles in controlling the hydrologic, chemical, and biological processes[21]. However, the effects of shallow-deep drainage ditch on soil fertility and quality are scarcely known. The aim of this research was to assess the effect of shallow-deep drainage ditch on the soil chemical properties (pH, available phosphorus, available potassium, total nitrogen and soil organic matter) and soil enzyme (phosphatase, urease and sucrase) activities. And find the relationships among soil chemical properties and enzyme activities.

2. Materials and Methods

The experiment was carried out by planting winter wheat which grows from October to May in Sanhu Farm (latitude 30°10' N, longitude 112°29' E, elevation 27.3—31.1 m above sea level, slope 0.011°), located in Hubei Province, China. The experimental site belongs to north subtropical humid monsoon region with rainy spring and summer. The mean annual precipation and air temperture are 1100 mm and 16.7°C, respectively. The soil is yellow brown paddy soil and its texture is loam. The topsoil (0-20 cm) particle size is summarized as 21.96% clay (0–2 μm), 75.00% silt (2–50 μm) and 3.04% sand (50–2000 μm), with following properties: pH 7.1(1:2.5 soil : water ratio), available phosphorus (AP) 7.02 mg kg^{-1}, available potassium (AK) 152.10 mg , total nitrogen (TN) 0.8g $kg^{-1}kg^{-1}$, available nitrogen (AN) 118.42 mg kg^{-1}, soil organic matter content (SOC) 28.8 g kg^{-1}.

Soil was collected with a soil sampler from experimental site. Deep (1m deep, 0.5 m wide, 20 m apart) and shallow (0.3 m deep, 0.3 m wide, 2 m apart) drainage ditch system was designed as Figure 1. Soil samples (0-20cm depth) were collected from five plots that were 1m, 3m, 5m, 7m, 9m away from deep drainage ditch, respectively, and air-dried before determination. To improve the accuracy, each soil sample was analzyed with three replicates.

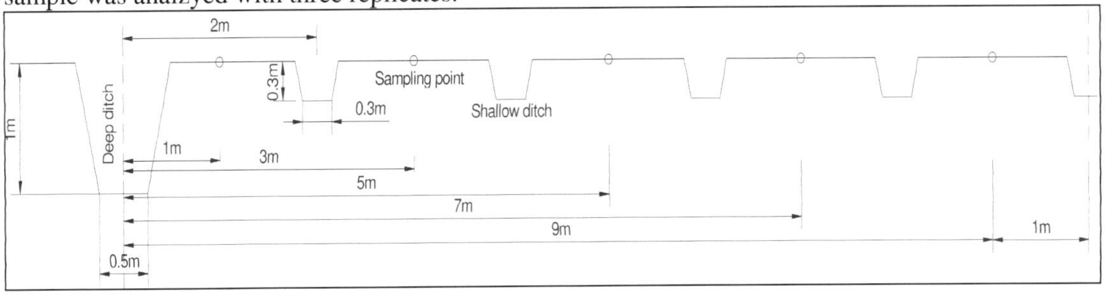

Figure 1. Arrangement plan of shallow-deep drainage ditch and sampling positions.

Soil was sampled for three times respectively at the wheat growth stages of jointing (March), heading, and grain filling (May). Gravimetric soil water content (SWC) was determined by an oven at 105°C, soil organic matter content (SOC) was analyzed with wet combustion method by titration, and soil pH was measured by a pH meter (FG3-ELK, Mettler - Toledo international trading (Shanghai) Co., Ltd, China) in 1:2.5 soil-water extract. Soil total nitrogen content (TN) was determined by automatic kjeldahl apparatus (K9840, Hanon Instrument, Jinan, China), available phosphorus content (AP) was assayed spectrophotometrically (UV-5500PC Specrtophotometer, Shanghai Metash Instrument Co., Ltd. Instrument Co., Ltd., China) at 700 nm, available potassium content (AK) was determined on flamephotometer (FP640, Shanghai INESA Scientific Instrument Co., Ltd., China). Urease, phosphatase, and sucrase activities were determined according to Guan's methods[22]. Urease activity was assayed using a 10% aqueous urea as substrate and incubated at 37°C for 24 h. The released NH_4^+ was determined by a spectrophotometer at 578 nm, and the urease activity was expressed as mg NH_4^+ g^{-1} soil 24 h^{-1}. Alkaline phosphatase activity was measured using disodium phenyl phosphate substrate solution, and incubated for 24 h at 37 °C. The formation of phenol was determined spectrophotometrically at 600 nm, and enzyme activity was expressed as mg phenol g^{-1} soil 24 h^{-1}. Sucrase activity was assayed using sucrose solution as the substrate. After incubating at 37 °C for 24 h, the filtrate was boiled in a water bath for 5 min with 3 mL 3, 5- dinitrosalicylic acid (DNS). Absorbance of the reducing sugars was measured at 508 nm, and sucrase activity was expressed as mg glucose g^{-1} soil 24 h^{-1}. The determination procedures of all soil properties were according to standard methods and instrucment instruction manuals[22], [23].

The significance of differences among treatment means was tested by one-way ANOVA, using SPSS 21.0 (SPSS Inc. Chicago, USA). When the F value in the ANOVA was statistically significance, a least significant difference test (significance level of P=0.05) was used for seperation of the means. Pearson correlation was applied to analyze the relationship of soil physcal-chemical properties and enzyme activities, and the level of statistical significance was set at P=0.05.

3. Results and Discussion

3.1. Soil Chemical Properties

Changes of SWC and soil chemical properties influenced by different distances away from deep ditch were shown in Figure 2. SWC was higher in jointing stage and lower in filling stage, and increased with distance away from deep ditch in different growing stage owing to the drainage effects of deep ditches.

pH decreased with time and distance away from deep ditch. With the decrease of soil water content, soil pH decreased because of reduced groundwater evaporation and subsequently salt accumulation in surface soil [24]-[26].

Due to natural environment factors such as temperture, illumination intensity, wind and so on, it was higher AP in jointing stage, higher AK and SOC in heading stage than other two growing stages[27]. AP, AK, TN and SOC increased, then decreased with distance away from deep ditch in all growing stages, owing to the effects of SWC on soil chemical properties[28]. High level of SWC altered the oxygen and nutrient availabilities to block respiration, and retarded the growth of plant root which is closely associated with nutrient transformation and absorption[4], [29]. Drainage is generally required to facilitate respiration of plant roots[30]. That's why AP, AK and TN contents were little lower at 9m compared to those at 5m away from deep ditch. However, soil nutrient may be decomposed and subsided quickly if groundwater level is lower than 5m, which would accelerate nutrient loss and further reduce plant productivity[31]. That may lead to decrease in soil nutrition cotents near the deep ditch. Appropriate SWC could provide good soil porosity that promotes dynamic of N and C cycle which is positively correlated with root growth and soil enzymatic activity[32].

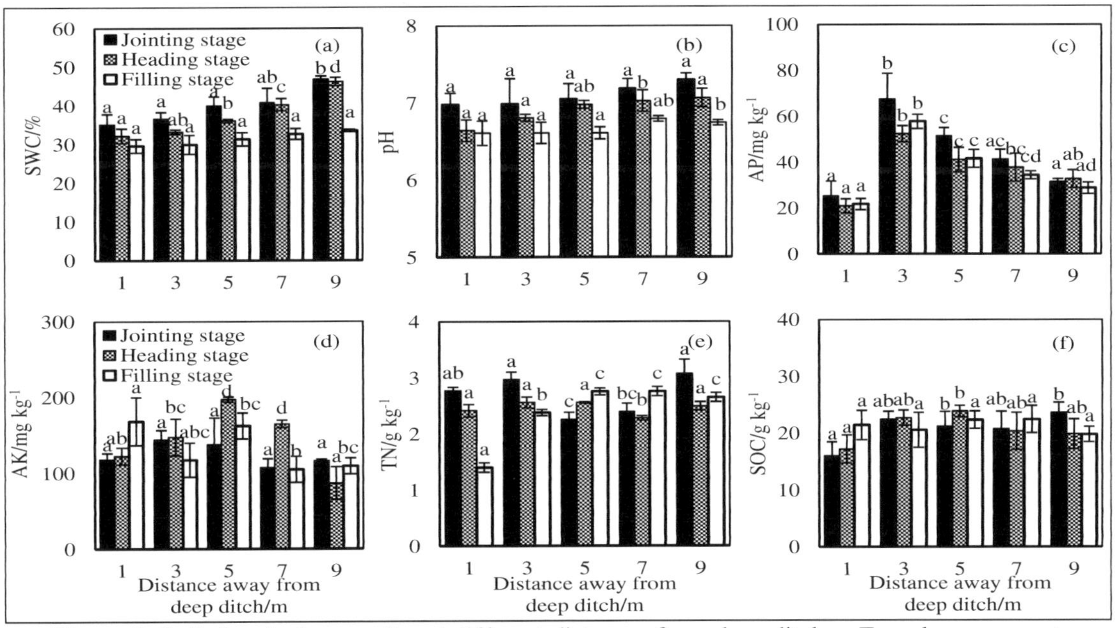

Figure 2. Soil chemical properties at different distances from deep ditches. Error bar represent standard error of the mean. Letters above error bars donote significant differences at $p < 0.05$ levels.

3.2. Soil Enzyme Activities

Variation of soil enzyme activities including phosphatase, urease and sucrase was shown in Figure 3. Similar to previous studies, phosphatase urease and sucrase activities were lower in heading stage and higher in filling stage [33]. Geisseler and Horwath have found similar reault that β-glucosidase activity and β-glucosaminidase activity are lower in April than those in May, due to environmental factors especially temperature[18], [34]. Temperature can influence enzyme activity by direct modification of enzyme kinetics and indirect influence on microbial proliferation[35]. Also, closely related to nutritional conditions, enzyme activity is regulated by soil chemical property in different growth periods[36]. For example, ammonium availability is much lower in spring due to active uptake of ammonium by microorganisms and plants[37].

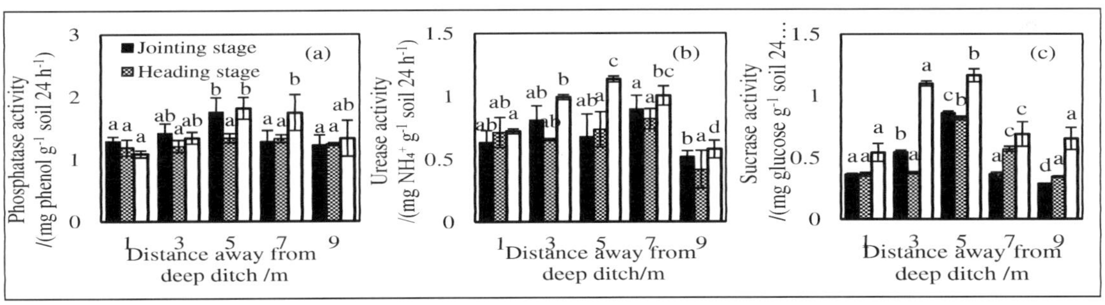

Figure 3. Soil enzyme activities at different distances from deep ditches. Error bar represent standard error of the mean. Letters above error bars donote significant differences at $p < 0.05$ levels.

Fluctuations of soil enzyme activities were similar that enzyme activity increased then decreased with distance away from deep ditch, except urease. Peaks of phosphatase and sucrase activities appeared at 5m, by 11.31% – 21.49% and 42.95% – 64.04% increase compared to other sampling points, respectively. Urease activity at 7m was 0.91 mg NH_4^+ g^{-1} soil 24 h^{-1}, by 6.24% – 44.29% increase compared with other sampling points. Minimums of phosphatase and sucrase activities were 1.19 mg phenol g^{-1} soil 24 h^{-1} and 0.42 mg glucose g^{-1} soil 24 h^{-1} at 1m by 6.71% – 37.92% and 26.80% – 79.50% decrease against rest sampling points, respectively. Urease activity was 0.50 NH_4^+ g^{-1} soil 24

h^{-1} at 9m, lower by 0.66% – 178.11% than other sampling points. Different distances away from deep ditch represented different levels of SWC. Nearer deep ditch, SWC was lower which limited soil enzyme activities by water stress[30]. Related to root activity closely, microbial and enzyme activities could be retarted due to the obstacles of soil anaerobic environment resulting from waterlogging or of water deficit by drought[7], [38], [39]. Furthermore, high groundwater level had significant effects on enzyme activity that decreased with increasing SWC[40]. Because waterlogging soil markedly affected the reaction rates of soil enzymes, impeded the decomposition, substrates cycling and enzymatic potential tied to substrate availability[41].

3.3. Pearson's Correlation Analysis

Correlations among soil chemical properties and soil enzyme activities were shown in Table 1. SWC was positively correlated with pH, TN, SOC and phosphatase. pH had negative correlation with AP, AK, TN and sucrase. AP was positively correlated with TN, SOC and soil enzyme activities. AK was positively correlated with phosphatase ($P < 0.05$), urease ($P < 0.05$), sucrase ($P < 0.01$) significantly. TN was positively correlated with SOC, phosphatase and sucrase instead of urease. The correlations among these three soil enzymes turned out to be positive.

Mostly, soil enzyme activities were negatively correlated with SWC in result of waterlogging[40], [42]. Soil enzyme activities were positively correlated with SOC due to promoted soil enzyme activity by increasing soil microorganism[43], [44] Phosphatase increased with soil chemical properties because of the positive effects of soil C, N, P and K on phosphatase[45]. Urease had positive correlations with soil chemical properties except TN due to higher microbial biomass and greater stabilization via humic substances as the same results of Zhang and Zhou[38]. Urease is offered an abundant substrate with nitrogen, and also can transfer organic N into mineralized N, then may produce more N_2O to cause TN loss[45]. Sucrase decreased with pH significantly may due to the suppression of pH [45], [46]. Influenced by soil microbial biomass, sucrase which involves in C transformation was positively correlated with TN, AP, AK and SOC[47]. The way to enhance sucrase activity has been suggested to enhance microbial activity and substrate concentrations in soil[48].

Table 1. Pearson's correlation coefficients between soil chemical properties and soil enzyme activities.

	AP	AK	TN	SOC	Phosphatase	Urease	Sucrase
SWC	-0.228	-0.514*	0.614*	0.203	0.081	-0.42	-0.179
pH	-0.253	-0.033	-0.182	0.150	0.055	0.145	-0.485**
AP	1	0.325	0.467	0.555*	0.368	0.496	0.511
AK		1	-0.289	0.238	0.556*	0.586*	0.767**
TN			1	0.597*	0.196	-0.193	0.11
SOC				1	0.529*	0.174	0.493
Phosphatase					1	0.572*	0.819**
Urease						1	0.502
Sucrase							1

* Correlation is significant at the 0.05 level (2-tailed).

** Correlation is significant at the 0.01 level (2-tailed).

This study demonstrated that enzyme activities not only involved in mineralization of C, N, P and K, but correlated with each other [33]. Further studies found the significant linear correlations between phosphatase and protease activities, acid phosphatase and cellulase activities in rhizosphere soil due to the produce and release processes by physiological activities of root and microorganism in rhizosphere soil[49]. However, Zhang indicated that no significant differences in invertase and alkaline phosphatase activity were detected among different treatments[50]. The mechanism of the interaction among soil enzymes still need further research.

4. Conclusions

Soil chemical properties and enzyme activities were influenced by shallow-deep drainage ditch obviously in Sanhu farm. We found that SWC and pH decreased with time and distance away from deep ditch. It was higher for AP and TN in jointing stage, for AK in heading stage contrary to soil enzyme activities. The change of SOC by winter wheat growth stage was not significant. Soil chemical properties as AP, AK, TN, SOC and soil enzyme activities were significantly lower near the deep ditch and the midpoint of two deep ditches compared to other sampling points. The result of pearson's correlation analysis showed that soil chemical properties and enzyme activities were correlated with each other. TN, SOC and phosphatase were positively correlated with SWC. Phosphatase was negtively correlated with pH. Except the relationship of TN and urease, the correlations of soil nutrient contents and enzyme activities turned out to be positive. So did the correlations among enzyme activities. The results indicated that shallow-deep drainage ditch could improve soil quality and fertility by lower groundwater table. Our suggestions were based on the limited condition and more research on the mechanism should be conducted in the future.

5. References

[1] Goyal V C, Jane S K and Pareek N 2005 *Journal of the Indian Society of Remote Sensing* 22 7–15
[2] Choudhury B U, Sharma B D, Mukhopadhyay S S and Verma B C 2015 *Proceedings of the National Academy of Sciences, India Section B: Biological Sciences* 86 835–45
[3] Zhang X, Shabala S, A.Koutoulis, Shabala L, Johnson P, Hayes D, Nichols D S and Zhou M 2015 *Plant Soil* 394 355–72
[4] Ghobadi M E, Ghobadi M and Zebarjadi A 2017 *International Journal of Biometeorology* 61 635–45
[5] Bansal R and Srivastava J P 2016 *Indian Journal of Plant Physiology* 22 130–34
[6] Zhang X, Jiang D, Zheng C, Dai T and Cao W 2011 *Journal of Agronomy & Crop Science* 197 31–39
[7] Ni L, Xu J, Chu X, Li S, Wang P, Li Y, Li Y, Zhu L and Wang C 2016 *Bulletin of Environmental Contamination & Toxicology* 97 131–37
[8] Xiao M, Yu S E, She D, Hu X and Chu L 2014 *Arabian Journal of Geosciences* 8 4411–20[9] Wang S, Xu D, Chen H, Han S and Jiao P 2014 *Journal of Drainage and Irrigation Machinery Engineering* 32 343–49 (in chinese)
[10] Yu S, Liu J, Li Z, Liu H and Tan L 2013 *Chinese Journal of Eco-Agriculture* 20 1664–72 (in chinese)
[11] Wang S 2010 *SHUILI XUEBAO* 41 697–702 (in chinese)
[12] Kannan N 2007 *Water Resources Management* 22 1113–25
[13] McArthur S A, Allen D M and Luzitano R D 2010 *Environmental Earth Sciences* 63 581–93
[14] Darzi-Naftchally A, Mirlatifi S M and Asgari A 2013 *Paddy and Water Environment* 12 103–11
[15] Yang J, Cheng J, Shi N, Jiang S and Tong Y 1978 *Today Science & Technology* 2 44–56
[16] Dick R P, Rasmussen P E and Kerle E A 1988 *Biology and Fertility of Soils* 6 159–64
[17] Shi W, Dell E, Bowman D and Iyyemperumal K 2006 *Plant Soil* 288 285–96
[18] Geisseler D and Horwath W R 2009 *Biology and Fertility of Soils* 46 65–72
[19] Allison S D 2006 *Biogeochemistry* 81 361–73
[20] Kang Y H, Liu S H, Wan S Q and Wang R S 2011 *Paddy and Water Environment* 11 87–95
[21] Barua G and Sarmah R 2016 *Acta Geophysica* 64 169–205
[22] Guan S 1986 *Book Methods of Soil Enzymology* ed J H Chen (Beijing: China Agriculture Press) pp 274–339 (in chinese)
[23] Bao S 2008 *Book Soil Agro-chemistrical Analysis(III)* (Beijing: China Agriculture Press) pp 25–114 (in chinese)
[24] Wang Y, Liu X, Butterly C, Tang C and Xu J 2013 *J Soil Sediment* 13 654–63
[25] Abliz A, Tiyip T, Ghulam A, Halik Ü, Ding J, Sawut M, Zhang F, Nurmemet I and Abliz A 2016 *Environmental Earth Sciences* 75 260–75
[26] Haddadi B S, Hassanpour H and Niknam V 2016 *Acta Physiologiae Plantarum* 38 119–30
[27] Kanda K, Miranda C H and Macedo M C 2002 *Soil Science & Plant Nutrition* 48 179–84
[28] Fromin N, Pinay G, Montuelle B, Landais D, Ourcival J M, Joffre R and Lensi R 2010 *Ecohydrology* 3 339–48
[29] Bailey-Serres J and Voesenek J 2008 *Annual Review of Plant Biology* 59 313–39
[30] Liu J, Zhang R, Zhang G, Guo J and Dong Z 2016 *Journal of Forestry Research* 28 255–63

[31] Miyamoto E, Matsuda S, Ando H, Kakuda K, Jong F and Watanabe A 2009 *Nutrient Cycling in Agroecosystems* 85 157–67
[32] Azouzi R, Charef A, zaghdoudi S, Khadhar S, Shabou N, Boughanmi H, Hjiri B and Hajjaj S 2015 *Arabian Journal of Geosciences* 9 3–16
[33] Hu W, Jiao Z, Wu F, Liu Y, Dong M, Ma X, Fan T, An L and Feng H 2014 *Ecotoxicology* 23 2069–80
[34] Wang C, Lü Y, Wang L, Liu X and Tian X 2013 *Journal of Forestry Research* 24 683–89
[35] Kang H, Kang S and Lee D 2009 *Ecological Research* 24 1137–43
[36] Dilly O and Nannipieri P 2001 *Biology & Fertility of Soils* 34 64–72
[37] Vetanovetz R P and Peterson J C 1992 *Communications in Soil Science & Plant Analysis* 23 379–88
[38] Zhang B C, Zhou X B and Zhang Y M 2014 *J Arid Land* 7 101–09
[39] Spohn M and Kuzyakov Y 2014 *Plant Soil* 379 67–77
[40] Hassan W, Chen W, Cai P and Huang Q 2013 *Journal of Thermal Analysis and Calorimetry* 116 969–88
[41] Sinsabaugh R L, Lauber C L, Weintraub M N, Ahmed B, Allison S D, Crenshaw C, Contosta A R, Causack D, Frey S, Gallo M E *et al* 2008 *Ecology Letters* 11 1252–64
[42] Anuradha M, Sivaraju K and Krishnamurthy V 2013 *Indian Journal of Plant Physiology* 18 67–70
[43] Lal R 2004 *Science* 304 1623–27
[44] Bhattacharyya R, Kundu S, Srivastva A K, Gupta H S, Prakash V and Bhatt J C 2010 *Plant Soil* 341 109–24
[45] Wang Z, Bao X, Li X, Jin X, Zhao J, Sun J, Christie P and Li L 2015 *Plant Soil* 391 265–82
[46] Wang R, Filley T R, Xu Z, Wang X, Li M H, Zhang Y, Luo W and Jiang Y 2014 *Plant Soil* 381 323–36
[47] Kandeler E, Luxhøi J, Tscherko D and Magid J 1999 *Soil Biology and Biochemistry* 31 1171–79
[48] Parthasarathi K and Ranganathan L 2000 *Biology & Fertility of Soils* 30 347–50
[49] Xiao S, You H, You W, Liu J, Cai C, Wu J, Ji Z, Zhan S, Hu Z, Zhang Z *et al* 2017 *Journal of Forestry Research* 28 521–28
[50] Zhang M, Cheng G, Feng H, Sun B, Zhao Y, Chen H, Chen J, Dyck M, Wang X, Zhang J *et al* 2017 *Environmental Science and Pollution Research* 24 10108–20

Acknowledgments
Project supported by the Special Fund for Agro-scientific Research in the Public Interest (201203077)

Crop Water Demand for Rain-Fed Maize in Northeast of China

Hanqing Xu[1,2,*], Zhan Tian[1,2], Minlan Wang[1], Dongli Fan[1], Biao Hu[1,2], Xiangyi Wang[1,2]

1 Shanghai Institute of Technology, Shanghai, China
2 Shanghai Climate Center, Shanghai, China
E-mail: xuhanqing2015@gmail.com

Abstract. Drought risk is one of the main constraint for stabling high maize production in the Northeast of China (NEC), where about 30% of the national maize is produced. Maize in the NEC is especially vulnerable to climate change and extreme weather. Previous studies on water demand of maize are based on field experiments by crop model, but water demand of different maize growth stages is rarely studied. Given the importance of NEC in China's food security, it is crucial to optimize the irrigation schedule to mitigate the negative effects of drought. In this study, we use the AEZ model to examine the climate change impacts on maize water demand in NEC. This model is employed to simulate the future maize water demand on climate scenario from NorESM1-M model driven by 2 Representative Concentration Pathways (RCP2.6 and RCP8.5). Results indicate that climate change will affect water demand of mazie in NEC. The increasing frequency and rising amount of water demand in the western part of NEC where should be given priority for soil moisture monitoring and irrigation.

1. Introduction

Drought risk has been one of the main limiting factors for maize production in the Northeast of China (NEC). The total maize planting area in the NEC takes more than 36% of the total crop area in NEC and accounts for 30% of the nation's total maize production [1]. But this region has experienced a warming trend about 0.38 °C per decade in the last 50 years [2]. The warmer climate accelerates crop growth, reduces crop growing length and causes yield loss [3]-[5]. Although maize yield has shown an increasing trend during recent decades in NEC, maize production has been highly affected by climate change in this region [6], and the projected less precipitation and more frequent drought events in the coming decades [7], [8], and may put maize production in NEC under greater risk in the expanded cropping area in recent years [9].
Previous studies on assessing the impact of climate change on maize yield and water demand based on field experiments in NEC, without considering that water demand of maize at a regional scale [10]. Therefore, in this study, we estimate the spatial variation of crop water demand under climate change in NEC, which could provide a better understanding of water demand under climate change.

2. Study Region

The Northeast Farming Region is located in the Northeast China (118° 50′ −135° 05′ E, 38° 43′ −53° 24′ N) (Figure 1), it includes Heilongjiang, Jilin and Liaoning Provinces. The total area of NEC is 787,300 km^2 with a population of 112 million (in 2010). During the maize growing season, the average temperature is 24.7 ℃ and annual average precipitation ranged from 308 to 657 mm.

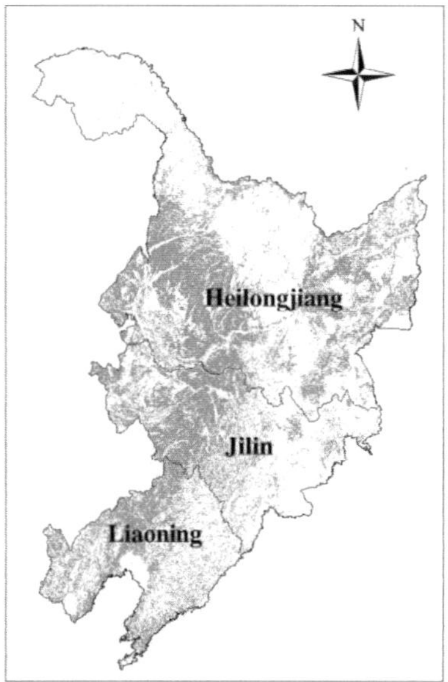

Figure 1. The map of maize cropping area in the NEC.

3. Datasets and Methods

The data used in this research include climate/weather data, land and soil information, and cropland data in the NEC. Climate change in the NEC and its impact on maize production in 2050s is obtained with NorESM1-M climate model under 2 RCP scenarios. Figure 2 summarizes the procedure of our work.

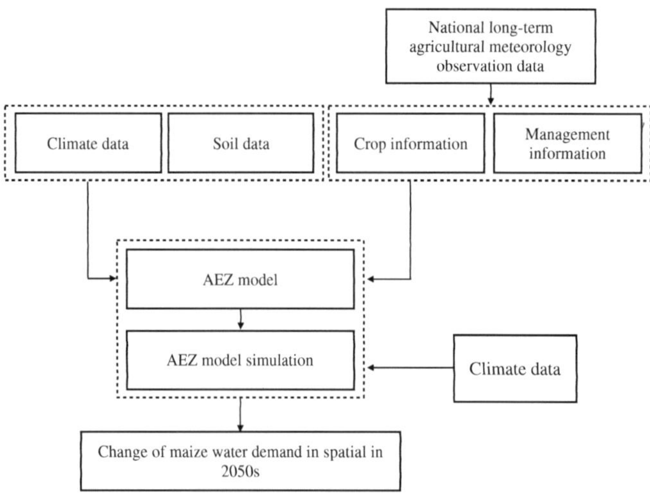

Figure 2. Flowchart to examine the climate change impacts on the crop water demand of maize.

3.1. AEZ Model

The Agro-Ecological Zone (AEZ) model was jointly developed by IIASA and FAO. AEZ model employs simple and robust crop models and provides standardized crop-modeling and environmental matching procedure to identify crop-specific limitations of prevailing climate, soil and terrain resources under assumed levels of inputs and management conditions. The standardized crop-modeling and environmental matching procedure in the AEZ makes it well suited for crop productivity assessment at regional, national and global scales.

4. Results

4.1. Crop Water Demand of Rainfed Maize

The crop water requirement (CWR) which is the amount of moisture that is expected to be supplied from the irrigation system to replenish the moisture loss through evapotranspiration in order to sustain the growth of the crop in different climate scenarios. The regions with the highest CWR of maize was western Liaoning, while the regions with the lowest CWR were northern and eastern Heilongjiang province. Under the baseline condition, the average CWR of rainfed maize were 415 mm. Our simulation showed the spatial variation in average crop water requirement under rainfed condition of maize across the study area, with highest in the middle part of Liaoning province and some parts of western Jilin province (Figure 3).

The changes in crop water demand are calculated using the results from climate variables of temperature and solar radiation. Under future climate projection, our results highlight spatial CWR patterns of rainfed maize in increasing significantly over NEC from baseline to 2050s. Under the scenario of RCP2.6, the CWR of rainfed maize decreased in the west of Jilin and southwest part of the Heilongjiang province. CWR increases across by 2050s compared to the baseline (1981-2010).

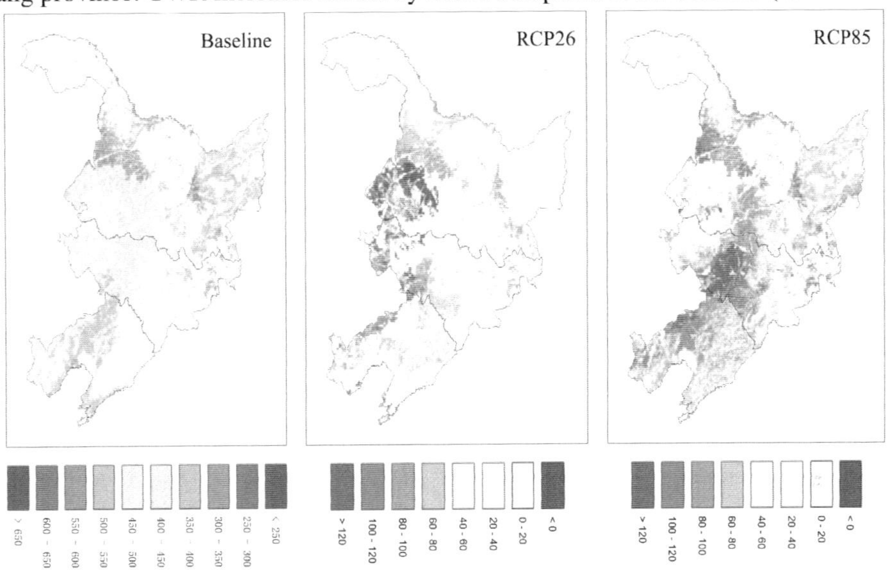

Figure 3. The crop water requirements change for rainfed maize at the middle in the century (2050s)

5. Conclusion

In this research, we combined outputs with observational data to produce more detailed climate scenarios of local climate suitable for assessing the impact of climate change on water demand of rainfed maize across NEC. Under the baseline, crop water demand is higher in the western NEC. Climate change will affects water demand of mazie in NEC. We found that the uncertainty of the impact of climate change on water demand of maize increases. The future climate scenarios indicate an increasing frequency and rising amount of water demand in the western part of NEC where should be given priority for soil moisture monitoring and irrigation.

6. References

[1] Z. Liu, K. G. Hubbard, X. Lin, and X. Yang, "Negative effects of climate warming on maize yield are reversed by the changing of sowing date and cultivar selection in Northeast China," *Glob Change Biol*, vol. 8, no. 2, pp. n/a–n/a, Aug. 2013.

[2] Z. Liu, X. Yang, K. G. Hubbard, and X. Lin, "Maize potential yields and yield gaps in the changing climate of northeast China," *Glob Change Biol*, vol. 18, no. 11, pp. 3441–3454, Aug. 2012.

[3] J. R. Porter, "Rising temperatures are likely to reduce crop yields," *Nature*, vol. 436, no. 7048, pp. 174–174, Jul. 2005.
[4] J. E. Olesen, "Climate Change and CO2 Effects on Productivity of Danish Agricultural Systems," *Journal of Crop Improvement*, vol. 13, no. 1, pp. 257–274, Sep. 2008.
[5] F. N. Tubiello, M. Donatelli, C. Rosenzweig, and C. O. Stockle, "Effects of climate change and elevated CO2 on cropping systems: model predictions at two Italian locations," *European Journal of Agronomy*, vol. 13, no. 2, pp. 179–189, Jul. 2000.
[6] Z. Liu, X. Yang, F. Chen, and E. Wang, "The effects of past climate change on the northern limits of maize planting in Northeast China," *Climatic Change*, vol. 117, no. 4, pp. 891–902, 2012.
[7] B. I. Cook, T. R. Ault, and J. E. Smerdon, "Unprecedented 21st century drought risk in the American Southwest and Central Plains," *Science Advances*, vol. 1, no. 1, pp. e1400082–e1400082, Feb. 2015.
[8] X. Yu, X. He, H. Zheng, R. Guo, Z. Ren, D. Zhang, and J. Lin, "Spatial and temporal analysis of drought risk during the crop-growing season over northeast China," *Nat Hazards*, vol. 71, no. 1, pp. 275–289, Nov. 2013.
[9] X. Xu, Q. GE, J. ZHENG, E. Dai, X. Zhang, S. He, and G. Liu, "Agricultural drought risk analysis based on three main crops in prefecture-level cities in the monsoon region of east China," *Nat Hazards*, vol. 66, no. 2, pp. 1257–1272, 2013.
[10] X. Song, L. Li, G. Fu, J. Li, A. Zhang, W. Liu, and K. Zhang, "Spatial–temporal variations of spring drought based on spring-composite index values for the Songnen Plain, Northeast China," *Theoretical and Applied Climatology*, vol. 116, no. 3, pp. 371–384, Jul. 2013.

Acknowledgments

This work was supported by the National Natural Science Foundation of China (Grant Nos. 41371110 and 41601049).

Evaluation of Land Ecological Security in Liaoning Province based on Grid

Hongfei An and Shuqin Zhao*

College of Land Resources and Rural-Urban Planning, Hebei GEO University, Huai'an East Road, Shijiazhuang, China
Email: shuqinzhao999@163.com

Abstract. Nowadays, green development has become an important concept of economic and social development. Ecological security issues have gradually been the concern of the general public in the rapid economic development. Based on the statistical data and remote sensing of Liaoning Province as the main data source, this paper extracts three groups of 12 indicators according to the principle of establishing the ecological security evaluation index, and the practical circumstance and data of the study district based on remote sensing technology. The weight is determined by the analytic hierarchy process and the entropy method. Using the comprehensive index method carries out land ecological security evaluation in Liaoning Province from 1000m*1000m grid and the municipal administrative of evaluation unit. The result shows that the level of land ecological security is mainly medium-security and medium-high-security, and the middle part is gradually reducing to the east and west in Liaoning Province. The conclusion of this study is in line with the actual situation, which can provide the theoretical basis and valuable experience for the ecological construction of Liaoning Province.

1. Introduction

With the rapid development of urbanization, the problems of ecological and environmental degradation are becoming more and more obvious, and people's awareness of the protection of ecology and environment is gradually enhancing. Land, as a special material, is a prerequisite for our production. With the increasing of human activities and the sharp contradiction between human and land, the land problem is becoming more and more significant, such as land desertification, increasing land pollution, and reducing level of quantity and quality of arable, which leads to a lack of land resources ([1]). With the concept of green development rising in the people's mind, the land ecological security issue has received wide public attention. In recent years, many scholars have used different evaluation methods to study land ecological security from multi-level and multi-scale. Because of the complexity of the land ecosystem and the impact of human activities on the land, there is no clear and comprehensive evaluation system to evaluate the land ecological security. There are many methods for evaluating land ecological security, such as ecological model, mathematical model, landscape ecological model and GIS technology ([2];[3]). Ecumenical used evaluation models are mainly "Pressure - State - Response" Model (PSR), the "Drivingforce - Pressure - State - Exposure - Effect - Action" model (DPSEEA), the matter-element model and projection pursuit model and so on ([4];[5];[6];[7];[8]). Yang Chunhong and other scholars evaluated Shantou by collecting relevant data based on PSR model to build index system, using principal component analysis method to determine the weight of the index system, and calculating the land ecological security composite index ([9]). Han Lei and other scholars evaluated ecological security of XiuShan County by using GIS technology combined with digital elevation model, remote sensing image and the present situation of land use map data, and establishing the land ecological security evaluation model with topographic index as the dominant factor ([10]). Li Mingyue and other scholars assessed the land ecological security in

Guangzhou based on BP neural network approach from nature - economy - society ([11]). All the above scholars have used various methods to make relevant research from the city scale and put forward reasonable opinions, and there are relatively few cases of this kind of researches and analysis based on remote sensing technology. This paper expandes the research scope and respectively evaluates the ecological security in Liaoning Province in 2015 from the 1000 m*1000 m grid and the municipal administrative unit, two kinds of evaluation unit based on PSR model, application of RS and GIS technology, which provides reasonable basis for the management and restore of ecological environment in Liaoning Province.

2. Establishment of index system and data processing

2.1. *Establishment of Index System*

Ecological security evaluation is a complex and multidisciplinary cross process, there are many different kinds of evaluation indexe. Therefore, we should be in line with the scientific, operational, comparability and the principle of universality and regional in selecting indicators ([12]). Combined with the availability of research methods and data, this paper establishes three groups of 12 evaluation index systems based on the P-S-R model.

2.2. *Data Sources and Processing*

According to the above index system, this article selects the statistical data and remote sensing image in Liaoning Province in 2015, which comes from Geospatial Data Cloud. In order to completely cover the area of Liaoning Province, there are 13 scene remote sensing image. Digital elevation model is provided by Data Center for Resources and Environmental Sciences, Chinese Academy of Sciences (RESDC) (http://www.resdc.cn). Combined with the “current land use classification” and the requirements of this article, the land use type of Liaoning Province can be divided into: vegetation, water, construction land and other land ([13];[14]). This article performs geometric correction, stitches and crops the remote sensing image, and then supervises classification. The overall classification accuracy of remote sensing image is higher than 90%, and the Kappa coefficient exceeds 0.85, which indicates that the interpretation result is accurate (Figure 1).

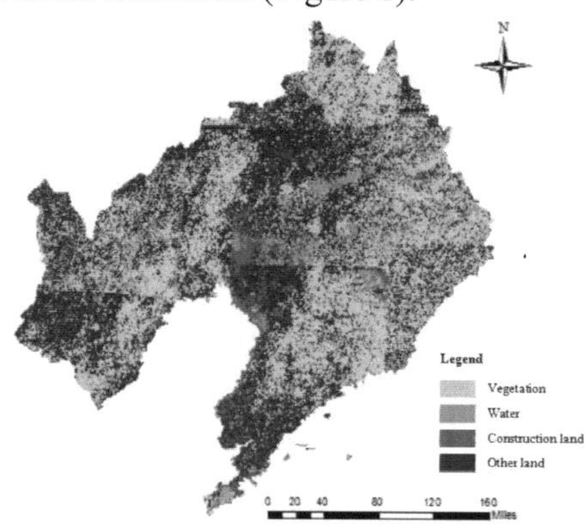

Figure 1. Land use map in Liaoning Province

3. Ecological security evaluation in Liaoning Province

3.1. *Divide Evaluation Unit*

This paper evaluates the ecological security in Liaoning Province in 2015 respectively from the 1000 m*1000 m grid and the municipal administrative unit. Due to the different data sources in this article, the steps implemented under the ArcGIS platform are inconsistent and the processing methods are slightly different.

3.2. *The Index Value*

3.2.1. Vegetation fraction (f). F is made by Normalized Difference Vegetation Index (NDVI) through modeling, which reflects the vegetation index of case ([15]). The formula is as follows:
In the formula 1, NIR is near infrared band, R is red band. In the formula 2, $NDVI_{min}$ is the minimum value of NDVI in the study area, $NDVI_{max}$ is the maximum value of NDVI in the study area.

$$NDVI = \frac{(NIR\text{-}R)}{(NIR+R)} \quad (1)$$

$$f = \frac{NDVI - NDVI_{\min}}{NDVI_{\max} - NDVI_{\min}} \quad (2)$$

3.2.2. Slope and aspect. Slope is the degree of steepness of the surface units, while the aspect is the direction of the terrain slope. These two indexes have an important influence on the land use patterns and play a greater role in the mountain ecology.

3.2.3. Ecosystem *services value system (ESV).* Most of the coefficients used in China are proposed by Costanza, but this method has a higher assessment of wetlands and a lower assessment of arable land. Therefore, Xie Gaodi puts forward corresponding ecological service value formula for China's ecosystem. The formula is as follows:
In the formula 3, ESV is the total value of ecosystem services in the study area, A_i is the area of the i-st land use type, P_i is the ESV of each land use type i per unit area (¥ / (km^2 * a)) ([16];[17]) (Table 1).

$$ESV = \sum_{i=1}^{5} P_i \times A_i \quad (3)$$

Table 1. The value of ecosystem services for each land use type

Land use type	Construction land	Water	Vegetation	Other land
ESV	377	56133.4	6203	521

3.2.4. Human impacts *index (HAI).* The initial state characteristics of landscape components are influenced by human activities, which are constantly decreasing. Artificial changes or exploitation represent different landscape components in varying degrees. HAI shows the impact of human activities on the landscape in this study area ([18]). The formula is as follows:
In the formula 4, HAI is human impacts index, A_i is the area of the i-st landscape component, TA is the total landscape area, P_i is the anthropogenic influence intensity parameter reflected by the i-st landscape component (Table 2).

$$HAI = \sum_{i=1}^{5} \frac{A_i P_i}{TA} \quad (4)$$

Table 2. The human impacts index for each land use type

Land use type	Construction land	Water	Vegetation	Other land
HAI	0.95	0.115	0.55	0.075

3.3. Ascertainment of Weights

Because the data level of each index is relatively large, the selected indexes cannot be directly applied to the model in the back, and the data should be normalized ([19]). The calculation model is:

$$R_{ij} = \frac{x_{ij} - \min(x_{ij})}{\max(x_{ij}) - \min(x_{ij})} \quad (5)$$

$$R_{ij} = \frac{\max(x_{ij}) - x_{ij}}{\max(x_{ij}) - \min(x_{ij})} \qquad (6)$$

Equation 5 is the positive index calculation formula, and equation 6 is the negative index calculation formula. R_{ij} is the standard value of the indicator. When it is 1, each indicator reaches the best condition. When it is 0, each indicator is in the lowest condition, and X_{ij} is the actual value of each indicator.

In this paper, analytic hierarchy process (AHP) method is used to determine the weight of the criterion layer factor, and the weight of the indicator layer factor is calculated according to the entropy method ([20];[21]). The final result is shown in Table 3.

Table 3. Land ecological security assessment index system in Liaoning Province

The target layer A	The criterion layer B	Weight	The indicator layer C	Weight	Trend
Ecological security assessment A	Natural state of the environment B_1	0.325	F C_{11}	0.0847	+
			Slope C_{12}	0.0834	-
			Aspect C_{13}	0.0851	-
			ESV C_{14}	0.0840	+
	Regional system pressure B_2	0.440	Population density C_{21}	0.0844	-
			HAI C_{22}	0.0850	-
			Industrial wastewater discharge C_{23}	0.0738	-
			Industrial sulfur dioxide emissions C_{24}	0.0844	-
			Industrial smoke (powder) emission C_{25}	0.0832	-
	Humanistic response B_3	0.235	Per unit of GDP C_{31}	0.0842	+
			Environmental protection expenditure as a percentage of GDP C_{32}	0.0835	+
			General industrial solid waste comprehensive utilization rate C_{33}	0.0843	+

4. Ecological security Evaluation in Liaoning Province

4.1. Ecological Security Index Calculation

After the normalization of the original data, according to the weight of each index, the eco-security index of each grid of the 1000m*1000m evaluation unit in the study area is calculated by the comprehensive index method, and the calculation model is:

$$Y_j = \sum_{i=1}^{n} X_i \times W_i \qquad (7)$$

In the formula 7, Y_j is the ecological security index of grid element j, X_i is the normalized result of the i-st index, W_i is the weight of the i-st index, and n is the number of indicators.

4.2. Ecological Security Classification

After the above results are standardized, the integrated value of land ecological security is divided into five levels on average from low to high.

4.3. Evaluation and Analysis of Ecological Security in Liaoning Province

4.3.1. Comprehensive Analysis of Ecological Security in Liaoning Province. As can be seen from Figure 2, the ecological security index in Liaoning province is generally in a good state, but moderate security and the following areas occupy a larger area, which needs more attention. Among them, the low-security area is 1.8 thousand hm^2, accounting for 1.26% of the total area of Liaoning Province; the light-security area is 30.7 thousand hm^2, accounting for 21.48% of the total area; the largest is medium-security area, with 61.8 thousand hm^2, accounting for 43.25% of the total area; medium-high-security area account for 19.31% of the total area, with 27.6 thousand hm^2; the high-security area is 2.1 hm^2, accounting for 14.7% of the total area. Different degrees of ecological security are classified according to the size of area: low-security < high-security < medium-high-security < light-security < medium-security. The high-security area is small and needs to be upgraded to a medium-high-security and high-security area through the improvement of light-security and medium-security areas, which enhances the ecological security of the entire area.

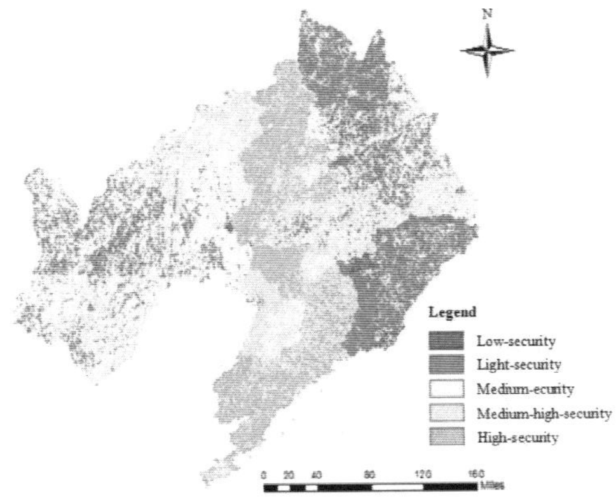

Figure 2. Ecological composite index rating in Liaoning Province

4.3.2. Ecological Security Analysis based on the Administrative Unit. The ecological security integrated value of each city is extracted, and the ecological security situation of each city is visualized and expressed, according to the ecological security integrated result of 1000m*1000m grid and the Zonal statistics tool in ArcGIS.

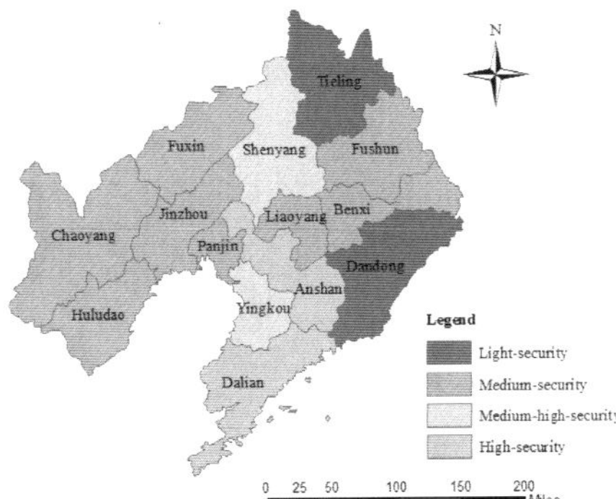

Figure 3. Ecological security level of various cities in Liaoning Provinc

As can be seen from Figure 3, the ecological security level of Dalian and Anshan is high-security, which mainly dues to the large area of the Yellow River and Bohai Sea, there are large sea area and good air quality. At the same time, Dalian has a higher vegetation coverage and per unit of GDP, with a lower population density. The ecological security value of Dandong and Tieling is the lowest. Tieling has low vegetation coverage and relatively serious soil erosion. At the same time, the per unit of GDP in this area is the lowest and its overall economic strength is weak. Dandong has a strong human disturbance index, and the ecosystem is damaged. There is a big difference in the distribution of ecological grade areas in various cities, which can reflect the differences in economic conditions, natural conditions and policy support among cities in Liaoning Province. As can be seen from Table 4, the low-security and light-security are mainly distributed in Chaoyang, Dandong and Fushun, with the accumulative area percentages of 26.42%, 76.53% and 35.33% in two levels; the medium-security is mainly distributed in Chaoyang, Huludao, Jinzhou and Panjin, with the accumulative area percentages of 71.51%, 71.67%, 76.14% and 82.46%; the medium-high-security and high-security are mainly distributed in Anshan, Dalian and Shenyang, with the accumulative area percentages of 97.95%, 95.90% and 93.25%. From the above analysis, the ecological security of Liaoning Province shows a vertical ribbon-like distribution. The ecological safety index of the middle area is high and mainly locates at medium-high-security and high-security, gradually decreasing from the middle to the east and the west.

Table 4. Ecology level area distribution of various cities in Liaoning Province

City	I	II	III	IV	V
Anshan	0	19	174	3860	5368
Benxi	4	496	4576	3023	28
Chaoyang	519	4903	14678	426	0
Dalian	0	1	459	3246	7507
Dandong	610	10028	3241	19	3
Fushun	185	3797	6746	542	1
Fuxin	0	379	6448	3831	9
Huludao	15	960	6886	1747	0
Jinzhou	4	1284	7076	914	10
Liaoyang	0	462	2035	2430	21
Panjin	60	442	2431	10	5
Shenyang	0	9	850	3519	8354
Tieling	258	7478	5164	8	9
Yingkou	0	81	2012	3133	8

5. Conclusions

This paper evaluates the ecological security in Liaoning Province in 2015 respectively from the 1000 m*1000 m grid and the municipal administrative unit, two kinds of evaluation unit based on PSR model and application of RS and GIS technology. The result shows that the level of ecological security in Liaoning Province are mainly distributed in medium-security and medium-high-security in 2015, the overall distribution is a vertical ribbon-like, and the ecological security status in the middle area is good, which is gradually reduced from the middle to the two sides. These indicators have an inhibitory effect on the ecological security of land, including population density, HAI, industrial wastewater discharge, industrial sulfur dioxide emissions and industrial smoke (powder) emission, and these indicators are generally lower in areas of light ecological security levels. Combined with the evaluation index, we can see that the rapid population growth, prompt economic development and human disturbance have brought great pressure on the ecosystem in Liaoning Province. Therefore, controlling the population growth rate and publicizing the family planning policy will reduce the pressure on land resources; we should also further develop our economy and shorten the gap between the rich and the poor, so as to provide a guarantee for the healthy development of the ecosystem.

6. References

[1] Quanjing, Zhang and Wanmao, Wang, 2003. China's Land Problems Classification, Reasons and Countermeasures, Journal of Qufu Normal University. 29(2), pp.97-101.

[2] Lei, Han, Xiangbin, Kong, Hongfeng, Guo and Zhuo, Luo, 2015. Study on Land Eco-security Evaluation of Ecological Sensitive Area in Southwest Mountains - A Case Study in Xiushan County, Research of Soil and Water Conservation. 22(1), pp.229-240.

[3] Mingyue, Li and Xiaojuan, Lai, 2011. Evaluation on Ecological Security of Urban Land Based on BP Neural Network - A Case Study of Guangzhou, Economic Geography. 31(2), pp.289-293.

[4] Huaiying, Wang, 2016. Evaluation Research of Regional Land Use Change and Ecological Security Based on Remote Sensing and GIS - by the Case of Leshan City, Sichuan Normal University.

[5] Baiming, Chen and Xiaoping Zhou, 2007, Explanation of Current Land Use Condition Classification for National Standard of the People's Republic of China, Journal of Natural Resources. 22(6), pp.994-1003.
[6] Pengfeng, Xiao, Shunxi, Liu, Xuezhi, Feng and Guangfa, Lin, 2006. A Land UseCover Classification System Based on Medium Resolution Remote Sensing Data, China Land Science. 20(2), pp.33-38.
[7] Yuandong, Zhang, Xiaohe, Zhang and Shirong Liu, 2011. Correlation Analysis on Normalized Difference Vegetation Index (NDVI) of Different Vegetations and Climatic Factors in Southwest China, Chinese Journal of Applied Ecology. 22(2), pp.323-330.
[8] Gaodi, Xie, Yili, Zhang, Du, Zheng and Shengkui, Cheng, 2001. Study on Valuation of Rangeland Ecosystem Services of China, Journal of Natural Resources. 16(1), pp.47-53.
[9] Zhiyou, An, 2011. The Urban Ecological Risk Assessment Based on GIS - A Case Study of Shanghai, Shanghai Normal University.
[10] Danhua, Zhou, 2013. Ecological Vulnerability of Wanquan River Basin in Hainan Province, Central South University of Forestry and Technology.
[11] Zhijuan, Qi, 2015. Small Town Residential Environment Livability Evaluation Under the Background of New Urbanization - A Case Study in Handan City Linzhang. Hebei Normal University.
[12] Peipei, Chu and Meichen, Fu, 2014. Research Progress on Ecological Security and Evaluation of Regional Land in China, Chinese Agricultural Science Bulletin. 30(11), pp.160-164.
[13] Xiaoshuang, Wang, Xuehua, Zhang and Ji, Lei, 2013. Tianjin Ecological Livable City Construction Index and Evaluation Research, China Population, Resources and Environment. 23(5), pp.19-22.
[14] Fang, Gong and Xiuming, Chen, 2016. Comprehensive Evaluation of Livable City Based on the Analytic Hierarchy Process, Journal of Changchun Normal University. 35(12), pp.88-94.
[15] Changcheng, Liu and Chun, Chen, 2013. Appraisal of Land Ecological Security Based on Ecological Footprint, Research of Soil and Water Conservation. 20(1), pp.193-201.
[16] Ruhai, He, Peng, Sun, Dianzhou, Xu and Cailie, Gao, 2017. Evaluation of Land Ecological Security in Hefei Based on PSR, Journal of Changchun University of Science and Technology (Social Sciences Edition). 30(3), pp.118-122.
[17] Tao, Lu, 2016. Evaluation of Land Ecological Security Based on the Variable Weight TOPSIS-DPSIR Model - A Case Study of Hebei City, China University of Geosciences (Wuhan).
[18] Dun, Yu and Wenbo, Chen, 2011. Land Ecological security in Poyang Lake Eco-economic Zone An Evaluation Based on Matter-element Model, Chinese Journal of Applied Ecology. 22(10), pp.2681-2685.
[19] Wei, Pei, Qiang, Fu, Dong, Liu, Tianxiao, Li and Kun, Cheng, 2016. Assessing Land Resource Ecological Security in the Heilongjiang Province Based on Improved Projection Pursuit Model, Journal of Northeast Agricultural University. 47(7), pp.92-100.
[20] Xuefeng, Xie, Tao, Wu, Cui, Xiao, Guojun, Jiang, Huajing, Bian, Yong, Ma and Jianhua, Chen, 2014. Ecological Security Assessment of the Dongyang River Watershed Using PSR Modeling, Resources Science. 36(8), pp.1702-1711.
[21] Chunhong, Yang, Zhengdong, Zhang, Nannan, Tian, Shenfeng, Wu and Wumei, Zhang, 2012. Evaluation on Land Ecological Security in Shantou Based on P-S-R Model, Research of Soil and Water Conservation. 19(3), pp.209-214.

IOP Conf. Series: Earth and Environmental Science **178** (2018) 012048 doi:10.1088/1755-1315/178/1/012048

Analysis of Population Genetic Diversity of *CynoglossusSemilaevis* in Three Groups from Northern China Sea

ZHANG Bo[1], LIU KeFeng[2], ZHENG DeBin[3], JIA Lei*

Tianjin Bohai Sea Fisheries Research Institute, Bohai Sea Fisheries Research Center of Chinese Academy of Fishery Sciences,Tianjin300457, China
E-mail: jialei@163.com

Abstract. Taking the Cultured and Wild Populations of Cynoglossussemilaevis in Tianjin Coast and cultured Populations in Haiyang as study objects, there were 32746 SNP locis screened by 2b-RAD sequencing. The genetic relationship, genetic structure, genetic diversity in these three groups were also analyzed by SNP markers. The genetic variation within populations (Fis), the genetic variation between groups (Fst), and total variation (Fit) in each SNP locus in these three groups were estimated , which shown that Fstvalue was ranged from 0.0731~0.1635. The genetic differentiation between Tianjin culture population and Tianjin wild population is the smallest, and the genetic differentiation between Tianjin and Haiyang culture populations, Tianjin wild population and Haiyang culture population is great. The results of selective sweep showed that the polymorphism of wild population was highest in Tianjin, followed by Tianjin culture population, and lowest in Haiyang breeding population. The polymorphism information content （PIC） statistics of Haiyang, Tianjin wild and Tianjin cultured populations respectively in each SNP site shown that Tianjin wild was up to 0.220, followed by Haiyang cultured and Tianjin breeding, the lowest is 0.175. The differentiation between Tianjin wild and Tianjin culture populations is not significant. Principal component analysis was carried out by using software GCTA, and PCA showed that the contribution rates of the first two principal components PC1 and PC2 were 16.25% and 9.54%, respectively, and the cumulative contribution was 25.79%. Two dimensional clustering results show that a small part of the region in Tianjin wild population had a small intervals with that of Tianjin cultured population, but the two principal components generally can be clearly divided into two categories, and these two populations had a clear boundaries with haiyang cultured one. Based on the above SNP locus diversity, polymorphism information content, genetic diversity comparison, the genetic diversity of Tianjin wild population is higher than Haiyang farming groups , but not significant difference between wild and cultured populations in tianjin revealed that Tianjin populations may have a certain extent mixture, which means the possibility of population mixture between wild and culture individuals may exist.

1. Introdution

The molecular marker assisted breeding technology is the reseach focus and hotspot in recent years in the field of molecular breeding of fish. With the development of biotechnology, different molecular markers, such as RPAD, RFLP, SSR and SNP, have been used to assist molecular breeding, and the methods of developing new markers have been further developed. The development of complete genome sequencing has brought huge amounts of DNA marker, and the group diversity analysis, population structure analysis, genetic relationship and estimation of linkage disequilibrium degree can be explored by using these molecular markers, all these explorations can make the breeding ideas more clearly and the breeding goal easier to achieve （[1]）

2b-RAD method is a streamlined restriction site-associated DNA (RAD) genotyping method based on sequencing the uniform fragments produced by type IIB restriction endonucleases,which (for example, BsaXI and AlfI) cleave genomic DNA upstream and downstream of the target site, producing tags of uniform length that are ideally suited for sequencing on existing next-generation platforms([2])The method provides a cost-effective and flexible genotyping platform for aquaculture species lacking sufficient genomic resources.The simplicity of the 2b-RAD protocol makes it particularly suitable for high-throughput genotyping as required for linkage mapping and profiling genetic variation in natural populations. 2b- RAD can be used in population genetic diversity analysis, population evolution research, genome assisted assembly, genome selection breeding and other fields([3]). In this study, three populations of *C.semilaevis* have been screened and 32746 SNPs have been found, further more， diversity analysis and SNP polymorphism analysis have also been Investigated.

2. Material and Methods

2.1. Animals

Here we picked 10 healthy C.semilaevis weighing 25.2 ± 3.4 g were cultured in the Huanghai fishery company in haiyang city in Shandong Province ,defined as 'haiyang culture group'and 10 healthy C.semilaevis weighing 22.1 ± 2.4 g were cultured in the XINGSHENG fishery company in Tianjin ,defined as 'Tianjin culture group'， also, 10 healthy C.semilaevis weighing 21.1 ± 5.7 g were captured from Bohaisea,which are considered as 'Tianjin wild group'. All the fish were kept in three 70 L tanks at 22 °C in aerated seawater. Before the experiment, three fish were euthanized and sampled for the presence of DNA extraction. The animal experiments were conducted in accordance with the "Regulations for the Administration of Affairs Concerning Experimental Animals" from the State Science and Technology Commission of Tianjin. The study was approved by the ethics committee of Bohai Sea Fisheries Research Institute, Chinese Academy of Fishery Sciences([4]).

2.2. Extraction of genomic DNA

For DNA extraction, the larvae or fins fixed in anhydrous ethanol were placed in 1.5 mL centrifuge tubes. The minced tissues were homogenized in 400 μL TNES lysis buffer (10 mMTris-HCl, pH 7.5; 400 μMNaCl; 100mM EDTA; 0.6% SDS, and freshly added 10μL proteinase K, 10mg/mL), then the homogenate was lysed at 55 °C. After that, 140 μL saturated NaCl solution was added, and the mixture was centrifuged at 12000 rpm for 30 min at 4 °C. The DNA was precipitated by two volumes of ethanol which had been refrigerated at -20 °C. Then the DNA pellets were washed twice with 70% ethanol, dried and dissolved in 50μL TE buffer (10 mMTris-HCl, pH 8.0, 10 mM EDTA).

2.3. 2b-RAD library preparation

The concentrations of all DNA samples were adjusted to 50 ng/μL. The DNA (200 ng) was digested in 10 μL reaction volume of NEB Buffer 4 with 1 U BsaXI at 37°C for 2 h. An additional DNA was digested simultaneously to detect the digestion efficiency by 1% agarose gel electrophoresis. The primary DNA band disappeared and became disperse, indicating a successful digestion. Then the ligation reaction was completed in the same tube as the digestion, combining the remaining digested DNA with 2 μL of the adapter1, 2 μL of the adapter2, 2 μL ligase buffer and 400 U of the T4 ligase. BsaxI could not be inactivated by heating, so the digested productions was recommended performing the ligations in 4°C for 1 hour, then hold on ice .

2.4. Genotyping and quality analysis

30 samples were sequenced on the HiSeq X-Ten platform by Paired-end sequencing.In the procedure, the standard protocol features adaptors with fully degenerate cohesive ends (5′-NNN-3′). Eliminate the sequences containing no BsaXI recognition sites, excluding low quality sequences, keeping the high quality original reads whose ratio was above 75% on average, and the label depth was between 14-25, the average depth was 20, all these showed the Library of good quality. After genotypes procedure, totally 32,746 SNPs yielded by SOAPsnp were confirmed. All of the markers information was showed in Additional filethe results in the original reads filter is in Table 1:

Table 1. Constuction Library and sequencing of 2b-RAD tags

NO.	Original reads	High qulity reads	Percentage	Unique	Depth
HY-1	6,396,349	5,054,386	79.02%	143,337	20
HY-2	6,396,349	5,587,054	87.35%	145,862	21
HY-3	6,396,349	5,798,571	90.65%	145,740	22

HY-4	6,396,349	5,545,037	86.69%	146,059	21
HY-5	6,396,349	4,045,274	63.24%	144,101	16
HY-6	7,065,887	5,445,752	77.07%	144,571	21
HY-7	7,065,887	5,972,173	84.52%	144,777	23
HY-8	7,065,887	6,462,181	91.46%	145,127	25
HY-9	7,065,887	6,444,063	91.20%	143,720	25
HY-10	7,065,887	4,748,605	67.20%	143,837	19
TJcl-1	6,646,800	4,688,066	70.53%	145,915	18
TJcl-2	6,646,800	5,804,242	87.32%	146,395	21
TJcl-3	6,646,800	5,728,174	86.18%	145,486	21
TJcl-4	6,646,800	5,573,184	83.85%	146,024	21
TJcl-5	6,646,800	4,244,575	63.86%	145,211	16
TJcl-6	7,027,055	5,482,653	78.02%	145,998	21
TJcl-7	7,027,055	6,021,305	85.69%	146,142	22
TJcl-8	7,027,055	6,082,922	86.56%	145,434	23
TJcl-9	7,027,055	5,766,252	82.06%	146,083	22
TJcl-10	7,027,055	4,655,306	66.25%	146,354	18
TJYS-1	5,866,332	4,209,487	71.76%	143,403	16
TJYS-2	5,866,332	5,149,297	87.78%	143,053	20
TJYS-3	5,866,332	5,138,678	87.60%	142,469	20
TJYS-4	5,866,332	4,993,461	85.12%	143,425	19
TJYS-5	5,866,332	3,645,928	62.15%	142,983	14
TJYS-6	7,404,738	4,473,240	60.41%	145,430	16
TJYS-7	7,404,738	5,977,990	80.73%	143,673	22
TJYS-8	7,404,738	5,375,319	72.59%	143,216	20
TJYS-9	7,404,738	6,021,680	81.32%	143,343	22
TJYS-	7,404,738	4,184,500	56.51%	143,167	15

For the species which have reference genomes, according to the genotyping strategy of RAD-typing, the markers are classified as followsConstruction of reference sequences: extract the tags includesBsaXI restriction site label as a reference sequence from the Cynoglossussemilaevisreference genomes. *Cynoglossussemilaevis* *(Cynoglossussemilaevis)* genome reference: ftp://ftp.ncbi.nlm.nih.gov/genomes/all/GCF_000523025.1_Cse_v1.0/GCF_000523025.1_Cse_v1.0_genomic.fna .gz

For Individual SNP genotyping: Mapping the high quality reads to the reference sequence of the individual by using SOAP software (the parameter is: -M 4 - V 2 - R 0). Site genotyping was performed using the maximum likelihood ML. In order to ensure the accuracy and preciseness of typing of SNP loci, the following conditions were filtered: in individuals, the tag depth was below 500; Leave only 3 SNP tagged sites within the all tags; the tags can be typed in 80% of the individuals. The minimum allele frequency MAF $\geq$ 0.05; There were a total of 32746 genotyping have been typed.

3. Results and Analysis

According to the 32746 SNP loci obtained, the genotying data have been analysed of each SNP at the individuals in 3 groups to get the value of Fis, Fst, Fit by Genepop software (version 4.2.2) ([5]) . $\theta\pi$ and Tajima 'D have been calculated by GD software(version: 0.8) to evaluate the nucleotide diversity according to the genotying data and mismatch analysis.

3.1. The value of Fst among threegroups

Table 2 . The value of Fst

Groups	HY	TJCL
TJCL	0.1635	
TJYS	0.1778	0.0731

The statistical data from table 2 showed that the value of Fst among three groups ranged from 0.0731~0.1635. The genetic differentiation between Tianjin cultured group and Tianjin wild group was the smallest, and the genetic differentiation between Tianjin cultured group and Haiyang group, Tianjin wild group and Haiyang group were great. As we all know the coefficient of genetic differentiation (Fst) was less than 0.05, indicating that the variation was mainly from within the group and no differentiation among the group. The value of Fst was greater than 0.05, indicating that genetic variation occurred mostly among groups, but there was low differentiation among the groups

3.2. Tajima's D test
Tajima's D is a statistical test s to distinguish between a DNA sequence evolving randomly ("neutrally") and one evolving under a non-random process.A randomly evolving DNA sequence contains mutations with no effect on the fitness and survival of an organism. The randomly evolving mutations are called "neutral", while mutations under selection are "non-neutral".

The purpose of Tajima's test is to identify sequences which do not fit the neutral theory modeatequilibriumbetween mutation and genetic drift.

Table 3. θπ and Tajima'D among three groups

	HY	**TJCL**	**TJYS**
θπ	0.118343	0.185802	0.206763
Tajima'D	1.457026	0.448594	0.59124

In the standard neutral evolution model, the theoretical value of 'Tajima' s D is zero,which means the randomly evolving mutations happened. If the actual 'Tajima' s D value deviates significantly from zero, the actual allele frequency is found to be biased against the expectation of the neutral evolution model, which means mutations under selection are "non-neutral". In Table 3,theTajima'D in HY group was highest, which shown that the group deviated from the neutral pattern and may be affected by population expansion and natural selection. θπ is an important index to measure the population polymorphism. Table 3 showd that there was the highest degree of polymorphism among the wild population in Tianjin, followed by the Tianjin culture group and the lowest in Haiyang group.

3.3. SNP polymorphism analysis
According to the 32746 SNP loci were obtained, respectively, TJCL, TJYS and HY three groups of each SNP locus were analysed by using PowerMarker software (version 3.25) ,the polymorphism information content (PIC), heterozygosity observed (Ho) and heterozygosity expected (He) were shown in table 4.

Table 4. SNP polymorphism analysis

Group		He	Ho	PIC
HY-	range	0~0.5	0~1	0~0.375
	average	0.2212	0.3115	0.1752
TJCL	range	0~0.5	0~1	0~0.375
	average	0.2531	0.2877	0.2066
TJYS	range	0~0.5	0~1	0~0.375
	average	0.2473	0.2805	0.220

The polymorphism information content of HY, TJCL and TJYS respectively in each SNP site (PIC)were be statistically analysed, TJYS up to 0.220, followed by HY and TJCL, the lowest is 0.175, the differences between TJCL andTJSY were not significant.

3.4. Principle Component Analysis, PCA
Principal component analysis was carried out using software GCTA (version, 1.25.0). According to the first two principal components, each individual is depicted as a scatter diagram. In general, the individual can be clustered by the first two principal components. The result is shown below(.Fig1.)

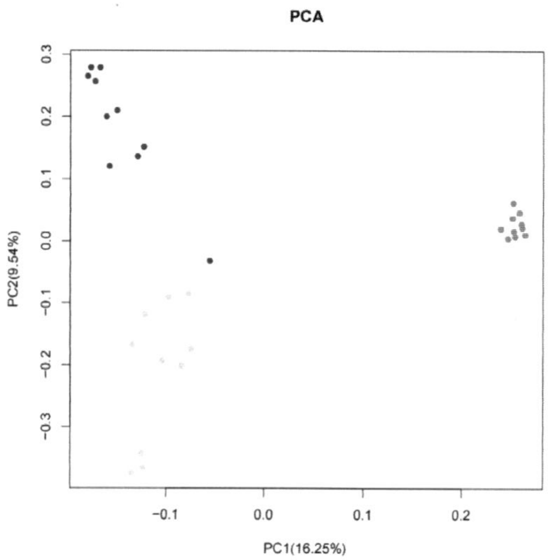

Figure 1. Scatter plot of PCA

(Dark blue dots indicate TJYS group;Light blue dots indicate TJCL group; Pink dots indicate HY group)
PCA results showed that the contribution rates of PC1 and PC2 in the first two principal components were 16.25% and 9.54%, respectively, and the cumulative contribution was 25.79%(Fig1). Two dimensional clustering results shown that a small part of TJYS had a smaller intervals with TJCL, but the two principal components generally can be clearly divided into two categories, and the two groups in Tianjin had clear boundaries with the Haiyang cultured group

3.5. Construction of phylogenetic tree based on individual shared tag sequences

Select the SNPs shared among 30 individuals, and if the corresponding loci are missing, use "-" instead. The obtained sequence were used to construct NJ (Neighbor-joining) tree by treebest (Version: 1.9.2), and the NJ (Neighbor-joining) tree was tested by bootstrap method (repeated 1000 times)(Fig2); the percentage of bootstrap were represented as the value at internal nodes in the graph, as the object name is at leaf node.

4. Discussion

2b-RAD method is a streamlined restriction site-associated DNA (RAD) genotyping method based on sequencing the uniform fragments produced by type IIB restriction endonucleases, 33-36bp enzyme digesting tags were produced after enrichment for high-throughput sequencing reaction downstream. Through genome-wide high-throughput SNP screening and genotyping analysis, SNPs can be found for the genetic diversity analysis of population ([6]) The principal component analysis found that the genetic differentiation between Tianjin cultured population and Tianjin wild populations were small, genetic differentiation of Tianjin wild population and Haiyang cultured population was great. The genetic diversity of half smooth tongue sole in Tianjin population is higher degree of polymorphism than that of the population of in Haiyang.The population from Tianjin wild and Tianjin cultured had not a significant difference, revealed Tianjin wild and cultured populations may be mixed to a certain extent.

In this study, the same geographical conditions, differences between wild and cultured populations of genetic diversity and differentiation in three groups of *Cynoglossussemilaevis* were examined by SNP genotying([7]). In fact, this difference is not significant, because the implementation of all the restocking activities, make breeding groups may part after releasing into the wild population, resulting in two in the morning was statistically significant difference. Here are some suggestions of protection of germplasm and utilization of the wild population in *C.semilaevis*: 1) As wild resources of *C.semilaevis* were reduced sharply, protection and conservation of the spawning field should be paid more attention.2) *Cynoglossussemilaevis* wild populations deviated from the Hardy-Weinberg equilibrium,Therefore, the artificial restocking and releasing in the sea should pay attention to its genetic structure, to avoid man-made accelerated half smooth tongue natural germplasm resources ;3) The genetic diversity of populations should be analysed before the artificial restocking and releasing in the sea, and the population genetic structure can not be too simple, causing disturbance to the natural population structure

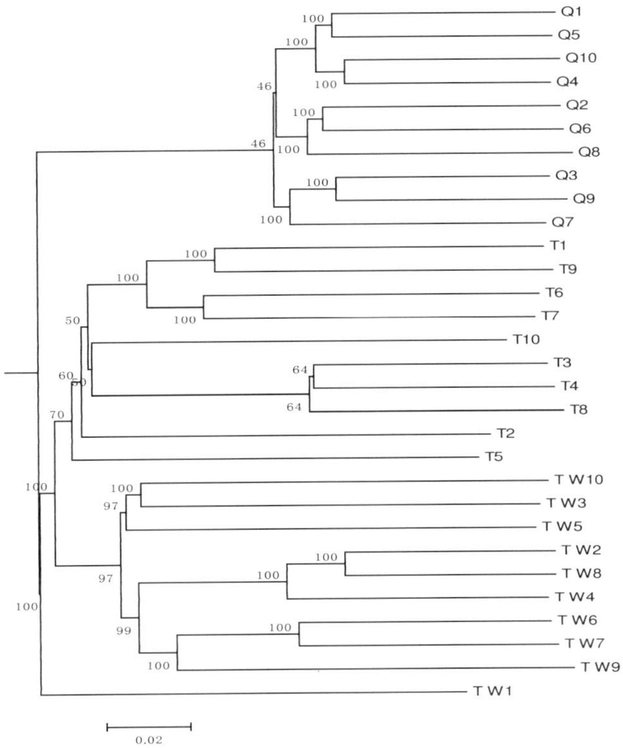

Figure 2. Neighbor-joining tree of 30 individuals in three different groups of *Cynoglossussemilaevis*
（Q1——Q10 indicate HY1-HY10; T1——T10 indicate TJCL1-TJCL10； TW1-TW10 indicate TJYS1-TJSY10）

5. References

[1] V I,Delphine E M,ALBRECHT L, CLAUDE ET AL. 2010, POPULATION STRUCTURE AND GENETIC DIVERSITY IN A COMMERCIAL MAIZE BREEDING PROGRAM AS-SESSED WITH SSR AND SNP MARKERS[J]. THEORETICAL AND APPLIED GE-NETICS. 120 (7): 1289 -1299.

[2] X,FU J, DOU J,MAO ET AL. 2013,RADTYPING: AN INTEGRATED PACKAGE FOR ACCURATE DE NOVO CODOMINANT AND DOMINANT RAD GENOTYPING IN MAPPING POPULATIONS. PLOSONE ,8(11): E79960

[3] J T,Gerdes C F, BEHR J G, COORS ET AL. 1994.COMPILATION OF NORTH AMERI-CAN MAIZE BREEDING GERMPLASM[M]. MADISON, WI: CSSA,

[4] L J,LU A J,MA X A，WANG ET AL. 2013, POLYMORPHISMS ANALYSIS OF FIVE POPULATIONS OF TAKIFUGURUBRIPES WITH MICROSATELLITE.[J]. PROGRESS IN FISHERY SCIENCE, 34(4): 27-34.

[5] S K，MAO,A J，MA F H，, DING ET AL. 2016, ANALYSIS OF GENETIC STRUCTURES OF FOUR WILD GEOGRAPHOIC POPULATIONS OF MULLET LIZA HAEMATOCHEILA BY USING MICROSATELLITE MARKER TECHNIQUE[J]. PROGRESS IN FISHERY SCIENCE, 37(2): 67-75.

[6] S, Wang E, MEYER JK, MCKAY MV,MATZ2012, 2B-RAD: A SIMPLE AND FLEXIBLE METHOD FOR GENOME-WIDE GENOTYPING. NAT METHODS. 9: 808-810. 10.1038/NMETH.2023.

[7] W J , ZHANG Y H，YU Y F，SHEN ET AL. 2003, ADVANCES ON GENETIC ANALYSIS OF MICROSATELLITE DNA IN PROTOZOOLOGY. [J].ACTA HYDROBIOLOGICA SINICA, 27(2) : 185-190.